Introduction to the
DESIGN AND ANALYSIS
OF BUILDING ELECTRICAL SYSTEMS

Introduction to the
DESIGN AND ANALYSIS
OF BUILDING
ELECTRICAL
SYSTEMS

John H. Matthews

John Matthews & Associates, Inc.
Atlanta, Georgia

VNR VAN NOSTRAND REINHOLD
New York

Library of Congress Catalog Card Number 92-26279
ISBN 0-442-00874-0

Printed in the United States of America

Van Nostrand Reinhold
115 Fifth Avenue
New York, New York 10003

Chapman and Bell
2-6 Boundary Row
London, SE1 8HN, England

Thomas Nelson Australia
102 Dodds Street
South Melbourne 3205
Victoria, Australia

Nelson Canada
1120 Birchmount Road
Scarborough, Ontario M1K 5G4, Canada

16 15 14 13 12 11 10 9 8 7 6 5 4 3 2 1

Library of Congress Cataloging-in-Publication Data

Matthews, John H., 1943-
    Introduction to the design and analysis of building electrical systems / John H. Matthews.
        p. cm.
    Includes bibliographical references and index.
    ISBN 0-442-00874-0
    1. Buildings--Electric equipment. 2. Interior lighting.
3. Electric engineering. I. Title
TK4001.M37 1992
621.319'24--dc20   92-26279
                                                                          CIP

To

R. P. Derrick, P.E. (1931-1992)
Good friend, great shipmate, and one fine engineer.

Contents

Preface

This book discusses the design and analysis of electrical systems for buildings. It evolved from two courses taught by the author, a consulting engineer, at Georgia Institute of Technology in Atlanta. The first course introduces building electrical systems and the second, illumination engineering. The school has offered both courses for a number of years, and each has enjoyed good student interest and acceptance. Electrical engineering students normally take the courses as technical electives in their senior year; however, they are also open to students from other engineering disciplines.

Although excellent texts exist for utility-level power systems, very little has been written to support these specific courses. Even professionals often forget that utility systems exist solely to serve the needs of residential, commercial, and industrial power-system customers. Surely these utilization level systems deserve the careful design and analysis accorded the systems that serve them.

Excellent resource material does exist on building electrical systems, due largely to the efforts of organizations such as the Institute of Electrical and Electronic Engineers, Inc., (IEEE), and the Illuminating Engineering Society of North America, (IESNA), as well as numerous technical reports, papers, and application guides by various electrical manufacturers. There also exists a significant body of local and national construction standards such as those prepared by the National Fire Protection Association, NFPA, whose publication NFPA-70 is commonly referred to as the National Electrical Code or simply the *NEC*.

The goal of the present text is to provide an introduction to both the theoretical considerations and practical challenges associated with the design and analysis of building electrical systems. Emphasis is placed on system planning, design and protection. It was also felt important to include sufficient reference material within the text to provide a reasonably stand alone treatment. References are provided at the end of each chapter for those who wish to probe deeper into the available literature on a given subject.

Several topics are not treated in the present work. These include topics such as variable speed control of *ac* induction motors, building signal and communication systems and building control and automation systems. These are of great interest but have been omitted from the present edition in the interest of presenting a compact treatment of essential concepts.

The following teaching sequence might be considered. For a one quarter introductory course in building power systems, Chapters 1 through 8. For a one quarter course in illumination systems, Chapters 9 through 12. As presently taught at Georgia Tech the illumination course also requires a student research paper and class presentation on an illumination engineering topic selected by each student. The class presentations are grouped together on several consecutive class periods at the end of the quarter. This provides each student the opportunity to research a topic of personal interest and to share his/her findings with the class. These presentations, limited to about 10 minutes each and followed by a short question period, have been one of the most popular aspects of the course and always seem to prompt lively, thought-provoking, and often surprising, class discussion.

Finally a word about the author—a consulting engineer with more than twenty years experience in the design of building power and lighting systems. He also is a part-time faculty member at Georgia Institute of Technology.

Acknowledgments

The author gratefully acknowledges the support and encouragement of several members of the building industry. Robert Frost, Rick Leeds, and Cheryl English provided technical information on luminaire photometrics. Al Meredith provided much information on the early days of lighting as well as technical data on light sources. Ken Beckworth, P.E., provided useful information on the development of the utility industry and also arranged for access to Georgia Power's corporate library. Kenneth Box provided technical data on low-voltage fuses. Bill Shellman, Wendell Carter, P.E., Bill Moncrief, P.E., and Tom Derby provided technical information on power-quality matters. Chuck Newcombe and Doug Severance provided information on measuring instruments. Alexander McEachern of Basic Measuring Instruments gave permission to use several graphs from his excellent book on power quality. Robert Atkinson, P.E., of Atkinson & Associates permitted the use of his electrical design software, ECALC. Russell Ohlson, P.E., offered suggestions on the fault-current analysis sections.

Thanks also go to my friends in the academic community. Dr. Roger Webb, P.E., and Dr. Bill Sayle, P.E., have offered unflinching support for the building electrical system courses at Georgia Institute of Technology. Colonel Harold Askins, P.E., of the Citadel and Dr. Marshall Molen, P.E., of Mississippi State University have offered much encouragement. Special thanks go to Dr. Robert Broadwater, who first encouraged preparation of this book.

After completing the manuscript, I realize why my friends say that you never finish a book, only abandon it. I may well have abandoned it without the support and active participation of Mary Matthews, whose diligence and professionalism made its organization and preparation possible.

Many students have participated in the proofing of this manuscript. Their assistance is much appreciated. Special thanks go to Tammy Gammon, who devoted considerable time and energy to this task.

Introduction to the
DESIGN AND ANALYSIS
OF BUILDING
ELECTRICAL
SYSTEMS

1

Introduction to Building Electrical Systems

This book introduces the design and analysis of building electrical systems. The building electrical system is the final link in the power generation and delivery process. Broadly speaking, we can divide this process into four phases or levels: the *generation, transmission, distribution,* and *utilization* level systems.

As shown in Figure 1.1, the power-delivery process begins at utility generating facilities. Here power is generated using nuclear fission, fossil fuels (such as oil or coal), natural gas, or hydroelectric power.

The voltage from these generators is immediately stepped up to transmission levels as high as 500 kV to 1 MV at substations adjacent to the generation facility. This higher voltage facilitates transmission of large quantities of power over long distances. This is because the high voltage results in lower current flow, thereby requiring smaller conductors. In addition, the lower current flow also results in lower transmission-line losses. These transmission lines are often integrated into large power networks which, in turn, provide increased system reliability in case of the unexpected outage of one generation unit.

The transmission lines deliver power to cities and towns, where the voltage is stepped down at substations to distribution levels of perhaps 25 kV. It is at the distribution level that overhead and underground lines are extended to residential neighborhoods, industrial parks, shopping centers, and office complexes.

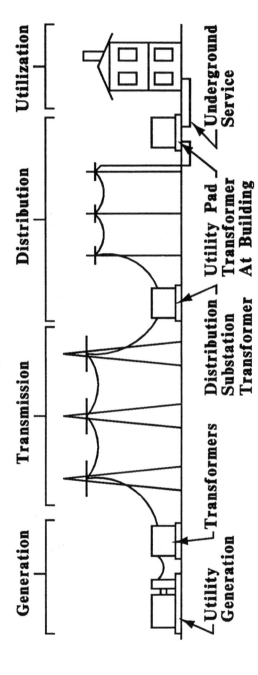

Figure 1.1 The four areas of power system activity (generation, transmission, distribution, and utilization).

2

At these locations, utility transformers are provided to once again step the voltage down to less than 1 kV for connection to the customer's electrical service. It is at this point that the utilization level power system commences. It is the utilization-level power system that distributes the power within buildings for lighting, air conditioning, electric heating, computers, and hundreds of other items of equipment. These building power systems are the last step in a complex power generation and delivery system. They are, in fact, the reason for the existence of utility generation, transmission, and distribution systems.

Consider for a moment the crucial role that utilization level power systems play. Building power systems provide the lighting, air conditioning, computer, and process system power for our work environments. These systems provide the power within a wide variety of other buildings in which we play, shop, worship, and live. They play crucial roles in hospitals, hotels, airports, schools, and colleges. Each type of building has unique electrical requirements—a personality. Building power systems are also the point at which people are in closest proximity to the potential dangers associated with electricity.

Building power systems have grown in complexity and size since the earliest days, when building electrical systems existed primarily to serve incandescent lighting systems pioneered by Edison in the 1880s. By the turn of the century, electrical power was being used in many applications, ranging from pumps to fans to machine tools—in fact, almost any application where power had previously been supplied by other means. As the applications of electric power grew, so did the complexity of building electrical systems. It soon became apparent that only careful engineering would optimize the efficiency and safety of building electrical systems.

Therefore, this book introduces the major design and analysis aspects associated with the building, or utilization level, power system.

During the first half of this century, young electrical engineers were limited to a choice of careers in two areas, power and electronics. Those who opted for careers in power often found themselves in either the utility industry or the building construction industry. Those involved in building construction played important roles in the consulting engineering community, and from their ranks came the principals of our leading professional design firms. Following World War II, electrical engineering education opened many more career paths, as solid state devices appeared and later as digital systems evolved in both power and capability. Today, young engineers have dozens of career paths.

Sadly, both the utility industry and those involved in utilization-level power systems often find it difficult to attract the engineering talent to deal with the sophistication of today's power systems. To illustrate: recently I returned to Atlanta on a late-night flight from a tiring, one-day inspection trip to a construction project. The passenger in the next seat asked about my work. I briefly explained about the design of building electrical systems. Coincidentally, the other passenger was also an electrical engineer with a consumer electronics company. After listening to my description, he said that he thought building design was "shake and bake" without real engineering challenge: "Just look it up in a table and draw a sketch."

For those with similar misconceptions, consider the following. Engineers in automotive, aircraft, and consumer electronics develop their products during months or years of extensive development and design. Once designed, the product is duplicated hundreds or thousands of times. Nevertheless, you only have to read the newspaper to see that this process does not always produce the desired results. Contrast this with the results of a building design project. Every building is unique and must be designed to meet the requirements of the owner as well as to meet local and national construction and energy-code requirements, using hundreds of specialized products. The design must not only meet these functional requirements, but must also result in the desired architectural appearance as well as meet construction-cost constraints. And the final product is expected to have a useful life of more than a generation! This is somewhat like asking Detroit to design a car, build a prototype, and expect it to perform satisfactorily for years without refinement.

The reader should realize by the end of this book that building design presents innumerable engineering challenges and that each design project presents an opportunity to be creative, which provides an enormous sense of professional pride and satisfaction. After all, buildings provide the environment for so many of the important activities that contribute to the quality of life. I will never forget the pride I felt, many years ago, when I visited my first design project. Almost thirty years and several thousand projects later, I feel exactly that same thrill. No, my fellow passenger on that late night flight, our profession is not "shake and bake." It's alive and well—full of challenges and opportunities for those with the technical education, creativity, and determination to succeed in a very demanding environment.

In this chapter, we will discuss the building design process, meet its major participants, and see how design projects are organized. We will also learn something of the role of professional registration and of several of the major technical organizations that keep our industry on the leading edge of technology. Finally, in the last section, we will briefly outline the organization of the material covered in the remaining chapters.

DESIGN AND CONSTRUCTION

The building design and construction process involves numerous participants, each filling a part. It is therefore important to understand these roles and how the entire group interacts as the project moves from concept through construction.

The Owner and Building Tenant

It may be surprising that the building owner and tenant are not necessarily the same person or organization. In fact, many buildings are constructed by an owner or developer for lease to a specific tenant. Examples include shopping centers constructed by developers for lease to a tenant, such as K-Mart, Kroger, or Belk's. In many cases, such building tenants have well-developed ideas about what they require in the way of building space and even construction details. These requirements are often made a part of their lease with the owner as criteria drawings for the architectural/engineering team in developing final building plans. Similarly, industrial parks are developed and buildings are developed for lease to companies for manufacturing or warehouse operations. In either case, the owner or developer works with the prospective building tenant to develop the space requirements of the facility. At this stage the owner may approach a lending institution for initial financial arrangements. Often the terms of such loans are for periods of 10 to 20 years.

In other cases, the owner might also be the building tenant. For example, a company might have outgrown its present facility or wish to relocate from one part of the country, or world, to another. Such owners vary widely in understanding their building requirements. Some experienced companies may have developed quite refined building requirements, while others require significant design-team support in determining their needs.

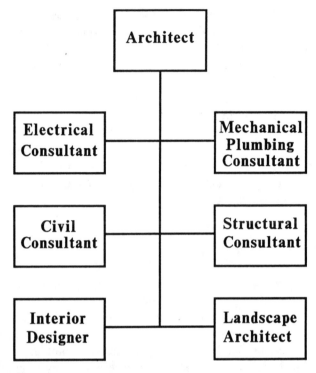

Figure 1.2 Typical design team makeup and organization.

The Architect/Engineer Team

The next participants are the members of the actual design team, the architects and engineers. The architect/engineer, or A/E, team can be members of a single organization with all required design disciplines. In many cases, however, the team consists of an architectural firm working with consulting engineering firms representing the major design disciplines. Such an organization is shown in Figure 1.2.

The major team members are as follows:

1. The *Architect*: The architectural firm generally acts as the overall design coordinator; from a contractual standpoint it is employed directly by the building owner or developer. The architect is responsible for building-space planning, architectural design, and

coordination of the work of the consulting engineers. The architect often participates with the owner in preparing early design-development plans used in the financial planning process. Many architectural firms specialize in specific areas of construction, such as office buildings, medical facilities, religious buildings, or commercial shopping centers. These firms develop a high degree of competence in their specialty. The architect often shoulders much of the inspection responsibility during the construction process as well as acts as the owner's agent in approving progress payments to the contractor. The architect's design activities must also comply with a variety of national, state, and local construction codes.

2. The *Electrical Engineer*: The electrical consultant is responsible for planning the building's electrical system, which consists of power distribution, lighting, telephone/data, emergency power, and lighting and fire alarm systems. These design activities are governed by national codes, such as those prepared by the National Fire Protection Association (NFPA) as well as state and local codes. The mechanical engineer, architect, and electrical engineer must closely coordinate a number of items such as the location of ceiling-mounted air diffusers and lighting fixtures in order to avoid conflict and also to present the best architectural appearance. The electrical engineer will also participate in the inspection process during the construction phase.

3. The *Mechanical/Plumbing Engineer:* The mechanical and plumbing engineer plans the building's heating, ventilating, air conditioning, and plumbing systems. Much of the mechanical and plumbing equipment requires significant physical space and can be both bulky and heavy. As a result, close coordination is necessary between the mechanical engineer and the architect and structural engineer. During the process of selecting the proper capacity for heating and cooling equipment, the mechanical engineer must work closely with the architect to determine the level of building insulation, exterior glass area, and other factors affecting this equipment. The mechanical engineer must also supply the electrical engineer with detailed electrical data on the mechanical equipment, which requires a significant part of the building's electrical load. The work of the mechanical engineer is guided by stringent national, state, and local code requirements.

4. The *Civil Engineer*: The civil engineer plays a role in the design of the building site itself, including features such as grading requirements and site drainage systems. The civil engineer works very closely with the architect in planning the overall site layout, including building placement and location of drainage facilities. The civil engineer must also see that site drainage does not endanger nearby property. Many sites require significant grading to level the necessary building area, while in other cases the area must be filled to bring it up to the desired building level. Some sites require grading of some areas and filling of others. The planning for optimal grading and filling to achieve the minimum transportation of earth to or from the site is quite a challenge.

5. The *Structural Engineer*: The structural engineer designs the all-important structural system supporting the building itself. This task varies in complexity from simple one-story structures to multi-story office buildings requiring sophisticated structural systems. In many areas of the country, consideration must be given to potential earthquake damage. In coastal areas, the dangers of high winds caused by hurricanes impose special structural-design requirements. In these areas, tall buildings are often designed to permit deflection at the upper levels of as much as 18 inches or more during the 150 knot "breeze" of a hurricane. Since the architect conceives the basic building layout and interior details, there must be close coordination with the structural engineer to ensure adequate space for columns, cross-bracing, and other structural members. The structural engineer also designs the support foundation necessary to carry the weight of the building and its contents. In various parts of the country, this can be challenging, due to widely varying soil conditions. Again, the structural design must meet a variety of national, state, and local building-code requirements.

6. The *Interior Designer*: The interior designer plans much of the interior finishing details of the building and often selects carpet, drapery, and room-surface colors, finishes, and even furniture. The interior designer is often a member of the architect's staff who specializes in this area of design. Obviously, close coordination between the architect, interior designer, and building owner/tenant is very important. It is also important for the electrical engineer and interior designer to coordinate any special lighting requirements.

7. The *Landscape Architect:* The landscape architect plans the building's exterior landscaping, including placement (or removal) of trees, shrubs, and flowers. This work can enhance the architectural appearance of the site and can be used to emphasize site-design features. Anyone who has visited a facility that has benefitted from the design activities of a landscape architect will agree on the impact such site-landscape planning can achieve. Many corporate office complexes achieve their "campus" atmosphere by creative landscape planning. Obviously, the landscape architect must work closely with the building architect to achieve the desired results. In some cases the landscape architect is a member of the architectural design firm, but in many cases he is a separate consultant.

From the above discussion, the reader probably suspects that the key to the successful design of any building is the very close coordination between the various members of the design team. This coordination involves careful communication between architects and engineers on hundreds or even thousands of details during the design phase.

The Construction Team

Once the building is designed, the project moves into the preconstruction phase. During this phase, a general contractor is selected. The *general contractor* acts as overall construction coordinator and normally performs much of the actual work with his forces. Certain areas of construction, such as electrical and mechanical systems, are normally subcontracted to specialist contractors. Depending on the nature of the project, there may also be subcontractors for specialties such as steel erection, grading, carpentry, drywall construction, and roofing.

The entire construction effort is coordinated by the general contractor, who is responsible for the orderly progression of construction from start to finish—a major undertaking. For example, grading must be completed before foundation work can proceed. Steel must arrive on site and must be installed, walls must be erected, and the building made weathertight before interior construction can begin. A major construction project may require 18 to 24 months and must be planned and coordinated like a military operation if the building is to be kept on schedule. Before construction begins, elaborate construction scheduling charts are developed so that each member of the construction team understands the critical dates at which various stages of his work must be completed to avoid delay for other members of the team. The subcontractors must then coordinate the timely arrival of the hundreds of system components, such as switchgear, roof air-conditioning units, door hardware, glass, steel, carpet, and paint. Many of these items must be ordered weeks or months earlier with specific shipping dates identified. Since site storage space is often very limited, most equipment and material must arrive just in time to be off-loaded from trucks and placed in final locations in the building. On any construction site, the general contractor is a very busy person, orchestrating these varied activities and dealing with the inevitable errors and delays.

The Vendors

The vendors are the hundreds of individual companies who supply products used in the construction of the building, including lighting fixtures, lamps and panelboards, doors, water heaters, air-conditioning units, ceiling tile, nails, and pipe. These products are detailed as part of the architectural and engineering plans and specifications. Because most buildings are unique, the architect and engineering team often contacts the technical representatives of the major vendors and requests them to supply technical and application information. Once the plans are completed and a contractor selected, the vendors are asked to prepare shop drawings of their specific products for approval by the A/E team. After the products are approved, they are ordered by the general contractor or subcontractors as a part of their construction contracts. As you can imagine, the vendors maintain close liaison with both the A/E team members who specify their products and with the contractors and subcontractors who purchase the products.

As part of this liaison process vendors normally supply architects, engineers, and contractors with catalogs on their products and periodically visit architectural and engineering offices to update these and apprise the design team of new products. Many consultants maintain libraries of several hundred vendor catalogs.

Building Inspection and Code Authorities

It is the responsibility of local building-inspection authorities to review building plans before construction, issue construction permits, and periodically inspect the building during the construction process. These authorities, usually a part of the local city or town government, review the plans for code compliance, although such compliance is ultimately the responsibility of the design and construction team. The inspection authorities have no responsibility for whether the building will have adequate space or even meet the owner's requirements. They are tasked only with assuring the building meets applicable building codes and is safe for public use.

Building inspection departments maintain copies of approved construction plans and schedule site inspections at appropriate intervals. Typically, there will be separate inspectors for electrical and mechanical systems, while a single building inspector might review structural and other building construction details.

Normally, the state fire marshal's office will also review building plans for compliance with the relevant fire safety codes, such as the presence of adequate building egress paths, emergency lighting, and fire-alarm systems.

At the conclusion of the construction process, these authorities make a final building inspection, resolve any outstanding questions, and issue a certificate of occupancy for the building, making it officially available for use by the owner or tenant. Again, these inspections do not indicate that the building will perform any of its desired functions, only that it is safe for public occupancy. The architect and engineers also inspect for code compliance as well as verify that the building will meet its design objectives as set forth in their plans and specifications.

Typical Project Sequence

A design and construction sequence might proceed as follows. The project owner or developer initiates the project in response to the need for the facility. Preliminary financing arrangements may be made with a lender for construction funds or permanent financing. The owner might then contact an architect who has an established reputation in the type of construction contemplated. The architect will assemble the engineering team based on its experience and ability. The architect will then prepare a design contract covering the scope of architectural and engineering services as well as the total fee requirements and method of payment. After contract approval, the architect will enter into individual contracts with the various consultants. Once the paperwork is completed, the architect may engage in preliminary design activities aimed at accurately determining the owner's requirements and expressing these in terms of preliminary floor plans and building elevations. During this phase the owner and architect work very closely.

It is also common to solicit the early design input of each of the consultants on matters dealing with that consultant's activities. For example, the electrical engineer might be consulted regarding the required space for switchboards and panelboards. The mechanical engineer might be consulted regarding the space requirements for air conditioning and other mechanical system components. The result of this preliminary design phase is an initial building floor plan that will meet the owner's needs and also provide the space for the various building systems.

This preliminary design process can require several weeks or several months, but the time is well spent because at its conclusion the project will be ready to enter the working-drawing or design phase. During the early stage of the working drawing phase, the architect refines his architectural plans and finally issues architectural plans that are used as the basis of final design by the various engineers. Each engineer prepares plans based on the architect's work, showing only the specific systems under the control of that engineer. The assembled plans are then coordinated by the architect to eliminate conflicts. Each engineer might consult a number of individual vendors for technical data during the design phase. The engineers also transfer relevant data among themselves in order to coordinate items such as mechanical-system equipment locations and power requirements.

During the working-drawing phase, several coordination meetings are normally held to discuss problems or resolve conflicts. This phase of the process can also require several weeks or several months, depending on the scope of the project. During this phase, each discipline will also prepare a set of written specifications detailing materials and workmanship requirements not included on the drawings.

The completed package of plans and specifications becomes the design documents for the project.

After the completion of the plans and specifications, the owner may distribute the design documents to several general contractors for preparation of competitive bids. The general contractors then issue sets of plans and specifications to selected subcontractors for preparation of their respective bids. The subcontractors assemble their bids based on the plans and by soliciting bids from the vendors whose products have been chosen. On a specified day, the sealed bids from the general contractors are publicly opened. The construction contract is normally then awarded to the lowest qualified bidder. The key word here is *qualified*. Ideally, only competent contractors are allowed to submit bids, but mistakes will occur and even competent companies can err. It is therefore in everyone's interest to see that a construction agreement based on obvious errors be avoided. Once the owner is satisfied that the contractor's proposal is acceptable, a contract is prepared and signed. The terms of the contract stipulate key items such as total cost, construction-time period, and any penalty associated with late completion of the work.

At this point, the construction phase begins. The plans are reviewed by the building inspection department and building permits issued. The general contractor contracts with the subcontractors. The vendors prepare shop drawings for submittal to and approval by the engineers and architect.

Once construction begins, the architect and engineer visit the site in accordance with the requirements of their contract with the owner to assure that the material and workmanship meet the requirements of the plans and specifications as well as the applicable codes. The local building authorities also visit the site to assure that code requirements are met.

At the conclusion of construction, the architect and engineer visit the site and prepare a final inspection report covering any discrepancies. Once these items have been rectified by the contractor, the local authorities make their final inspection and the contractor addresses any final items noted during their visit. At this point, a certificate of occupancy is issued and the owner receives a set of keys to the building along with a sizeable final invoice from the general contractor.

This process, of course, almost never goes smoothly. Problems inevitably seem to occur. Errors may be made, material may not arrive on time, labor problems can occur, code conflicts might arise, and weather can intervene to cause delays. The owner, architect, engineers, and contractors must attempt to resolve these problems and continue forward to meet the necessary construction date.

Amazingly enough, this process is normally carried out successfully. The few cases where irreconcilable problems occur are not the subject of this book and are more appropriate for a legal text!

Other Construction Methods

The typical team organization discussed above reflects the traditional design and construction process. There are, however, several alternative approaches.

For example, the owner might determine the need for the facility and contact a general contractor directly. The owner and general contractor refine the scope of the project and enter into a design build contract. This contract requires that the contractor retain the architect and engineers, who then perform the design services for him rather than the owner. Theoretically, this process offers several advantages. First, the owner pays only one person, the general contractor. Secondly, the general contractor acts as the owner's representative in dealing with the design team. Unfortunately, under this arrangement the design team cannot always offer their advice as effectively to the owner as when they report directly to him or her.

Another common approach involves a negotiated contract between the owner and general contractor. In this approach, the owner retains the A/E team, identifies a preferred general contractor, and at the conclusion of the design process negotiates a contract with this contractor. One value to this approach is the opportunity to take advantage of the experience of the contractor during the design phase when alternative construction suggestions can be evaluated to reduce cost and increase quality. At the conclusion of the design phase, the general contractor already possesses sufficient data to assemble his negotiated cost amount. This process can be effective provided the A/E team and general contractor are able to coordinate their knowledge and experience.

PROFESSIONAL REGISTRATION

Architects and engineers who plan buildings must be registered. Professional registration dates back many years and results from concern for public safety. Around the turn of the century, our country was rapidly developing new industries requiring larger buildings. At the same time, larger bridges and tunnels were being built. Inevitably, buildings failed, bridges collapsed. In response to public concern, a movement developed to require registration of engineers in order to limit practice to those with demonstrated competency. The first state to enact registration requirements was Wyoming, which passed its first such law in 1907. Today, all states have laws governing the practice of engineering, architecture, and surveying. Those registration laws vary from state to state, but the common goal is safety.

The Evolution of Registration Laws

During the early days of professional registration, each state enacted laws and established licensing procedures and qualification requirements. After several years, however, it became clear that intra state registration problems were rapidly developing. In 1920, seven of the thirteen states with registration laws formed an organization to coordinate information among state boards of registration. This organization was the forerunner of today's National Council of Examiners for Engineers and Surveyors (NCEES). Today, NCEES represents all states and U.S. territorial jurisdictions. One of the primary functions of NCEES is the preparation of the engineering examinations.

Professional registration is a multi-step process beginning at the college level:

Step 1. Graduation from a four-year engineering program accredited by the Accreditation Board for Engineering Technology, Inc. (ABET). Today, 27 major engineering organizations make up ABET. ABET periodically sends inspection teams to accredited engineering schools to evaluate their programs. Graduates of non-ABET programs must be approved by the state board.

Step 2. Successful passing of the Fundamentals of Engineering (FE) examination. This examination is offered twice a year, in April and October. College seniors are generally permitted to take the exam as well as those who have already graduated. The FE exam is frequently administered on campus at most engineering universities, making it far more convenient for busy students. The multiple-choice exam takes 8 hours; there is a morning and afternoon session. The morning portion contains 140 questions and the afternoon 70 questions. The exam tests basic knowledge in several areas of undergraduate engineering education. Among these are science, general chemistry, and physics. Life science, earth science, advanced chemistry, and physics might also be included. Other topics covered are differential and integral calculus and differential equations. Probability and statistics, linear algebra, numerical analysis, and advanced calculus might also be included. Finally, the test includes questions on engineering science involving mechanics, thermodynamics, electrical and electronic circuits, material science, and transport phenomena. The FE examination is an open-book test. All questions are worth the same number of points. Because no points are deducted for incorrect answers, all questions should be answered. NCEES reports the grades to the individual state boards on a scale of 100. The state board then issues a pass or fail grade, or, in some cases, the numerical score. Those successfully completing the FE examination are awarded the Engineering-in-Training (EIT) certificate.

Step 3. Professional Experience: Following the successful completion of Steps 1 and 2, the engineer must work satisfactorily for four years. Some states also require the supervision of a professional engineer. Generally, the engineer must progress, demonstrating increasing responsibility.

Step 4. The Principles and Practice of Engineering (PE) Examination: After completion of the required professional experience, the next step is the Principles and Practice of Engineering Examination in a specialty. The exam is 8 hours and contains both multiple-choice and essay questions. Each of the 8 examination items counts as ten points for a total of 80. A passing score requires at least 48 points. NCEES prepares and scores these examinations and returns the score to the state board, which issues the final grades.

There are presently 14 different areas for professional registration: chemical, civil (sanitary and structural), electrical, mechanical, aeronautical/aerospace, agriculture, control systems, fire protection, industrial, manufacturing, metallurgical, mining/mineral, nuclear, and petroleum.

After completing these four steps, the engineer receives registration as a professional engineer.

Why You Should Take the FE Exam While in College

If you are still in college, there are several reasons why you should take the FE exam before graduation. In the stress of completing your degree, these points may elude you:

First, even if you don't plan to enter an area of the profession presently requiring registration, you should take the FE exam because registration laws may be extended to include your area.

Second, you are probably as well prepared today as you will ever be to take the examination. Remember that the examination covers several areas of your undergraduate program, and you may fail to retain some of this material months or years after graduation.   Stop and think for a moment. Do you remember all of your freshman course-work even now?

Third, the examinations are probably offered on campus and the application process is relatively straight forward.

Fourth, an EIT certificate in your resume indicates that you take your career seriously. You will already have the entry step to the profession behind you; this point is not lost on most employers.

PROFESSIONAL DEVELOPMENT

Over the years, I have had the opportunity to interview, and hire many engineers as they complete their undergraduate degree and enter the professional world. Far too many fall into one of several tempting traps. First, immediately after graduation, a well-deserved sense of accomplishment predominates in the thinking of many young engineers. I have heard many say, "I'll never go near a book again" or "Now I'll have a chance to spend time with my family or significant other." The danger here is in letting down your guard and forgetting that engineering education never stops. Your technological half-life is about four years! You simply must continue your professional development or you will wake up one day and find yourself obsolete and unmarketable.

Second, many young engineers become so immersed in their office activities that they forget the importance of the professional organizations they supported in school. You may well be keeping abreast of activities in your area, but you may be ignorant of advances in other areas of electrical engineering.

How do you keep this from happening? Continue your involvement in organizations such as IEEE. If professional registration is in your plans, join the National Society of Professional Engineers (NSPE). If you are involved in the building industry, join the Illuminating Engineering Society of North America (IESNA). As your career develops and you become eligible, seek a certificate from the National Council of Examiners for

Engineering and Surveying (NCEES). If you enter private practice, join the American Consulting Engineers' Council (ACEC). These are some of the professional organizations that will help you maintain the professional and technological edge you have worked so hard to attain. Don't become obsolete by the time you are thirty.

Below are the names and addresses of some of these organizations:

- Institute of Electrical and Electronic Engineers
  445 Hoes Lane
  Piscataway, New Jersey 08855-1331

- National Society of Professional Engineers
  1420 King Street
  Alexandria, Virginia 22314
- American Consulting Engineers Council
  1015 Fifteenth Street, NW
  Washington, D.C. 20005

- Illuminating Engineering Society of North America
  345 East 47th Street
  New York, New York 10017

- National Council of Examiners for Engineering and Surveying
  P. O. Box 1686
  Clemson, South Carolina 29633-1686

HOW THIS BOOK IS ORGANIZED

This book is in two parts. The first part, Chapters 2 through 8, deals with building power systems. The second part, Chapters 9 through 12, deals with illumination engineering. A brief discussion of the material presented in each chapter follows.

Chapter 2: Fundamental Power Concepts

If it has been some time since you finished school, or if you need a refresher course on some basic power concepts, the first part of Chapter 2 should be reviewed carefully. Here the fundamental power concepts of time-varying voltages, currents and power are presented, along with *rms* values of current and voltage. Power is composed of two parts. One part has an average value of zero and represents power that simply flows back and forth in the system doing no useful work. The second part, however, has a positive average value and represents the useful power delivered by the system. Formulas for both single- and three-phase systems are developed. In the final sections, common building voltages are discussed along with their standard tolerance ranges. We will see that standards have been developed that define the permissible voltage range for utility systems as well as building power systems. These standards guide utilities in the voltage ranges supplied to customers, engineers who design building systems so that all equipment will receive appropriate operating voltage, and finally, equipment manufacturers, who must design their equipment for operation within the proper voltage range.

Chapter 3: System Components

In Chapter 3 we will introduce the basic components we will use in designing building electrical systems. We begin with conductors and raceways—the arteries and veins of the power system. We will review the *dc* and *ac* characteristics of conductors as well as the effects of temperature on conductor performance. We will see how current-carrying capabilities were developed and how various conductor insulation types evolved for specific applications. The raceway systems that have been developed to provide physical protection for conductors are then presented. We will also discuss the important characteristics of circuit breakers and fuses. These are circuit protective devices used to protect system components from overcurrent and short-circuit damage. We next discuss step-down transformers beginning with the ideal transformer model and then extending this to a more realistic model. We will discuss panelboards and switchboards, the electrical equipment that acts as the nerve center of the distribution system and houses the circuit breakers and fuses. Finally, we will review the performance characteristics of the single- and three-phase induction motor. By use of an induction motor model and performance equations, we will see that this versatile machine is capable of operation over a range of load conditions.

With this background, we will then learn how these components are integrated into an operating building power system.

Chapter 4: Introduction to System Design

In Chapter 4 we will begin our discussion of system design, including system planning, equipment placement, utility coordination, load studies, branch circuit and feeder design, panel schedules, and voltage-drop analysis. In this chapter, we will see how the components first presented in Chapter 3 are integrated into an operating system.

Chapter 5: System Design Example

Chapter 5 consists of a design example, involving a small industrial building. The project criteria is presented in terms of architectural plans and owner criteria, as well as mechanical and plumbing data supplied by the mechanical engineer. From this criteria, the project is developed step-by-step using the methodology discussed in Chapter 4.

Chapter 6: Fault Current Analysis

In Chapter 6, we will discuss the basic concepts of short-circuit analysis. We will see that under short-circuit conditions the normal power system load impedance is replaced by a short-circuit impedance of essentially zero ohms. This results in a large increase in current flow, often several thousand times the normal value. This tremendous current flow can damage system components unless the system is properly designed and constructed. Short-circuit analysis thus provides us with the necessary tools to make informed engineering decisions on how to best protect the building power system.

Chapter 7: Utility Systems

The operation of the utility system is crucial to the planning of building electrical systems. The historical forces that have molded the utility industry are discussed in Chapter 7. We will see that utilities have several operating characteristics making them susceptible to events such as unexpected fuel-supply dislocations (e.g., those following the Arab-Israeli War of 1973). Following this war and the Arab Oil Embargo, the utility industry was forced to its knees and many utilities almost collapsed.

During this same time, increased environmental concerns made it difficult for utilities to plan and construct new facilities. These events helped shape today's utility environment and should be understood by all engineers.

Chapter 8: Power Quality

In recent years the susceptibility of building systems to even small power interruptions has increased. During this same time, the use of power electronic devices has also increased dramatically. These devices draw nonlinear currents which can cause a variety of building system problems, such as overheating of neutral conductors and transformer damage. We have also found that many commonly used measurement instruments yield inaccurate data when measuring nonlinear voltages and currents. These power-quality concerns affect the design of all new building systems and also pose significant design challenges for engineers remodeling older buildings. In Chapter 8 we will discuss the nature and source of some of the common power disturbances. We will also learn how distorted current and voltage waveforms are analyzed and how we predict their effects on building power-system components. Examples of various power-system disturbance and waveform distortions are presented in the form of graphical data recorded by disturbance recorders and power analyzers.

Chapter 9: Introduction to Illumination Engineering

In Chapter 9 we will begin our discussion of illumination engineering by reviewing light as radiant energy. We will also see how the eye converts this radiant energy into visual images. We will learn how we adapt to daylight and darkness, how we perceive colors, and how our vision changes with age. This chapter also introduces the special lighting terms used in this and succeeding chapters. In the concluding section, we will learn how we evaluate the level of illuminance required for various visual tasks.

Chapter 10: Light Sources

Chapter 10 introduces light sources. Our discussion begins with the incandescent lamp, the first commercially viable electrical light source. We will see how the incandescent source has evolved and how development promises to maintain the incandescent lamp as a major light source for many years. We will also learn about the fluorescent lamp, a discharge light source that produces light by passing current through a mercury gas. The collision of this current with mercury atoms releases ultra-violet energy that interacts with the phosphor coating inside the fluorescent lamp, thereby creating visible light. We will also study the high-intensity discharge (HID) light sources. These are the mercury-vapor, metal-halide, and high-pressure sodium lamps. We will learn how the light output of the various light sources varies over the life of the lamp. Finally, we will see that fluorescent and HID lamps require special equipment called ballasts to provide the necessary starting voltage and to limit the flow of current through the lamp itself.

Chapter 11: Luminaires and Photometrics

In Chapter 11 we will discuss lighting fixtures and see how to evaluate their performance based on photometric test data. We will see that luminaires are classified by their light source and method of mounting, as well as their light-distribution pattern. Finally, we will learn about luminaires designed for exterior lighting. We will see how they are classified in accordance with their distribution patterns and how their photometric data is presented.

Chapter 12: Lighting Design

In Chapter 12 we utilize the concepts presented in the previous chapters. We will learn how a lighting system is designed based on an understanding of light, human response to light, proper luminaire and lamp selection, and system design techniques. We will see that the design of a lighting system also involves an understanding of the factors that, over time, reduce the initial level of illumination produced by the system. We will see that this light loss can be divided into recoverable and unrecoverable light loss factors. Recoverable light losses can be periodically corrected by simple maintenance items such as relamping and luminaire and room-surface cleaning. Nonrecoverable factors, on the other hand, are a permanent part of the lighting installation. Such unrecoverable factors include excessively high or low ambient temperatures as well as depreciation of the luminaire's reflective surfaces.

In the concluding sections of this last chapter, we will see how exterior area lighting systems are designed based on luminaire distribution characteristics and point-by-point calculations. We will see that several key benchmarks are used to evaluate the performance of exterior lighting systems. These include the average and minimum level of illuminance as well as the ratios of the maximum to minimum and the average to minimum levels of illuminance.

2

Fundamental Power Concepts

In this chapter we lay groundwork for the material that follows. We first review the basic concepts of single- and three-phase electrical systems. This will include both time-varying and *rms* values of voltages and current as well as instantaneous and average power. In later sections we discuss the various standard single- and three-phase voltage systems with their operating tolerance characteristics.

REVIEW OF BASIC POWER CONCEPTS

DC Systems

For *dc* systems the current flow is given by the familiar

$$i \;=\; \frac{e}{R} \quad A \tag{2.1}$$

Where *e* represents the *dc* voltage and *R* the resistance in ohms. The power dissipated in the resistance, *R*, is given by:

$$p \;=\; e\,i \;=\; \frac{e^2}{R} \quad W. \tag{2.2}$$

For example, a *dc* circuit voltage of 12 volts applied to a 6 ohm resistor will produce 24 watts.

Since *e* is constant, i.e., it does not vary with time, *p* will also be constant.

Single-Phase *AC* Systems

In *ac* systems, the voltage varies with time as follows:

$$v = v_{max} \sin(\omega t) \quad V \tag{2.3}$$

where v_{max} is the peak value of the voltage and ω is the angular frequency in radians per second. *v* is sinusoidal in nature, and its period, *T*, is inversely proportional to the system frequency, *f*.

In the United States, power systems are operated at a frequency of 60 cycles/sec or 60 Hz. So:

$$T = \frac{1}{f} = \frac{1}{60} = 0.0166 \quad Sec.$$

and

$$\omega = 2\pi f = 2\pi \cdot 60 = 377 \quad Radians/Sec.$$

If we apply this sinusoidally time-varying voltage to a resistance, *R*, it gives rise to a sinusoidal current, *i*.

$$i = \frac{v_{max}}{R} \sin(\omega t) \quad A \tag{2.4}$$

Example 2.1: If a sinusoidal voltage with frequency of 60 Hz and v_{max} = 169.7 volts is applied to a resistance, $R = 3$ ohms, the current will be

$$i = \frac{v_{max}}{R} \sin(\omega t) = 56.57 \cdot \sin(377 \cdot t) \quad A$$

and Figure 2.1 shows a graph of *v* and *i* for one complete cycle of *v*.

Note that *v* and *i* are in phase, that is, both reach positive or negative peaks at the same time.

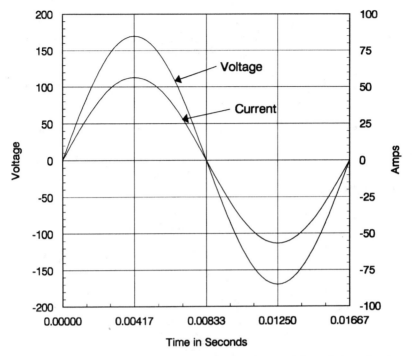

Figure 2.1 Instantaneous voltage and current in a purely resistive circuit.

The power dissipated in R is now a function of time and will be given by

$$p = vi = v_{max} \sin(\omega t) \frac{v_{max}}{R} \sin(\omega t)$$

$$= \frac{v_{max}^2}{R} \sin^2(\omega t) \quad W.$$

This can be simplified by use of the trigonometric identity

$$\sin^2(\alpha) = \frac{1}{2}[1 - \cos(2\alpha)]$$

yielding

$$p = \frac{v_{max}^2}{2R}[1 - \cos(2\omega t)] \quad W.$$

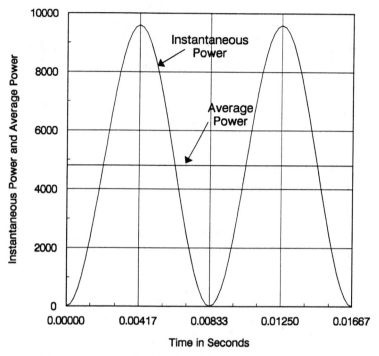

Figure 2.2 Instantaneous and average power in a purely resistive circuit.

Equation 2.5 gives us important insight into the nature of power. Let's apply this equation to the circuit discussed in Example 2.1. Figure 2.2 shows the resulting graph of the average and instantaneous power for one full cycle.

Notice that the power dissipated in R oscillates around a value of $v_{max}^2/2R$ at double the radian frequency of the voltage. Also note that this power is always positive. The average power dissipated is then

$$P_{avg} = \frac{v_{max}^2}{2R} \quad W. \tag{2.6}$$

Single-Phase Systems With Impedance

In most *ac* systems the load is actually composed of both resistive, R, and reactive, X, components. The resulting impedance, Z, is expressed as a complex number of the form

$$Z = R + jX \quad ohms.$$

Note that the units of both R and X are ohms. The current flow is now given by

$$i = \frac{v_{max}}{|Z|} \sin(\omega t - \varphi) \quad A \tag{2.7}$$

where

$$|Z| = \sqrt{R^2 + X^2} . \tag{2.8}$$

and φ represents the angle by which the current either *lags* or *leads* the voltage. If φ is positive, the current is said to *lag* the voltage. If φ is negative, the current *leads* the voltage.

The sign of φ is determined by the nature of X. If X is inductive in nature, φ is positive, and if X is capacitive in nature, φ is negative.

In many building systems, most loads are inductive in nature, and hence i generally lags v. The angle φ is given by

$$\varphi = \arctan\left[\frac{X}{R}\right] . \tag{2.9}$$

Let's look at an example.

Example 2.2: Given $v_{max} = 169.7$ V, $f = 60$ Hz, $Z = 4+j3$ Ω

$$v = 169.7 \cdot \sin(377 \cdot t) \quad V$$
$$\varphi = \arctan\left[\frac{3}{4}\right] = 36.9°$$
$$i = 33.94 \cdot \sin(377 \cdot t - 36.9°) \quad A.$$

A graph of v and i is shown in Figure 2.3. Notice that i lags v by 36.9 degrees.

Let us now examine the nature of power in loads involving both resistive and reactive components. Recalling the expression for power and letting

$$i_{max} = \frac{v_{max}}{|Z|} \quad A.$$

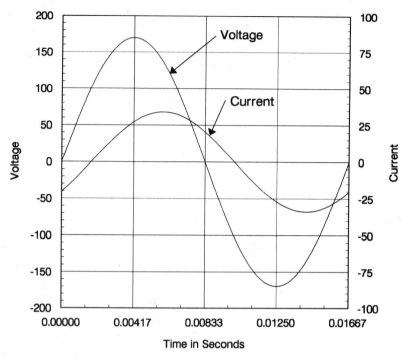

Figure 2.3 Instantaneous voltage and current in a circuit with both resistive and inductive reactance.

The expression for power will be given by

$$p = vi = v_{max} \sin(\omega t) i_{max} \sin(\omega t - \varphi) \quad W. \quad (2.10)$$

Using the trigonometric substitution

$$\sin(\alpha) \sin(\beta) = \frac{1}{2} [\cos(\alpha - \beta) - \cos(\alpha + \beta)]$$

yields

$$p = \frac{v_{max} i_{max}}{2} [\cos(\omega t - \omega t + \varphi) - \cos(\omega t + \omega t - \varphi)]$$

$$= \frac{v_{max} i_{max}}{2} [(\cos(\varphi) - \cos(2\omega t - \varphi)] \quad W. \quad (2.11)$$

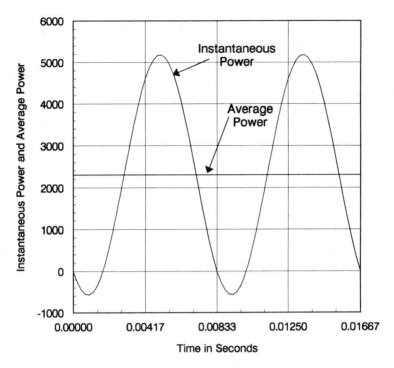

Figure 2.4 Instantaneous and average power in a circuit with both resistive and inductive reactance.

Equation 2.11 reveals that power now oscillates around a new average value and also that it is occasionally negative in value. Figure 2.4 shows a graph of the average and instantaneous power for the previous example.

Effective and *rms* Values of Current and Voltage

Let us return for a moment to the case of a purely resistive load and the expression for current given in Equation 2.4

$$i = \frac{v_{max}}{R} \sin(\omega t) \quad A. \tag{2.4}$$

As a practical matter this expression is a bit cumbersome for daily use in a design environment. We would like to reduce this expression to a single value, or effective value, to simplify calculations.

We would like for this effective value of current to produce exactly the same average heat in a resistance, R, as that produced by the time-varying current, i. We will call this value I_{eff}. Expressing this mathematically, we would like to find a value for I_{eff} such that

$$
\begin{aligned}
I_{eff}^2 \cdot R &= \frac{1}{T} \int_0^T i^2\, R\, dt \\[2ex]
&= \frac{1}{T} \int_0^T i_{max}^2 \sin^2(\omega t)\, R\, dt \qquad W.
\end{aligned}
$$

(2.12)

We can eliminate R since it appears on both sides of the equation and then simplify a bit further.

$$
I_{eff}^2 = \frac{i_{max}^2}{T} \int_0^T \sin^2(\omega t)\, dt.
$$

Using, the trigonometric substitution

$$
\int \sin^2(\alpha t) = \frac{1}{2} \cdot t - \frac{1}{4\alpha} \cdot \sin(2\alpha t)
$$

yields

$$
I_{eff} = \frac{i_{max}}{\sqrt{2}} \qquad A.
$$

So the effective value of i is related to its peak value by $\sqrt{2}$. Taking its name from the preceding mathematical development, the effective value of a sinusoidal voltage or current waveform is called its root mean square or simply *rms* value. We can now say that

$$
|V| = \frac{v_{max}}{\sqrt{2}} \quad V \quad \text{and} \quad |I| = \frac{i_{max}}{\sqrt{2}} \quad A.
$$

(2.13)

Real and Reactive Power in Single-Phase Systems

Returning now to the expression for power in circuits with both resistance and reactance

$$p = vi = v_{max} \sin(\omega t) \cdot i_{max} \sin(\omega t - \varphi)$$
$$= \frac{v_{max} \, i_{max}}{2} [(\cos(\varphi) - \cos(2\omega t - \varphi)] \qquad W. \qquad (2.11)$$

and realizing that $v_{max} = \sqrt{2} \, |V|$ and $i_{max} = \sqrt{2} \, |I|$, Equation 2.11 reduces to

$$P = |S| \cos \Theta \qquad\qquad |S| = |I| \cdot |V|$$

$$p = |V||I| [\cos(\varphi) - \cos(2\omega t - \varphi)] \qquad W \qquad (2.14)$$

where $|V|$ and $|I|$ represent the *rms* values of v and i.

We can simplify the second half of Equation 2.14 by the trigonometric substitution

$$\cos(\alpha - \beta) = \cos(\alpha) \cdot \cos(\beta) + \sin(\alpha) \cdot \sin(\beta)$$

With $\alpha = 2\omega t$ and $\beta = \varphi$, Equation 2.14 then reduces to

$$p = |V||I| [\cos(\varphi) - \cos(2\omega t) \cos(\varphi) - \sin(2\omega t) \sin(\varphi)]$$
$$= |V||I| \cos\varphi [1 - \cos(2\omega t)] - |V||I| [\sin(\varphi) \sin(2\omega t)] \quad W.$$
$$(2.15)$$

We see that the first part of Equation 2.15 oscillates around an average value of $|V| \cdot |I| \cdot \cos(\varphi)$ and never goes negative. On the other hand, the second half of Equation 2.15 oscillates around the zero axis and therefore has an average value of zero.

Simply put, $|V| \cdot |I| \cdot \cos(\varphi)$ represents the real power expended while $|V| \cdot |I| \cdot \sin(\varphi)$ represents power that travels back and forth in the system doing no useful work.

Complex Power

Based on the results developed in the last section, we will define complex power as follows:

$$S = P + jQ \quad VA \qquad (2.16)$$

$$= |V| \cdot |I|$$

where

$$P = |V||I| \cos(\varphi) \quad W$$

and

$$Q = |V||I| \sin(\varphi) \quad VAr.$$

This very important angle, φ, has been named the power-factor angle. The cosine of φ is employed to relate VA, or apparent power, to the W, or real power. The cosine of φ is referred to as the power factor, or PF. It is often expressed as a percentage and should properly be followed by either *leading* or *lagging*.

Finally, we should understand that the magnitude of typical building system loads is such that it is more convenient to express them in terms of thousands of watts, VA's or VAr's. For the remainder of the book, we will express these loads in terms of kW, kVA, and kVAr. It is not usually necessary to express amperes in terms of kA.

Three-Phase Systems

Almost all commercial and industrial power systems are now three-phase in nature. We will now review some of the more important characteristics of three-phase systems.

Three-phase systems can be viewed as having three separate voltage sources, equal in magnitude, but displaced by 120 degrees. The phases are referred to as a, b, and c. Expressions for these voltages are given by

$$
\begin{aligned}
v_a &= v_{max} \sin(\omega t) \quad V \\
v_b &= v_{max} \sin\left[\omega t - \frac{2\pi}{3}\right] \quad V \\
v_c &= v_{max} \sin\left[\omega t + \frac{2\pi}{3}\right] \quad V.
\end{aligned}
\qquad (2.17)
$$

Figure 2.5 shows a graph of v_a, v_b, and v_c for a typical three-phase system.

Figure 2.6 shows these same voltages in phasor diagram form. The *a* phase is normally selected as reference phase and is oriented along the positive x axis. Since

$$|V_a| = |V_b| = |V_c| = |V|$$

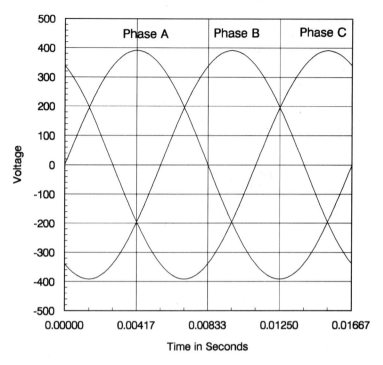

Figure 2.5. Instantaneous phase-to-neutral voltages in a typical three-phase system.

where $|V|$ is the *rms* value of v_a, v_b, and v_c.

We can express these voltages as

$$
\begin{aligned}
V_a &= |V| \angle\ 0° \\
V_b &= |V| \angle\ -\ 120° \\
V_c &= |V| \angle\ +\ 120°.
\end{aligned}
\tag{2.18}
$$

The values of V_{ab}, V_{bc}, and V_{ca} are given by:

$$
\begin{aligned}
V_{ab} &= V_a - V_b = |V| \angle\ 0\ -\ |V| \angle -120° = \sqrt{3}\,|V| \angle\ 30° \\
V_{bc} &= V_b - V_c = |V| \angle -120° - |V| \angle\ 120° = \sqrt{3}\,|V| \angle -90° \\
V_{ca} &= V_c - V_a = |V| \angle\ 120° - |V| \angle\ 0° = \sqrt{3}\,|V| \angle -210°.
\end{aligned}
\tag{2.19}
$$

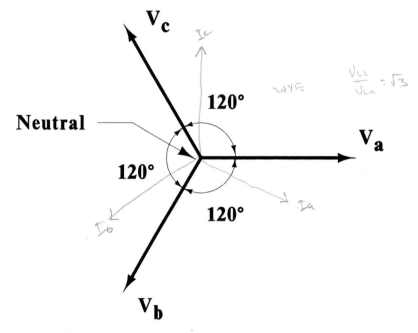

Figure 2.6. Phasor diagram for phase-to-neutral voltages in a three-phase system.

Note that V_a, V_b, and V_c are called the phase voltages or phase-to-neutral voltages. V_{ab}, V_{bc}, and V_{ca} are called line voltages or phase-to-phase voltages. Note that the phase voltages differ from the line voltages by a factor of $\sqrt{3}$.

We will restrict our discussion to balanced three-phase loads. These loads can be connected to the system in one of the following ways.

Wye Connected Loads

Wye loads are connected directly between each phase and neutral, as shown in Figure 2.7.

The line current equals the phase current in the wye connected load. The current for each phase is given by

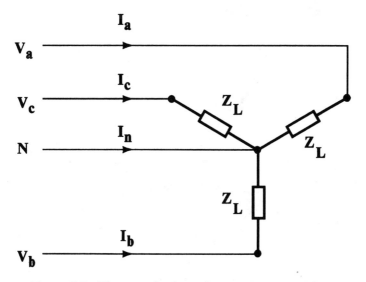

Figure 2.7. Diagram of a three-phase wye connected load.

$$I_a = \frac{V_a}{Z_a} \quad A$$

$$I_b = \frac{V_b}{Z_b} \quad A \qquad (2.20)$$

$$I_c = \frac{V_c}{Z_c} \quad A.$$

Realizing that

$$\left|V_a\right| = \left|V_b\right| = \left|V_c\right| = \left|V\right| \quad \text{and}$$
$$\left|Z_a\right| = \left|Z_b\right| = \left|Z_c\right| = \left|Z\right|$$

$$I_a = \frac{\left|V\right| \angle 0}{\left|Z\right| \angle \varphi} = \frac{\left|V\right|}{\left|Z\right|} \angle -\varphi \quad A$$

$$I_b = \frac{\left|V\right| \angle -120°}{\left|Z\right| \angle \varphi} = \frac{\left|V\right|}{\left|Z\right|} \angle -120° - \varphi \quad A \qquad (2.21)$$

$$I_c = \frac{\left|V\right| \angle 120°}{\left|Z\right| \angle \varphi} = \frac{\left|V\right|}{\left|Z\right|} \angle 120° - \varphi \quad A.$$

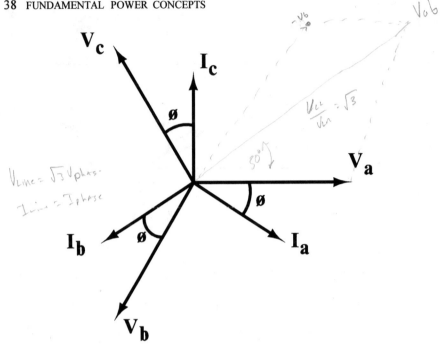

Figure 2.8 Phasor diagram for phase-to-neutral voltages and associated currents in a three-phase wye connected load.

Notice that in this balanced case

$$I_a + I_b + I_c = 0.$$

In other words, the neutral current is zero in the case of a balanced three-phase load. Figure 2.8 shows a phasor diagram for V_a, V_b, and V_c along with the corresponding I_a, I_b, and I_c, assuming an inductive load with a lagging power factor angle φ.

Example 2.3: Consider three equal impedances of $Z = 1.5 + j0.5$ ohms, which are wye-connected on a system with a voltage of 277 volts *rms* phase-to-neutral. Compute the currents.

$$
\begin{aligned}
V_a &= 277 \angle \ \ 0° \quad V \\
V_b &= 277 \angle -120° \quad V \\
V_c &= 277 \angle \ 120° \quad V
\end{aligned}
$$

Here,

$$\varphi = \arctan \left[\frac{0.5}{1.5} \right] = 18.43°$$

and

$$|Z| = \sqrt{(1.5^2) + (0.5^2)} = 1.581 \ \Omega$$

so that

$$I_a = \frac{277}{1.581} \angle (0 - 18.43°) = 175.2 \angle -18.43° \quad A$$

$$I_b = \frac{277}{1.581} \angle (-120° - 18.43°) = 175.2 \angle - 138.43° \quad A$$

$$I_c = \frac{277}{1.581} \angle (120° - 18.43°) = 175.2 \angle 101.57° \quad A.$$

The load has a power factor, *PF*, of

$$PF = \cos \left[\arctan \left[\frac{X}{R} \right] \right] = 0.949 = 94.9\%.$$

Delta Connected Loads

In this case the loads are connected directly between the phases. There is no common point of connection and hence no neutral. Figure 2.9 shows a typical Δ connected load.

Recall that

$$V_{ab} = \sqrt{3} \ |V| \angle \ 30° \quad V$$
$$V_{bc} = \sqrt{3} \ |V| \angle - 90° \quad V$$
$$V_{ca} = \sqrt{3} \ |V| \angle -210° \quad V.$$

Here $|V|$ is the *rms* phase-to-neutral voltage.

The currents in the Δ connected load are

$$I_{ab} = \frac{V_{ab}}{Z} = \sqrt{3} \ \frac{|V|}{|Z|} \angle (\ 30° - \varphi) \quad A$$

$$I_{bc} = \frac{V_{bc}}{Z} = \sqrt{3} \ \frac{|V|}{|Z|} \angle (\ -90° - \varphi) \quad A \qquad (2.22)$$

$$I_{ca} = \frac{V_{ca}}{Z} = \sqrt{3} \ \frac{|V|}{|Z|} \angle (-210° - \varphi) \quad A.$$

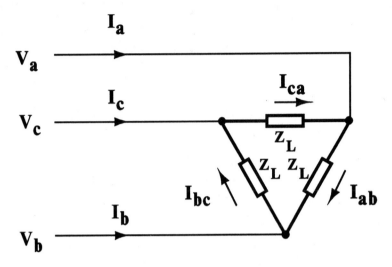

Figure 2.9 Diagram of a three-phase delta connected load.

The phase currents are given by

$$I_a = I_{ab} - I_{ca} = 3 \frac{|V|}{|Z|} \angle \quad -\varphi \quad A$$

$$I_b = I_{bc} - I_{ab} = 3 \frac{|V|}{|Z|} \angle (-120° - \varphi) \quad A \qquad (2.23)$$

$$I_c = I_{ca} - I_{bc} = 3 \frac{|V|}{|Z|} \angle (-240° - \varphi) \quad A.$$

Remember that $|V|$ is again the *rms* value of the phase-to-neutral voltage. Figure 2.10 shows a phasor diagram of the voltage and currents associated with a Δ connected load.

Example 2.4: Connect the three impedances of Example 2.3 in Δ and compute the Δ currents and phase currents. The phase-to-neutral voltage is again 277 V.
From our previous work

$$Z = 1.5\,t + j\,0.5\,\Omega \quad \varphi = \arctan\left[\frac{0.5}{1.5}\right] = 18.43° \quad |Z| = 1.581\,\Omega$$

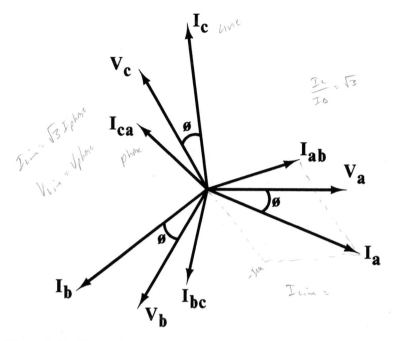

Figure 2.10 Phasor diagram for phase-to-phase voltages and associated currents in a three-phase delta connected load.

$$I_{ab} = \sqrt{3}\,\frac{|V|}{|Z|}\angle(\ 30° - \varphi) = \sqrt{3}\,\frac{277}{1.581}\angle(\ 30° - 18.43°) = 303.46\angle\ 11.57°\ A$$

$$I_{bc} = \sqrt{3}\,\frac{|V|}{|Z|}\angle(\ -90° - \varphi) = \sqrt{3}\,\frac{277}{1.581}\angle(-\ 90° - 18.43°) = 303.46\angle-108.43°\,A$$

$$I_{ca} = \sqrt{3}\,\frac{|V|}{|Z|}\angle(-210° - \varphi) = \sqrt{3}\,\frac{277}{1.581}\angle(-210° - 18.43°) = 303.46\angle-228.43°\,A.$$

The phase currents are

$$I_a = 3\,\frac{|V|}{|Z|}\angle\ -\varphi = 3\,\frac{277}{1.581}\angle\ -18.43° = 525.62\angle\ -18.43°\ A$$

$$I_b = 3\,\frac{|V|}{|Z|}\angle(-120° - \varphi) = 3\,\frac{277}{1.581}\angle(-120° - 18.43°) = 525.62\angle-138.43°\ A$$

$$I_c = 3\,\frac{|V|}{|Z|}\angle(-240° - \varphi) = 3\,\frac{277}{1.581}\angle(-240° - 18.43°) = 525.62\angle-258.43°\ A.$$

Power in Three-Phase Systems

Let us now extend the previous discussion of single-phase power to three-phase systems. Remember that our phase voltages are of equal magnitude and displaced by 120 degrees. Recalling Equation 2.17, and replacing v_{max} by $\sqrt{2}\,|V|$

$$
\begin{aligned}
V_a &= \sqrt{2}\,|V|\sin(\omega t) & V \\
V_b &= \sqrt{2}\,|V|\sin\left[\omega t - \frac{2\pi}{3}\right] & V \\
V_c &= \sqrt{2}\,|V|\sin\left[\omega t + \frac{2\pi}{3}\right] & V
\end{aligned}
\qquad (2.17)
$$

and the currents will *lead* or *lag* their respective voltages by φ so that

$$
\begin{aligned}
i_a &= \sqrt{2}\,|I|\sin(\omega t - \varphi) & A \\
i_b &= \sqrt{2}\,|I|\sin\left[\omega t - \frac{2\pi}{3} - \varphi\right] & A \\
i_c &= \sqrt{2}\,|I|\sin\left[\omega t + \frac{2\pi}{3} - \varphi\right] & A.
\end{aligned}
\qquad (2.24)
$$

For the total of all phases, the power will be

$$
\begin{aligned}
P_{3ph} &= P_a + P_b + P_c \\
&= \sqrt{2}\,|V|\sin(\omega t)\,\sqrt{2}\,|I|\sin(\omega t - \varphi) \\
&\quad + \sqrt{2}\,|V|\sin\left[\omega t - \frac{2\pi}{3}\right]\sqrt{2}\,|I|\sin\left[\omega t - \frac{2\pi}{3} - \varphi\right] \\
&\quad + \sqrt{2}\,|V|\sin\left[\omega t + \frac{2\pi}{3}\right]\sqrt{2}\,|V|\sin\left[\omega t + \frac{2\pi}{3} - \varphi\right] \\
&= 2\,|V|\,|I|\,[\sin(\omega t)\sin(\omega t - \varphi) \\
&\quad + \sin\left[\omega t - \frac{2\pi}{3}\right]\sin\left[\omega t - \frac{2\pi}{3} - \varphi\right] \\
&\quad + \sin\left[\omega t + \frac{2\pi}{3}\right]\sin\left[\omega t + \frac{2\pi}{3} - \varphi\right]\,].
\end{aligned}
\qquad (2.25)
$$

Again, using the trigonometric identity

$$\sin(\alpha)\sin(\beta) = \frac{1}{2}[(\cos(\alpha-\beta) - \cos(\alpha+\beta)]$$

$$
\begin{aligned}
P_{3ph} = |V||I| & [\cos(\varphi) - \cos(2\omega t - \varphi) \\
& + \cos(\varphi) - \cos\left(2\omega t - \frac{4\pi}{3} - \varphi\right) \\
& + \cos(\varphi) - \cos\left(2\omega t + \frac{4\pi}{3} - \varphi\right)].
\end{aligned}
\tag{2.26}
$$

We can simplify this expression further using the trigonometric identity

$$\cos(\alpha) + \cos\left(\alpha - \frac{2\pi}{3}\right) + \cos\left(\alpha + \frac{2\pi}{3}\right) = 0$$

so that Equation 2.26 simplifies to

$$P_{3ph} = 3|V||I|\cos(\varphi) \quad W. \tag{2.27}$$

Thus, three-phase power is constant and equal to three times the power delivered by each phase individually. Remember that $|V|$ is the *rms* value of the phase-to-neutral voltage and $|I|$ the *rms* value of the phase current.

Consider the three equal impedances of Example 2.3.

Example 2.5: Let three equal impedances of $Z = 1.5+j0.5\Omega$ be connected between the phase and neutral of a three-phase system with $|V| = 277$ V. Figure 2.11 shows a graph of p_a, p_b, p_c, and p_{3ph}, which confirms the results predicted by Equation 2.27.

Conventions and Summary of Power Formulas

A few words regarding conventions might be in order. As mentioned previously, it is generally more convenient to express the various power quantities in terms of kW, kVA, or kVAr. In the summary of formulas at the end of this section, this convention is followed. Another important point concerns the method of specifying building loads. Building loads such as electric heating and electric water heating are normally specified in terms of kW, voltage, and number of phases. Lighting and many other

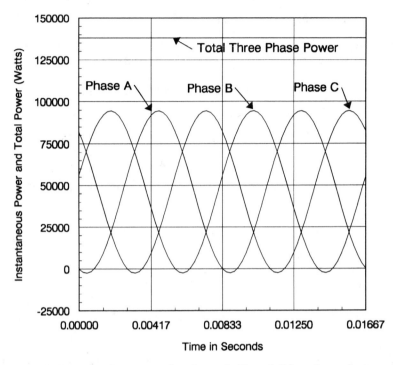

Figure 2.11 Instantaneous power for phases A, B, and C in a three-phase system along with total three-phase power.

equipment loads may be expressed in terms of kVA, voltage, and number of phases or kW, *PF*, voltage, and number of phases. Motor loads are often specified in terms of horsepower rating, voltage, and number of phases or alternatively, full load current, voltage, and number of phases. Quite often motor *PF* is not readily available and must be determined from standard references. In summary, we have three basic methods of specifying building system loads:

1. Current, voltage, and phase requirements, and Power Factor
2. kVA, voltage, and phase requirements, and Power Factor
3. kW, voltage, and phase requirements, and Power Factor

It is rare for equipment to be specified in terms of its impedance. In this sense, then, most of our calculations will be *load-driven* instead of *impedance-driven*.

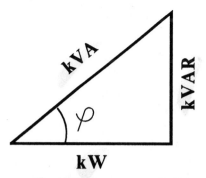

Figure 2.12. The power triangle.

As we will see in later chapters, the appropriate choice of load specification will depend on the purpose of the computation at hand. For example, the full-load current of the component or piece of equipment is necessary when determining overcurrent device rating or conductor size because these devices are rated according to current-carrying capacity. In other calculations it is more convenient to work in terms of kW or kVA.

From the above discussion, it should be clear that a good degree of facility in converting between the various methods of load specification is necessary.

A few of the more commonly used formulas developed earlier in this chapter follow. Note that we now express the *rms* value of voltage and current simply as V and I instead of $|V|$ and $|I|$.

Single-Phase Systems

$$
\begin{aligned}
I &= \frac{kVA}{kV} \quad \text{or} \quad \frac{kW}{kV \cdot PF} \quad A \\
kVA &= kV \cdot I \\
kW &= kVA \cdot PF
\end{aligned}
$$

(2.28)

Here kV represents the circuit voltage which could be either the phase-to-phase voltage or phase-to-neutral voltage depending on the load requirements.

Three-Phase Systems

$$I = \frac{kVA}{kV \cdot \sqrt{3}} \quad \text{or} \quad \frac{kW}{kV \cdot \sqrt{3} \cdot PF} \quad A$$
$$kVA = kV \cdot I \cdot \sqrt{3}$$
$$kW = kV \cdot I \cdot \sqrt{3} \cdot PF$$

(2.29)

Here kV represents the phase-to-phase voltage of the system.

Figure 2.12 shows the relationship between kVA, kW, kVAr, and φ in the form of the familiar power triangle.

SYSTEM VOLTAGE

The evolution of standard system voltages has been neither smooth nor swift, and changes continue through today.

The early systems developed by Edison in the late 1800s operated at 100 volts *dc*. Shortly, however, most utilities changed to *ac* systems. Today the building design engineer may find that a utility has several standard voltages available. A decision must be made very early in the design process as to the appropriate choice. This decision must be made before mechanical systems, lighting systems, or plant equipment is ordered or the power system to serve them is designed.

The Need for Standardization

The first electrical systems were generally stand-alone, in that generation, distribution, and utilization of power occurred literally on site. For example, an industrial plant might install its own stand-alone system to supply power for lighting. An electric trolley company might install a generation system to supply power for electric trolleys. These early systems were not interconnected and the concept of a central plant to generate power for an entire area was still in the future. There was no consistency in either voltage or system frequency between facilities.

It is not difficult to imagine the problems faced by the early manufacturers of equipment for the generation, distribution, or utilization of power. It was inefficient and costly to manufacture equipment specifically for the unique requirements of a given site. Mass production and the resulting economy of scale could not be realized without some degree of standardization.

In addition, during the early days it became clear that several different voltages and ranges of voltages would be necessary. Large amounts of power can be delivered more efficiently using smaller conductors if the voltage is increased. There were practical limits to the safe application of high voltages depending on the nature of equipment served and its proximity to operating personnel. Higher voltages could be safely used during transmission and distribution, while lower voltages were more appropriate within industrial or commercial facilities.

Finally, it became clear that voltage decreases along an electrical system could be a serious problem. For example, on a system operated from a 100 volt generator, loads near the generator itself would receive perhaps 95 to 100 volts, while equipment located further away might receive less than 90 volts. This presented a serious problem for manufacturers because equipment operation at substantially above or below the design operating voltage could result in performance degradation or outright failure.

The lack of standardized voltages, nominal voltage ranges and voltage tolerances had to be resolved in order for the infant utility industry to develop.

Standard Voltages

Today guidance is available from the American National Standard Institute (ANSI) Standard (ANSI C84.1, 1989) regarding preferred voltage systems. Table 2.1 lists some of the standard low voltages. The low voltage range is used in most commercial, residential, and industrial buildings. In a limited number of larger facilities, medium-voltage systems are used for distribution. High-voltage systems are used almost exclusively by the utility companies for transmission of power. Medium-voltage systems are widely used to distribute power from the last utility substation to the individual buildings where transformers step this voltage down to utilization level. The A and B voltage ranges shown in Table 2.1 will be addressed shortly.

A few definitions of some of the terms used in Table 2.1 might be in order.

Table 2.1 Preferred Nominal System Voltages and Voltage Ranges (Low Voltage Systems[1]).

NOMINAL SYSTEM VOLTAGE[3]		NOMINAL UTILIZATION VOLTAGE[7]	VOLTAGE RANGE A[4]			VOLTAGE RANGE B[4]		
			Max. Utilization & Service Voltage[5]	Min. Service Voltage	Min. Utilization Voltage	Max. Utilization & Service Voltage	Min. Service Voltage	Min. Utilization Voltage
Single-Phase Systems								
120/240	Three Wire	115/230	126/252	114/228	110/220	127/254	110/220	106/212
Three-Phase Systems								
208Y/120[6]	Four-Wire	200	218Y/126	197Y/114	191Y/110	220Y/127	191Y/110[2]	184Y/106[2]
480Y/277	Four-Wire	460	504Y/291	456Y/263	440Y/254	508Y/293	440Y/254	424Y/245
480	Four-Wire	460	504	456	440	508	440	424

1 Minimum Utilization Voltages for 120-600 V Circuits Not Supplying Lighting Loads (Preferred Values).

Nominal System Voltage	Range A	Range B
120/240	108/216	104/208
208Y/120	187Y/108	180Y/104
480Y/277	432Y/249	416Y/240
480	432	416

2 Many 220 V motors were applied on existing 208 V systems on the assumption that the utilization voltage would not be less than 187 V. Caution should be exercised in applying the Range B minimum voltages of Table 17 and Note (1) to existing 208 V systems supplying such motors.

3 Three-phase, three wire systems are systems in which only the three-phase conductors are carried out from the source for connection of loads. The source may be derived from any type of three-phase transformer connection, grounded or ungrounded. Three-phase, four wire systems are systems in which a grounded neutral conductor is also carried out from the source for connection of loads. Four-wire systems in this table are designated by the phase-to-phase voltage, followed by the letter Y (except for the 240/120 V delta system) a slant line, and the phase-to-neutral voltage. Single-phase services and loads may be supplied from either single-phase or three-phase systems.

4 The voltage ranges in this table are illustrated in ANSI C84.1-1989, Appendix B[2].

5 For 120-600 V nominal systems, voltages in this column are maximum service voltage. Maximum utilization voltages would not be expected to exceed 125 V for the nominal system voltage of 120, nor appropriate multiples thereof for other nominal system voltages through 600 V.

6 A modification of this three-phase, four wire system is available as a 120/208Y-volt service for single-phase, three wire, open wye applications.

7 Nominal utilization voltages are for low-voltage motors and control. See ANSI C84.1-1989, Appendix C [2] for other equipment nominal utilization voltages (or equipment nameplate voltage ratings).

Reprinted from IEEE Std. 541-1990, *IEEE Recommended Practice for Electric Power Systems in Commercial Buildings*, Copyright © 1991 by The Institute of Electrical and Electronic Engineers, Inc. with permission of IEEE.

- *Nominal system* voltage refers to the root mean square phase-to-phase voltage by which the system is designated and which is also near the normal operating voltage.
- *Service voltage* refers to the root mean square phase-to- phase voltage or phase-to-neutral voltage at the point in the system where the utility and the customer are connected.
- *Utilization voltage* is the root mean square phase-to-phase or phase-to-neutral voltage at the terminal of utilization equipment such as motors and lighting fixtures.
- *Maximum system* voltage refers to the highest phase-to-phase voltage occurring during normal operation. This is also the highest voltage which system equipment and components are designed to operate without the necessity of derating.

The voltage systems discussed above are supplied from utility transformers whose secondary connections are shown in Figure 2.13. This figure shows only the more commonly encountered low voltage connections.

The single-phase, three-wire connection is widely used in residential and small commercial or industrial applications. Lighting, receptacle, and small-appliance loads are often connected phase-to-neutral while larger items of equipment, such as electric heaters or air conditioners, are connected line-to-line.

The three-phase, three-wire connections are employed only where there is no requirement for phase-to-neutral loads.

By far the most common connection is the three-phase, four-wire wye. This versatile connection allows phase-to-neutral loads such as lighting, receptacles, and other small-equipment to be efficiently balanced on the three phases, while also allowing connections of single- or three-phase loads.

The three-phase, four-wire delta connection is used only where the proportion of phase-to-neutral load is very small compared with the three-phase load.

Evolution of Nominal Voltages

As mentioned previously, early electrical systems operated at 100 V. It soon became clear that, due to voltage decreases in the wiring system,

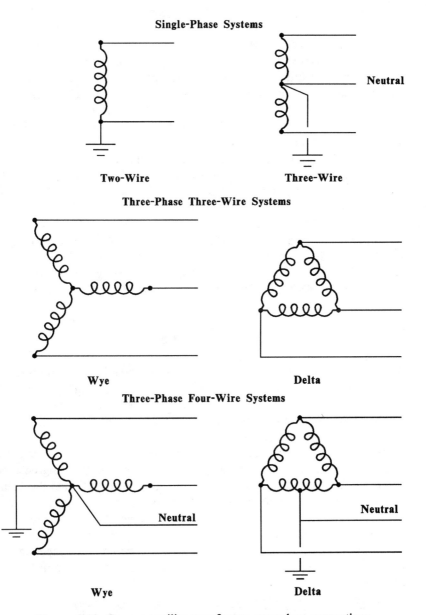

Figure 2.13 Common utility transformer secondary connections.

equipment near the generator might receive close to 100 V while equipment farther away might receive substantially less. When supply voltages were raised to 110 V to offset this decrease, equipment nearer the generator was subjected to an overvoltage condition. This in turn required manufacturers of utilization equipment to raise their design voltage to 110 V. At this same time higher voltages, multiples of 110, were also in common use. These were 220, 440 and 550 volts. Distribution system voltage followed suit and 2200 V, 4400 V, 6600 V, and 13200 V systems became quite common.

The next step in maintaining the supply voltage somewhat higher than the utilization voltage resulted in a change to 115 V and commensurate changes in other utilization voltages produced 230 V, 460 V, 575 V as well as 2300 V, 4600 V, 6900 V and 13000 V.

Later, when the popular 208Y/120 V system was developed, a new series of utilization system voltages were formed around a base voltage of 120 V. In addition to 208Y/120 V and 120 V, this included 240 V, 480 V, and 600 V. The corresponding distribution voltages became 2400 V, 4160Y/2400 V, 4800 V, 12000 V, and 12470Y/7200 V.

VOLTAGE TOLERANCE RANGES

Induction motors, which make up well over 90% of the motors in use today, represent such a significant percentage of a building's load that it is natural they play an important part in decisions regarding voltage tolerance levels. The need for standard voltage tolerances was felt by all levels of the electrical industry. Manufacturers of electrical equipment, ranging from coffee pots to washing machines, had to have accurate data on the system voltages to which their equipment would be connected. Building designers had to understand both the voltages supplied by the utility and the voltage requirements of the equipment manufacturers. Finally, the utilities needed guidance on the required building voltage ranges. The development of these voltage ranges was a somewhat arduous process which probably will continue to evolve for the foreseeable future. The common meeting point for all these groups is the American National Standard Institute (ANSI) Standard C84.1, 1989.

The voltage tolerance range for standard induction motors is 230 V and 460 V ±10%. Using the ratio of transformation between 480 V and 120 V we find that 460 V represents 115 V on a 120 V base. This 120 V base provides a very convenient way to present and compare tolerances for different system voltages. The voltage tolerance for an induction motor would then be 103.5 V to 126.5 V on the 120 V base. This results in a tolerance range of 23 volts.

Let us consider for a moment the distribution of the 23 V tolerance range. We know that there will be some decrease in voltage at every step in the process of producing and delivering power to the motor. Logically these are grouped into voltage decreases, or voltage drops, occurring in the generation, transmission, and distribution phase up to the utility transformer at the building, the voltage drop occurring in the transformer itself, and, finally, the voltage drop in the building wiring system between the service point and the utilization equipment.

ANSI C84.1, 1989 recognizes two ranges of voltage tolerances, the A and B ranges. Voltages within the A range will produce satisfactory performance in utilization equipment. The B range recognizes that short-term voltage excursions outside the A range will inevitably occur. It is generally understood that utilization equipment will give acceptable performance in the range between A and B, but corrective action to restore A range operation should occur promptly.

The tolerance of induction motors on a 120 volt base, 103.5 V and 126.5 V, were rounded to 104 V and 127 V respectively. This became the B range in the ANSI Standard. The 23 V tolerance was divided as follows:

Generation/Transmission/Distribution	13 V
Utility Transformer and Connection to Building	4 V
Building Wiring System	6 V
	23 V

The 6 V allowance for the building wiring system represents 5% on the 120 V base. The allocation of this 5% will be discussed as a part of voltage-drop analysis in a later chapter.

The A range criteria were developed by tightening the B range requirements from 127 V to 126 V and from 104 V to 108 V. The resulting 18 V tolerance was allocated as follows:

Generation/Transmission/Distribution	9 V
Utility Transformer and Connection to Building	3 V
Building Wiring System	6 V
	18 V

Table 2.2 Standard Voltage Profile for a Regulated Power Distribution System, 120 V Base.

	Range A	Range B
Maximum Voltage Allowed	126 (125*)	127
Voltage Drop in Generation/ Transmission/Distribution System	9	13
Minimum Primary Service Voltage	117	114
Voltage Drop Within Utility Transformer	3	4
Minimum Low Voltage at Service	114	110
Voltage Drop Within Building	6 (4**)	6 (4**)
Minimum Utilization Voltage	108 (110**)	104 (106**)

* For Utilization Voltage from 120 - 600 V.
** For Building Wiring Circuits Supplying Lighting Equipment.

Reprinted from IEEE Std. 241-1990, *IEEE Recommended Practice for Electric Power Systems in Commercial Buildings*, Copyright © 1991 The Institute of Electrical and Electronic Engineers, Inc. with permission of IEEE.

Finally, several additional adjustments were made to the ranges:

- The maximum A range voltage was reduced from 126 V to 125 V for systems rated 120 V through 600 V. This was partly in recognition of the sensitivity of incandescent light sources to overvoltages.

- The voltage drop within the building system was decreased to 4 V for lighting loads. This increased the minimum voltage from 108 V to 110 V in range A and 104 V to 106 V in range B.

Table 2.2 shows the resulting voltage profile for a power distribution system on a 120 V base. Figure 2.14 shows a graph of the results of the

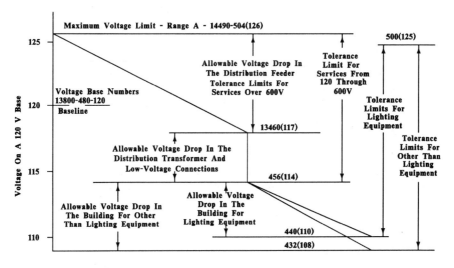

Voltage Profile of Limits of Range A, ANSI C84.1-1989

Figure 2.14 Voltage profile limits of range A, ANSI Standard C84.1-1989. (Reprinted from IEEE Standard 241-1990, *IEEE Recommended Practice for Electric Power Systems in Commercial Buildings,* Copyright © 1991 by The Institute of Electrical and Electronic Engineers, Inc. with permission of IEEE.)

preceding discussion which summarizes the requirements of ANSI C84.1, 1989.

SELECTION OF SYSTEM VOLTAGE

As mentioned earlier, selecting the optimal system voltage for a project is not always easy. The problem of selecting the best voltage for a given building was expressed well by consulting engineer Walter Hibble, P.E.: "Every type of building has different requirements, a different personality if you will. The challenge is to understand the unique personality of the building you are designing." Understanding this personality takes considerable experience, but here are a few guidelines.

When the total building load is small, perhaps on the order of 30 kVA, the 120/240 V single-phase system might be appropriate. In many locations, a three-phase utility distribution system is not available. Single-phase services are commonly found in residences, small commercial shops, and perhaps very small industrial facilities.

Building systems with larger power requirements normally require three-phase systems. Motors and other equipment, such as resistance heating and process loads, are almost always manufactured for use on three-phase systems. Recall the earlier discussion in this chapter of the uniform power-flow characteristics of three-phase systems. The two most widely used three-phase systems are the 208Y/120 V three-phase, four-wire system and the 480Y/277 V three-phase, four-wire system. Figure 2.15 shows the voltage characteristics of these two popular systems.

The 208Y/120 V system evolved first, and its popularity resulted from its versatility. Three-phase or single-phase 208 V circuits were available for motor and other loads. Single-phase 120 V circuits were available for

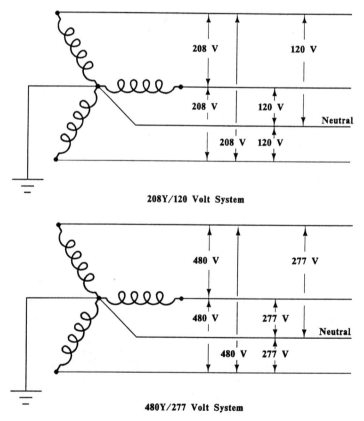

Figure 2.15 Common utility supply voltages. 208Y/120 V and 480Y/227 V.

lighting and receptacle loads. All of this from one system with no need for further voltage transformation! 208Y/120 V systems are widely available and are a good choice for medium-size buildings such as small shopping centers, food stores, small office buildings, and industrial facilities.

The 480Y/277 V system came into widespread use during the 1960s. It evolved from the need for a higher-voltage system capable of supplying larger commercial and industrial facilities using smaller and less expensive distribution equipment than would have been required for a comparable 208Y/120 V system. The 480Y/277 V system allows single-or three-phase 480 V circuits for motors or other equipment. In short order, manufacturers developed new ballasts for fluorescent and other discharge lighting sources to allow direct use of the 277 V phase-to-neutral voltage. The only problem was the lack of availability of 120 V single-phase circuits for receptacles and other miscellaneous loads.

The solution to the lack of 120 V circuits is the use of step-down transformers rated 480 V-208Y/120 V. Since such transformers represent additional equipment, and hence added cost, it is wise to specify as much equipment as possible for direct use at either 480 V three-phase or single-phase with as much of the lighting system at 277 V as possible.

Here the personality of the building comes into play. For example, office buildings tend to have substantial air conditioning and heating requirements along with widespread use of fluorescent or other discharge lighting-systems. The requirements for 120 V single-phase or 208 V three-phase power are normally substantially less than those for the mechanical system and lighting. Thus, the personality of this type of building suits the 480Y/277 V system well. On the other hand, retail food stores, which often have connected loads in the range of 1000 kVA, may not represent a good choice for 480Y/277 V. This is because a good portion of the store's load is in refrigeration systems for the many refrigerated and frozen-food cases. Traditionally such refrigeration systems are 208Y/120 V. The use of 480Y/277 V often requires a step-down transformer of several hundred kVA, offsetting the economic advantages of the higher voltage. Industrial facilities often have very heavy process loads as well as significant discharge-lighting loads. Such facilities are often excellent choices for 480Y/277 V, especially since the higher voltage lessens voltage drop concerns associated with the greater distances typically encountered in these buildings.

In a few facilities, the requirement for lighting and other phase-to-neutral loads is small or nonexistent. In these cases, a three-phase 240 V or 480 V system might be in order. An example is a remote pumping station employing only three-phase motors. Extreme care must be taken in making a decision that "no phase-to-neutral requirements exist in the facility" because such requirements have a way of appearing unexpectedly during or after construction.

In addition to the above engineering and economic considerations, the engineer also must work closely with the serving utility during the process of selecting the appropriate voltage system. Most utilities are extremely cooperative in this matter. Contact between the design engineer and the utility should be established early in the design process.

Finally, it should be clear that an understanding of the building's characteristics, including equipment voltage requirements, existing equipment brought from another facility, the magnitude of receptacle and other 120 V load, and the type of lighting and mechanical systems, provides important guidance for the engineer in the voltage selection process.

This voltage decision may seem bewildering, but after a few years of experience you will develop an appreciation for the personality of various types of buildings, making the choice easier.

3

System Components

A building electrical system is composed of hundreds of components, designed and assembled into a safe, functional power-delivery system. In this chapter, the primary components used in the distribution system, their design, and operational characteristics are introduced. In later chapters, how these components are integrated into an operating system is discussed.

Perhaps the best way to begin this review of system components is by understanding their respective functions. To do this, we need an electrical diagram of a typical building distribution system. Figure 3.1 shows a diagram of the major electrical components of a three-story building. This diagram is schematic in nature, but does convey the general relationship between the physical location of the various components shown. This is referred to as a *riser diagram*. We will refer to this diagram as we proceed through our discussion of system components.

The building system actually originates at the utility service, where the building's electrical system is connected to the utility system. In Figure 3.1, this is a pad-mounted transformer, but in other cases it might be a bank of transformers mounted overhead on a utility pole. The underground service connects the utility system to the building's main distribution panel. Located within the MDP is the main building overcurrent device, or main disconnect, as well as individual overcurrent devices for the system components connected to the MDP. The MDP may also contain provisions for utility metering, as well as instrumentation for the measurement of system voltage and currents. The main disconnect device can be either a circuit breaker or a fused switch. This main device often contains special circuitry for sensing low-level faults, which otherwise might escape detection. The branch devices, which also may be circuit breakers or fused

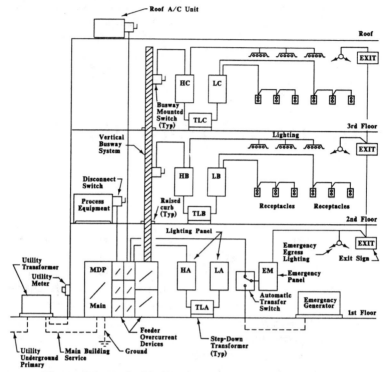

Figure 3.1 Typical building electrical system riser diagram.

switches, provide protection for the various feeders connected to the MDP. The MDP might be thought of as the electrical nerve center of the building. It is normally located near a building exterior wall and as close as possible to the utility transformer to minimize the cost of the main service feeder. From a safety standpoint, it should be remembered that the main disconnect device provides a single location at which the entire power system can be de-energized in case of emergency.

Feeders are installed from the MDP to other items of equipment such as panelboards, air-conditioning units, process equipment or motors. There are two basic means of conveying this power—individual insulated conductors or electrical busways. In the case of insulated cables, several individual cables or perhaps a multi-conductor cable are installed in a protective raceway, referred to as a *conduit*. For example, such conduit and cable feeders are shown connecting MDP to Panel HA and to the roof air-conditioning unit. The second method of conveying power, by means of an electrical busway system, often is used to convey energy over distances not economically feasible for cable and conduit feeders. The busway system utilizes insulated current-carrying bus bars installed in a protective metallic housing. The busway system also may include periodic tap locations along its length for easy connection of overcurrent devices, such as those shown at floors 2 and 3 of Figure 3.1. The panelboards, such as HA, LA, etc., provide locations for the connection of utilization equipment such as lighting fixtures and wall-mounted outlets called receptacles. Overcurrent devices to protect these circuits also are located in these panelboards.

The panelboards either can be connected directly to the MDP, in which case their operating voltage is the same as that of the main service, or alternatively, through transformers that allow changes in the voltage. In Figure 3.1, Panels HA, EM, HB, and HC are all operated at the same voltage as the main service. Panels LA, LB, and LC are connected through step-down transformers. A typical example is the 480Y/277 V system discussed in Chapter 2. In this system, major loads such as air-conditioning equipment, process equipment and even fluorescent or HID lighting is operated directly from either 480 or 277 volts. Receptacle loads and other equipment requiring lower voltages, such as 120 or 208 V, are operated through step-down transformers.

Finally, it is important to provide a safe means for building evacuation in times of emergency. In order to accomplish this, emergency lighting must be provided to mark the safe egress path and also to provide a safe level of lighting to facilitate building evacuation. The safe marking of building exit paths requires illuminated exit signs, and emergency luminaires located at strategic locations. This entire emergency system must function even in the event of total utility power failure. Thus, the emergency system is provided with two sources of power: normal building power and power from an emergency generator or battery system. Figure 3.1 shows an emergency panel (EM) connected to normal service and to an emergency generator through an automatic transfer switch. In the event of power failure, the generator is automatically started and, once stable operation is reached, the automatic transfer switch transfers Panel EM and its load to the generator. Since this process can require 5 to 10 seconds, battery-powered exit signs and emergency lighting units often are provided. The battery powered equipment provides temporary (90 minutes or so) operation and the generator provides longer-range operating capability, limited only by fuel supply. In cases where only short-term emergency lighting is required, the battery system alone often suffices.

All components of the system must be chosen carefully based on design requirements and must function safely, under normal operating conditions and also under abnormal conditions, such as short circuits.

As you may have now begun to suspect, this can be quite a design challenge.

CONDUCTORS AND CABLES

Cable installed in raceways or busways may be thought of as the arteries and veins through which the blood of the electrical system flows. As such, they are an extremely important part of the overall system planning process. Strictly speaking, the *conductor* is the actual metallic current-carrying component, while the *cable* is the complete assembly of conductor, insulation, and any required protective covering. Conductors are normally copper or aluminum, due to their excellent current-carrying properties. Figure 3.2 shows several typical cable configurations.

Before considering factors such as current-carrying capacity, insulation types and temperature levels, let us review some of the more important *dc* characteristics of conductor materials.

DC Characteristics of Conductor Materials

Conductors are typically of copper or aluminum. The *dc* resistance of a length (L) of conductor material of cross sectional area A and resistivity, ρ, is given by

$$R = \rho \frac{L}{A} \quad \Omega \qquad (3.1)$$

where
  R = *dc* resistance in ohms
  ρ = conductor resistivity in ohm-circular mils per ft.
  L = length in feet
  A = cross sectional area in circular mils.

The only unfamiliar concept involved in Equation 3.1 might be the measurement of cross sectional area in circular mils. A circular mil is defined as the area of a conductor of diameter 0.001 inch.

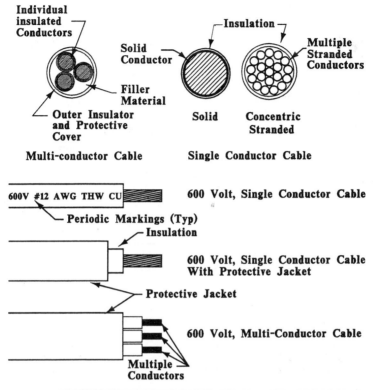

Figure 3.2 Typical single and multi-conductor cable construction.

To convert the area of a conductor of diameter, D, to circular mils, we must divide the conductor's area in in^2 by the number of in^2 per circular mil. The conductor area is given by

$$A_{Cond} = \pi r^2 = \pi \left[\frac{D}{2} \right]^2 = \frac{\pi}{4} D^2 \quad in^2. \tag{3.2}$$

The area of a circle of diameter 1 mil (0.001 inch) is given by

$$A_{cm} = \pi r^2 = \pi \left[\frac{D}{2} \right]^2 = \frac{\pi}{4} D^2 = \frac{\pi}{4} 10^{-6} \quad in^2. \tag{3.3}$$

Thus the area of an arbitrary conductor of diameter D will be

$$A = \frac{A_{in}}{A_{cm}} = \frac{\frac{\pi}{4}D^2}{\frac{\pi}{4} \cdot 10^{-6}} = D^2 \cdot 10^6 = (D \cdot 10^3)^2 \quad cm. \quad (3.4)$$

So the area of the conductor in circular mils will be simply its diameter in mils squared!

The student should verify that the units of Equation 3.1, resistivity, length, and area in circular mils indeed cancel to yield ohms.

Example 3.1: What is the cross sectional area of a conductor 0.5 inches in diameter?

for Cmil

Using Equation 3.4,

$$A = (D \cdot 10^3)^2 \qquad : \qquad (0.5 \, kio^3)^2 = 250 \, kCMIL$$

$$A = (D \cdot 10^3)^2 = (0.5 \cdot 10^3)^2 = 25 \cdot 10^4 \, cm = 250{,}000 \, cm \ or \ 250 \ kCMIL.$$

The above method of expressing the size of a conductor in terms of its cross sectional area in circular mils is common, especially for large conductors.

The resistivity, ρ, is different for copper and aluminum conductors and is normally expressed at a specified temperature. The approximate values of ρ are

$$\rho_{cu} = 10.371 \ \frac{\Omega \cdot cm}{ft} \ @ \ 20°C$$

$$\rho_{al} = 16.728 \ \frac{\Omega \cdot cm}{ft} \ @ \ 20°C.$$

Example 3.2: What is the dc resistance of a 1000 ft. length of copper conductor of cross sectional area 6530 circular mils?   Assume the temperature to be 20°C.

Using Equation 3.1,

$$R = \rho \frac{L}{A} = 10.371 \cdot \frac{1000}{6530} = 1.59 \ \Omega.$$

Note several important points regarding Equation 3.1. First, the resistance is directly proportional to the length, so longer conductors have greater resistance. Secondly, the resistance is inversely proportional to the cross sectional area. The latter provides a means of lowering the overall resistance of a given length of conductor by increasing its cross sectional area.

We now review how the temperature of a conductor affects its *dc* resistance. The value of R for copper and aluminum conductors is given by

$$R_{cu} = R_{T_1} \left[\frac{234.5 + T_2}{234.5 + T_1} \right] \Omega$$

$$R_{al} = R_{T_1} \left[\frac{228.1 + T_2}{228.1 + T_1} \right] \Omega.$$

(3.5)

Where T_1 is the initial temperature in °C at which the resistance was computed, and T_2 is the new temperature.

It is also useful to recall the relationship between °C and °F, which is given by

$$°C = \frac{5}{9} [°F - 32]$$

$$°F = \frac{9}{5} °C + 32.$$

(3.6)

As mentioned previously, resistivity is normally specified at 20°C, which according to Equation 3.6 is 9/5 · 20 + 32 = 68°F, about room temperature.

Example 3.3: What is the resistance of the copper conductor of example 3.2 if the temperature is 75°C?

Using Equation 3.5

$$R_{cu} = 1.59 \left[\frac{234.5 + 75}{234.5 + 20} \right] = 1.93 \ \Omega.$$

$T_1 = 20°$: (Almost Always use 20°C because that is Rated temp)

OR R at T_1

As we will see, 75°C (167 °F) is a common conductor operating temperature for certain classes of insulation. It should be understood that this conductor temperature occurs only under fully loaded conditions.

Let us now consider the *ac* resistance and impedance characteristics for conductors.

Resistance and Impedance

In the preceding discussion of resistance, we assumed *dc* operation of the conductors. In actual applications, however, the operating system is almost universally *ac* in nature. The *ac* impedance of conductors is different from the *dc* resistance due to factors such as skin effect. There also is an interaction of the magnetic field produced by current flow in the conductor with the magnetic characteristics of the raceway.

For example, conductors installed in steel raceway will have a higher overall impedance than similar conductors in non-magnetic raceway, such as PVC. Tests have been made using typical raceway materials and conductor types, with the resulting value of resistance, *R*, and inductive reactance, *X*, given in *NEC*[1] Table 9. A portion of this table is shown in Table 3.1.

Current-Carrying Capacity

The current-carrying capacity of a given conductor is determined by several factors. First, the I^2R heating that results from current flow will elevate the temperature of the conductor. The conductor's insulation naturally has thermal limits beyond which damage or shortened insulation life occurs. Second, the environmental conditions in which the conductor is installed must be considered. If, for example, a conductor designed to operate at 75°C in an ambient temperature of 30°C is installed in a higher temperature environment, the temperature difference between the conductor and its surroundings is less. As a result, the heat generated within the conductor is dissipated to the environment more slowly. Conversely, the same

---

[1]   *National Electrical Code*® and *NEC*® are registered trademarks of the National Fire Protection Association, Inc., Quincy, MA 02269.

Ohms to Neutral per 1000 feet

Size AWG/ kCMIL	X_L (Reactance) for All Wires		AC Resistance for Uncoated Copper Wires			AC Resistance for Aluminum Wires		
	PVC, Al. Conduit	Steel Conduit	PVC Conduit	Al. Conduit	Steel Conduit	PVC Conduit	Al. Conduit	Steel Conduit
14	.058	.073	3.1	3.1	3.1	-	-	-
12	.054	.068	2.0	2.0	2.0	3.2	3.2	3.2
10	.050	.063	1.2	1.2	1.2	2.0	2.0	2.0
8	.052	.065	0.78	0.78	0.78	1.3	1.3	1.3
6	.051	.064	0.49	0.49	0.49	0.81	0.81	0.81
4	.048	.060	0.31	0.31	0.31	0.51	0.51	0.51
3	.047	.059	0.25	0.25	0.25	0.40	0.41	0.40
2	.045	.057	0.19	0.20	0.20	0.32	0.32	0.32
1	.046	.057	0.15	0.16	0.16	0.25	0.26	0.25
1/0	.044	.055	0.12	0.13	0.12	0.20	0.21	0.20
2/0	.043	.054	0.10	0.10	0.10	0.16	0.16	0.16
3/0	.042	.052	0.077	0.082	0.079	0.13	0.13	0.13
4/0	.041	.051	0.062	0.067	0.063	0.10	0.11	0.10
250	.041	.052	0.052	0.057	0.054	0.085	0.090	0.086
300	.041	.051	0.044	0.049	0.045	0.071	0.076	0.072
350	.040	.050	0.038	0.043	0.039	0.061	0.066	0.063
400	.040	.049	0.033	0.038	0.035	0.054	0.059	0.055
500	.039	.048	0.027	0.032	0.029	0.043	0.048	0.045
600	.039	.048	0.023	0.028	0.025	0.036	0.041	0.038
750	.038	.048	0.019	0.024	0.021	0.029	0.034	0.031
1000	.037	.046	0.015	0.019	0.018	0.023	0.027	0.025

Reprinted with permission from NFPA 70-1990, *National Electrical Code*, Copyright © 1989. National Fire Protection Association, Quincy MA 02269. This reprinted material is not the complete and official position of NFPA on the referenced subject, which is represented only by the standard in its entirety.

Table 3.1 AC Resistance and Reactance for 600 V Cables, Three-Phase 60 Hz, 75°C (167°F) - Three Single Conductors in Conduit. (Based on Chapter 9, Table 9, *NEC*.)

conductor installed in an environment of less than 30°C is able to carry more current without danger of overheating.

Cable is available which is designed to operate at 60°C, 75°C, and 90°C.

Standard Conductor Sizes and Stranding

There are actually two systems for specifying conductor size. These are the *American Wire Gage (AWG)* and the *kCMIL* system. The earliest system was the AWG and conductor sizes were established with conductor cross sectional area increasing with decreasing wire-size numbers. For example, #14 AWG is smaller in cross sectional area than #12 AWG. In the early days, the requirement for wire sizes above #1 seemed unlikely. Normal building wire sizes were #14, 12, 10, 8, 6, 4, 3, 2, and 1 AWG. When it became clear that larger conductors would be required, 1/0, 2/0, 3/0, and 4/0, pronounced four-aught, were added. Finally, it became clear that even larger conductors would be required, and it was decided to express the conductor size in terms of its cross sectional area in circular mills. This change added 250 through 2000 kCMIL conductors. In actual practice, most building applications employ conductors no larger than 750 kCMIL, due to practical constraints posed by the physical handling and installation problems associated with larger conductors.

Finally, as you can imagine, larger solid conductors are difficult to bend. As a result, both smaller and larger conductors often are constructed with multiple smaller conductors or strands. These stranded conductors are far more flexible and the completed cable is much easier to handle and bend. Today almost all conductors above #8 AWG are stranded, as are many #12 and #10 conductors.

Conductor Insulation

A wide variety of conductor insulation types are available to meet the requirements of varying applications. These insulation types are designated in a standard fashion, and all cable is marked with information on its AWG or kCMIL size, voltage rating, and insulation type at intervals along its length.

Cable insulation is designated as follows:

A - Asbestos insulation
MI - Mineral insulation
R - Rubber insulation
SA - Silicon asbestos insulation
T - Thermoplastic insulation
V - Varnished cambric insulation
X - Cross-linked synthetic polymer

The cable is further designated according to its operating environment:

H - Heat resistant up to 75°
HH - Heat resistant up to 90°
    - No designation means 60°
W - Moisture resistant, for use in wet locations
UF - For use in underground direct burial applications

Many cables are designed and certified for use in several environmental conditions. Such multi-use cables are so marked. For example, a cable marked *TW* would indicate 60°C, thermoplastic insulation capable of use in wet environments. Type *THW* indicates 75°C, thermoplastic insulation for use in wet environments. Type *XHHW* represents cross-linked synthetic polymer insulation rated for 90°C.

Standard Ampacity and Temperature De-rating

The *ampacity* of a cable is its continuous current-carrying capacity under specified conditions. Data on the ampacity of various conductor sizes, conductor materials, and insulation types are summarized in Table 310-16 of the *NEC*, reprinted here as Table 3.2. Note that data is given for both copper and aluminum conductors and is based on an ambient temperature of 30°C. A table at the bottom of Table 3.2 gives corrections to be applied for actual ambient temperatures above or below 30°C. Note that if the actual temperature is below 30°C, the coefficients are greater than one, indicating that the conductor can safely carry more than its listed ampacity. Also note that the smallest practical conductor size is #14 AWG and that the maximum overcurrent protection for #14, #12, and #10 conductors is 15, 20, and 30 amperes respectively.

Table 3.2 Ampacities of Insulated Conductor Rated 0-2000 V, 60° to 90°C (140°F to 194°F). Not more than three conductors in Raceway or Cable or Earth (Directly Buried), Based on Ambient Temperature of 30°C (86°F).

Size	Temperature Rating of Conductor. See Table 310-13.								Size
	60°C (140°F)	75°C (167°F)	85°C (185°F)	90°C (194°F)	60°C (140°F)	75°C (167°F)	85°C (185°F)	90°C (194°F)	
AWG kcmil	TYPES †TW, †UF	TYPES †FEPW, †RH, †RHW, †THHW, †THW, †THWN, †XHHW †USE, †ZW	TYPE V	TYPES TA, TBS, SA SIS, †FEP, †FEPB, †RHH, †THHN, †THHW, †XHHW	TYPES †TW, †UF	TYPES †RH, †RHW, †THHW, †THW, †THWN, †XHHW †USE	TYPE V	TYPES TA, TBS, SA, SIS, †RHH, †THHW †THHN, †XHHW	AWG kcmil
			COPPER		ALUMINUM OR COPPER-CLAD ALUMINUM				
18	....	...	....	14	....	....	....	....	....
16	....	...	18	18	....	....	....	....	....
14	20†	20†	25	25†	....	....	....	....	
12	25†	25†	30	30†	20†	20†	25	25†	12
10	30	35†	40	40†	25	30†	30	35†	10
8	40	50	55	55	30	40	40	45	8
6	55	65	70	75	40	50	55	60	6
4	70	85	95	95	55	65	75	75	4
3	85	100	110	110	65	75	85	85	3
2	95	115	125	130	75	90	100	100	2
1	110	130	145	150	85	100	110	115	1
1/0	125	150	165	170	100	120	130	135	1/0
2/0	145	175	190	195	115	135	145	150	2/0
3/0	165	200	215	225	130	155	170	175	3/0
4/0	195	230	250	260	150	180	195	205	4/0
250	215	255	275	290	170	205	220	230	250
300	240	285	310	320	190	230	250	255	300
350	260	310	340	350	210	250	270	280	350
400	280	335	365	380	225	270	295	305	400
500	320	380	415	430	260	310	335	350	500
600	355	420	460	475	285	340	370	385	600
700	385	460	500	520	310	375	405	420	700
750	400	475	515	535	320	385	420	435	750
800	410	490	535	555	330	395	430	450	800
900	435	520	565	585	355	425	465	480	900
1000	455	545	590	615	375	445	485	500	1000
1250	495	590	640	665	405	485	525	545	1250
1500	520	625	680	705	435	520	565	585	1500
1750	545	650	705	735	455	545	595	615	1750
2000	560	665	725	750	470	560	610	630	2000

AMPACITY CORRECTION FACTORS									
Ambient Temp. °C	For ambient temperatures other than 30°C (86°F), multiply the ampacities shown above by the appropriate factor shown below.								Ambient Temp. °F
21-25	1.08	1.05	1.04	1.04	1.08	1.05	1.04	1.04	70-77
26-30	1.00	1.00	1.00	1.00	1.00	1.00	1.00	1.00	79-86
31-35	.91	.94	.95	.96	.91	.94	.95	.96	88-95
36-40	.82	.88	.90	.91	.82	.88	.90	.91	97-104
41-45	.71	.82	.85	.87	.71	.82	.85	.87	106-113
46-50	.58	.75	.80	.82	.58	.75	.80	.82	115-122
51-55	.41	.67	.74	.76	.41	.67	.74	.76	124-131
56-60	....	.58	.67	.71	....	.58	.67	.71	133-140
61-70	....	.33	.52	.58	....	.33	.52	.58	142-158
71-80	....	....	.30	.41	....	....	.30	.41	160-176

† Unless otherwise specifically permitted elsewhere in this Code, the overcurrent protection for conductor types marked with an obelisk (†) shall not exceed 15 amperes for 14 AWG, 20 amperes for 12 AWG, and 30 amperes for 10 AWG copper; or 15 amperes for 12 AWG and 25 amperes for 10 AWG aluminum and copper-clad aluminum after any correction factors for ambient temperature and number of conductors have been applied.

There are other design restrictions which relate to the temperature limits of cable termination equipment. These will be addressed in later design chapters.

Example 3.4: ✓(a) What is the ampacity at 30°C of #1 *THW* copper?
From Table 3.2—130 A

   ✓(b) What is the ampacity at 30°C of #400 KCMIL, *XHHW* copper?
From Table 3.2—335 A, wet location, 380 A, dry locations.

   ✓(c) What is the ampacity of 500 kCMIL *THW* copper at 40°C?
The normal ampacity is 380 A at 30°C. The correction factor for 40°C is 0.88, so the actual ampacity is 334 A.

$$= (380)(0.88)$$
$$= 334.4$$

Short Circuit Considerations

Short circuits can have a detrimental effect on insulated cables just as on any other components of the building electrical system. Although short-circuit analysis is discussed in a later chapter, it is appropriate now to discuss the effects of these high-level currents on conductors. In normal circuit operation, the flow of current is limited by the system impedance as well as the impedance of the load itself. The level of energy dissipated in the system components, such as cable, is proportional to the square of the current flow. We naturally wish to lose as little energy as possible in this I^2R heating, and also to minimize the attendant voltage drop. As a result, cables normally represent a very, very small part of the total impedance. The load impedance is many times greater than that of the cable. Under short-circuit conditions, the load's impedance is replaced by a far lower impedance and the resulting current flow can be several thousand times as great as normal. This current flow can produce tremendous thermal effects on the conductor insulation. Fortunately, overcurrent devices can reduce the time of this current flow to a few cycles, depending on the type of device (circuit breaker or fuse) employed. Finally, as seen in a later chapter, short-circuit currents are often asymmetrical in nature, especially during the critical first few cycles. This also must be considered.

Techniques have been developed to determine the short-circuit capability of insulated conductors based on the thermal limitations of the insulation, the length of time of current flow, and the material characteristics of copper and aluminum conductors.

Equations 3.7 give the relationship between the maximum asymmetrical fault-current flow and the maximum short-circuit temperature of copper and aluminum conductors.

$$\left[\frac{I}{A}\right]^2 t = 0.0297 \cdot \log\left[\frac{234 + T_2}{234 + T_1}\right] \quad Copper$$

$$\left[\frac{I}{A}\right]^2 t = 0.0125 \cdot \log\left[\frac{228 + T_2}{228 + T_1}\right] \quad Aluminum$$

(3.7)

Here I represents the maximum permissible asymmetrical current flow, A, the cross sectional area of the conductor in circular mils, and T_1, the normal maximum operating temperature. T_2 represents the maximum permissible conductor temperature under short circuit conditions. Both T_1 and T_2 are expressed in °C. The time, t, represents the time, in seconds, of short-circuit current flow before protective device operation. Typical times of interest vary from 1 cycle (1/60 second) to 90 cycles (1.5 seconds). For systems operating at less than 1,000 V, we would expect circuit breakers to operate in about 2 cycles and current limiting fuses in less than 1 cycle.

For THW insulated cable, T_1 is 75°C and T_2 is 150°C. For XHHW insulated cable, T_1 is 90°C and T_2 is 250°C.

With this information, Equation 3.7 can be solved to yield a family of curves relating conductor cross sectional area, insulation type, conductor material and duration of fault to the maximum asymmetrical fault current flow. Figures 3.3(a), (b), (c), and (d) show curves for THW and XHHW insulated copper and aluminum cable.

Once the maximum permissible asymmetrical short-circuit current flow for a given insulated cable is established, we convert this to an equivalent symmetrical current by use of a coefficient, K_0, as shown in Equation 3.8. For systems less than 1,000 V, the K_0 factor for circuit breakers is 1.3 and for current limiting fuses, 1.4.

$$I_{sym} = \frac{I_{asy}}{K_0}$$

(3.8)

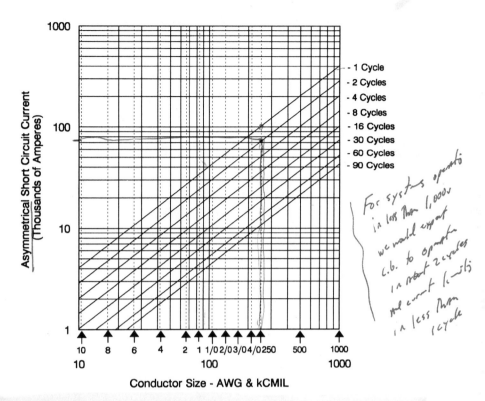

- 1 Cycle
- 2 Cycles
- 4 Cycles
- 8 Cycles
- 16 Cycles
- 30 Cycles
- 60 Cycles
- 90 Cycles

For systems operating in less than 1,000v we would expect c.b. to operate in about 2 cycles and current (a-sb) in less than 1 cycle

Conductor Size - AWG & kCMIL

Figure 3.3a Maximum allowable short circuit current for type THW, copper conductors.

Let's look at a few examples of how we determine the maximum permissible current flow for a given insulated conductor.

Example 3.5: (All examples are for systems less than 1,000 V.)
(a) What is the symmetrical short-circuit rating of 250 kCMIL, *THW* copper cable protected by a circuit breaker? = 2 cycles
From Figure 3.3(a), or Equation 3.7, we see that the maximum asymmetrical current flow is 72,500 amperes. Using $K_0 = 1.3$ for circuit breakers, the symmetrical current flow will be given by:

$$I_{sym} = \frac{I_{asy}}{K_0} = \frac{72,500}{1.3} = 55,800 \ A.$$

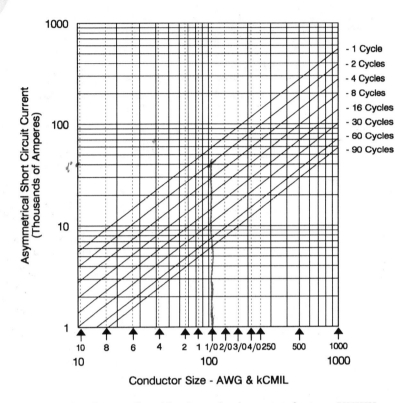

Figure 3.3b Maximum allowable short circuit current for type XHHW, copper conductors.

(b) What is the symmetrical short circuit rating of #1/0 *XHHW* copper cable protected by a circuit breaker?

From Figure 3.3(b) the maximum asymmetrical current flow is 41,600 amperes. Again, using $K_0 = 1.3$:

$$I_{sym} = \frac{41,600}{1.3} = 32,000 \ A.$$

(c) Repeat (b) except for an aluminum conductor.

Using Figure 3.3(d) the maximum asymmetrical current is 27,000 A, $K_0 = 1.3$:

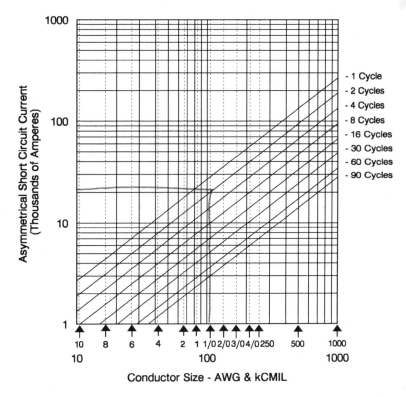

Figure 3.3c Maximum allowable short circuit current for type THW, aluminum conductors.

$$I_{sym} = \frac{27,000}{1.3} = 20,800 \ A.$$

Parallel Conductors

In many cases, it may be necessary to run several parallel sets of conductors. This situation occurs when the required current-carrying capacity of the circuit exceeds the capacity of the largest reasonable conductor size. It has been found that copper conductor sizes above 500 kCMIL or aluminum conductors above 750 kCMIL can become quite difficult to handle physically. As a result, many engineers regard 500 kCMIL copper and 750 kCMIL aluminum as maximum conductor sizes and resort to parallel sets for feeders exceeding the current-carrying

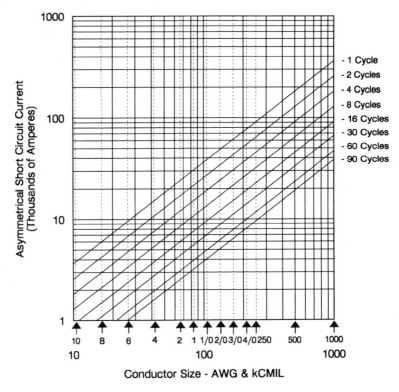

Figure 3.3d Maximum allowable short circuit current for type XHHW, aluminum conductors.

capacity of these two sizes. In fact, the *NEC* permits parallel conductors for #1/0 and larger conductors.

Figure 3.4 shows a parallel feeder served by a circuit breaker. In planning such parallel feeders, several points should be kept in mind.

1. All conductors must be of the same size and length in order to have the same impedance. This is important because each parallel set must carry its share of the total current in order to avoid overheating.
2. The conduits must be of the same type, size, and length.
3. The cable terminations must be the same in order to have equal connection impedances.
4. For economic reasons, the smallest feasible number of parallel sets should be used.

Circuit Breaker

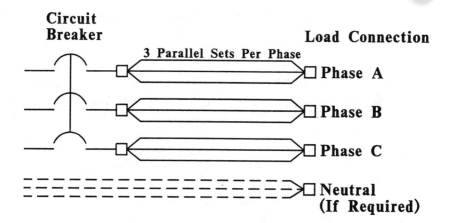

3 Parallel Sets Per Phase

Load Connection

Phase A

Phase B

Phase C

Neutral
(If Required)

(a) Parallel Feeder From Circuit Breaker

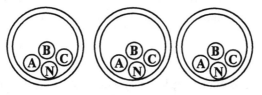

(b) Each Conduit Contains Phase A, B, C, and Neutral

Figure 3.4 Parallel sets of conductors in conduit.

Example 3.6: A given circuit requires 600 amperes. Using type *THW* copper conductor, what is the best parallel arrangement?

From Table 3.2, the current-carrying capacity of #500 kCMIL *THW* copper is 380 amperes, so parallel sets will be required. #350 kCMIL has a capacity of 310 amperes, so two parallel sets of #350 kCMIL will have a capacity of 620 amperes, which will meet our requirements.

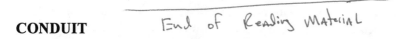

CONDUIT   End of Reading Material

Conduit provides physical protection for cable and, in some cases, acts as the required ground path. There are several types of conduit, each with its own specific application. In this section, we will discuss four common conduit types: *rigid, electric metallic tubing, intermediate metallic tubing,* and *polyvinylchloride tubing.*

Rigid Conduit (Rigid). Rigid galvanized steel (Rigid) conduit is perhaps the oldest type of electrical conduit. Its thick wall construction provides excellent physical protection for the cable and also provides the necessary grounding path. A variety of junction boxes and other fittings are available for the application of rigid conduit in wet environments or in applications where hazardous vapors are present, such as in petrochemical facilities. Rigid conduit is threaded much like ordinary water pipes, so the entire raceway assembly of conduit and fittings can be very effectively sealed. On the negative side, rigid conduit is quite heavy and, therefore, requires more installation time. Finally, the greater wall thickness results in the use of a significant amount of steel, adding to the overall cost. Like the other conduit types, rigid is available in a wide variety of diameters, varying from ½" to 6".

Electrical Metallic Tubing (EMT). The development of EMT resulted from applications requiring less physical protection than that afforded by rigid. The wall thickness of EMT is considerably less than that of rigid and the system is assembled using set-screw or compression type connectors instead of threaded fittings. Like rigid, EMT itself can be used for the necessary ground path. The reduced wall thickness results in lighter weight, less steel and hence, easier installation and lower costs. On the negative side, EMT cannot be used in hazardous areas and, despite the fact that special fittings are available, generally it is not viewed as being suitable for wet applications; Today, EMT is used widely in almost all interior applications.

Intermediate Metallic Conduit (IMC). Intermediate metallic conduit originated from a requirement for a new conduit system with many of the same physical protection characteristics of rigid, but with a somewhat thinner wall and lighter weight construction. IMC is approved for many of the same applications as rigid and was designed from its inception as an electrical conduit. Like rigid, IMC uses threaded fittings.

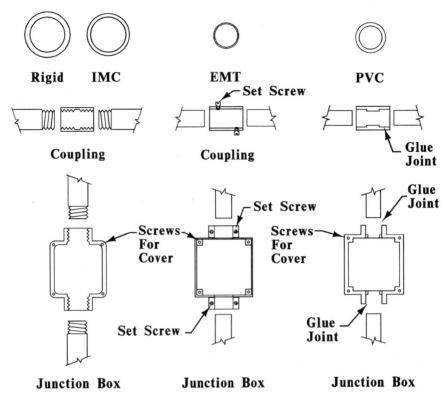

Figure 3.5 Typical fittings for rigid conduit, intermediate metallic conduit (IMC), electrical metallic tubing (EMT), and polyvinylchloride (PVC) conduit.

Polyvinylchloride Conduit (PVC). PVC conduit has gained increased acceptance during the past several years due to its light weight, ease of installation and great environmental protection. PVC conduit is, however, a petroleum-based material and is subject to the same cost variations that affect other similar products. PVC conduit is relatively easy to assemble, requiring only a saw and special adhesive glue. PVC is extremely weather resistant and is impervious to chemicals found in most soils, so it is widely used in underground direct burial applications. Due to the non-conducting nature of PVC, a separate ground conductor must be installed in all conduits.

Figure 3.5 shows these four conduit types, along with typical fittings.

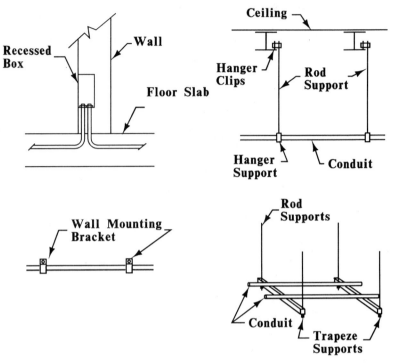

Figure 3.6 Typical conduit mounting and support details.

There are a number of methods of supporting conduit. Some of these are shown in Figure 3.6. In many applications, the conduit is installed in concealed locations, such as above the ceiling or in concrete slabs, where appearance is of little consequence. In these cases, conduit often takes a direct path instead of paralleling the building structure. In exposed applications, appearance generally dictates following the building structure. In industrial applications, conduit often is grouped together on a common support system.

Conduit Capacity

Table 3.3 shows the standard conduit sizes. The exact number of cables that can be safely installed in a given conduit size depends on a number of factors. Among these are the size of the conductor and its insulation. This is because the thickness of insulation determines, in part, the cross

Table 3.3 Standard Electrical Conduit Sizes.

½"	2½"
¾"	3"
1"	3½"
1¼"	4"
1½"	5"
2"	6"

sectional area of the finished cable. Another factor is the difficulty of pulling a given cable(s) into the raceway. The force cannot be too great or the insulation may be damaged. It has been found that a reasonable limit for the total cross sectional area of cable for a given conduit size is 40% of the conduit area. This limit also makes the pulling of cable around curves and bends far easier. Table 3 of the *NEC* contains data on the maximum number of insulated conductors in various conduit sizes. Table 3.4 summarizes some of the data from *NEC* Table 3.3.

Example 3.7: (a) What is the required conduit size for 4 #6, *THW* conductors?

From Table 3.4, the appropriate conduit size is 1 inch.

(b) Repeat (a) for *XHHW* conductors.

From Table 3.4, the appropriate conduit size is again 1 inch.

(c) How many #10 *THW* conductors can be installed in a 1 inch conduit?

From Table 3.4, 11 #10 *THW* conductors can be installed in 1 inch conduit.

(d) How many #250 kCMIL, *XHHW* conductors can be installed in 2-1/2 inch conduit?

From Table 3.4, 4 #250 kCMIL, *XHHW* conductors can be installed in a 2-1/2 inch conduit.

BUSWAY

It is frequently necessary to serve loads which are some distance from the point of service. In some cases, the cost of parallel sets of conduit and cable can be quite high. In other applications, such as large industrial areas

Table 3.4 Maximum Number of Conductors in Trade Sizes of Conduit or Tubing. (Based on Chapter 9, Tables 3A and 3B, *NEC*.)

Conductor Type	Conductor Size	½	¾	1	1¼	1½	2	2½	3	3½	4	5	6
THW	14	6	10	16	29	40	65	93	143	192			
	12	4	8	13	24	32	53	76	117	157			
	10	4	6	11	19	26	43	61	95	127	163		
	8	1	3	5	10	13	22	32	49	66	85	133	
	6	1	2	4	7	10	16	23	36	48	62	97	141
	4	1	1	3	5	7	12	17	27	36	47	73	106
	3	1	1	2	4	6	10	15	23	31	40	63	91
	2	1	1	2	4	5	9	13	20	27	34	54	78
	1		1	1	3	4	6	9	14	19	25	39	57
	1/0		1	1	2	3	5	8	12	16	21	33	49
	2/0		1	1	1	3	5	7	10	14	18	29	41
	3/0		1	1	1	2	4	6	9	12	15	24	35
	4/0			1	1	1	3	5	7	10	13	20	29
	250			1	1	1	2	4	6	8	10	16	23
	300			1	1	1	2	3	5	7	9	14	20
	350				1	1	1	3	4	6	8	12	18
	400				1	1	1	2	4	5	7	11	16
	500				1	1	1	1	3	4	6	9	14
	600					1	1	1	3	4	5	7	11
	700					1	1	1	2	3	4	7	10
	750					1	1	1	2	3	4	6	9
XHHW	14	9	15	25	44	60	99	142					
	12	7	12	19	35	47	78	111	171				
	10	5	9	15	26	36	60	85	131	176			
	8	2	4	7	12	17	28	40	62	84	108		
	6	1	3	5	9	13	21	30	47	63	81	128	185
	4	1	2	4	7	9	16	22	35	47	60	94	137
	3	1	1	3	6	8	13	19	29	39	51	80	116
	2	1	1	3	5	7	11	16	25	33	43	67	97
	1		1	1	3	5	8	12	18	25	32	50	72
	1/0		1	1	3	4	7	10	15	21	27	42	61
	2/0		1	1	2	3	6	8	13	17	22	35	51
	3/0		1	1	1	3	5	7	11	14	18	29	42
	4/0		1	1	1	2	4	6	9	12	15	24	35
	250			1	1	1	3	4	7	10	12	20	28
	300			1	1	1	3	4	6	8	11	17	24
	350			1	1	1	2	3	5	7	9	15	21
	400				1	1	1	3	5	6	8	13	19
	500				1	1	1	2	4	5	7	11	16
	750					1	1	1	2	3	4	7	10

For a complete listing of all types of cables, see *NEC* Chapter 9. Tables 1 - 7.

where the equipment being served is subject to frequent relocation, fixed wiring systems are impractical. In multi-story buildings, the use of conduit and cable may require additional internal cable support to avoid damaging the cable due to its own weight. These are the type of applications for which busway might be considered.

Instead of several parallel sets of conduits and cables, busway systems use insulated bus bars mounted within a protective metallic housing. These bus bars can be constructed with fairly large cross sectional areas, resulting in the ability to carry current levels equal to many parallel sets of cable. Busway is normally fabricated in lengths of about 10 feet to facilitate field handling and installation. The riser diagram in Figure 3.1 shows a vertical busway system serving the second and third floor areas. Note that the busway is provided with convenient locations at which disconnects (with fuses or circuit breakers) can be attached. In Figure 3.1, Panels HB and HC are connected to the busway system through such plug-in disconnect devices. In vertical applications, busway might offer the additional advantage of requiring less floor space, due to its reduced cross sectional area, than equivalent parallel runs of conduit and cable. In the case of an industrial system, the busway system might be installed near the ceiling and provided with tap locations at increments of several feet. Relocated equipment could be reconnected by moving the disconnect from the old to the new location. This type of busway is called *plug-in busway*. When frequent connections are not needed, such as in the case of a long run of busway to serve loads located at or near its end, feeder busway is used. *Feeder busway*, which does not contain tap locations, and plug-in busway, with tap locations, can be combined to meet the specific requirements of almost any application.

Busway is basically fabricated to meet the requirements of a specific project. Often, custom lengths are necessary to match the specific configuration of a given building. Figure 3.7 shows details of typical vertical and horizontal busway systems, along with a cross section of a typical busway.

Busway is available using either aluminum or copper conductors. Most busway systems are either three-phase or three-phase, four-wire, depending on the application. Three-phase busways frequently are used where only three-phase loads, such as motors or single-phase (phase-to-phase) loads such as welders, are to be connected. In applications where the busway serves loads which require neutral connections, such as the lighting panels in Figure 3.1, three-phase, four-wire busways are used. The required

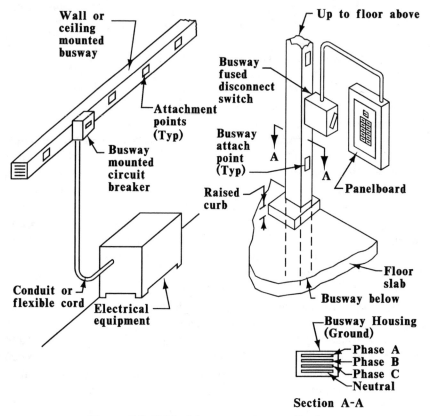

Figure 3.7 Typical busway system applications.

ground path can be provided by a separate ground bus within the busway or, alternatively, by the busway housing itself.

Busway is available in a variety of standard ampere ratings, including 100, 225, 400, 600, 800, 1000, 1350, 1600, 2000, 2500, 3000, 4000, and 5000 amperes.

In cases where a decision must be made between normal conduit and cable or busway systems, a simple economic analysis based on material and labor cost often provides the necessary guidance. Busway systems are highly flexible and are a valuable building component.

In the case of vertical busway systems, note the presence of raised curbs located at floor penetrations, as shown in Figures 3.1 and 3.7. Such curbs prevent the busway from being damaged by water or other liquids flowing across the floor following accidental spillage. The presence of such curbs has meant the salvation of many busway systems and its absence, the demise of many more.

— Begin Again —

CIRCUIT BREAKERS

In this and the following section we will discuss the two primary types of overcurrent protection devices normally used in building electrical systems, the *circuit breaker* and the *fuse*. In this section, we will discuss the characteristics of the circuit breaker and in the next, the fuse. Each type of device has its own unique advantages and disadvantages. The engineer must clearly understand these in order to make informed decisions regarding the appropriate type of device for a given application.

Both circuit breakers and fuses are overcurrent devices designed to protect the electrical system from overloads and short-circuits, but their method of providing this protection is quite different.

We begin by clarifying exactly what is meant by an overload and a short-circuit. An *overload* occurs when a circuit is required to carry more current than its normal rating. An example might be a circuit serving several receptacles in an office area where too many computers are connected. Such overloads are characterized by currents marginally above the circuit's protective device rating. On the other hand, *short-circuits* result from accidents that often reduce the normal load impedance to zero, leaving only the impedance of the conductors to limit current flow. The accidental dropping of a tool into energized components is an example of such an accident. In such cases, the current flow can reach levels of several hundred to several thousand times the overcurrent device rating.

As seen in the fault current analysis chapter, tremendous thermal stress as well as magnetically induced forces require that such short-circuits be removed as quickly as possible to minimize system damage. On the other hand, overloads represent less immediate danger to the system and can be tolerated for somewhat longer periods of time. Ideally then, we would like our overcurrent device to provide two discrete types of protection, namely from overloads and from short-circuits.

There are two basic types of circuit breakers used in most building electrical systems. These are the *molded case* circuit breaker and the *electronic trip* circuit breaker.

Molded Case Circuit Breakers

The molded case circuit breaker takes its name from the fact that the device is enclosed in a molded plastic case. This type of breaker is also often referred to as a *thermal-magnetic* device because of the method it uses to detect overloads and faults. Figure 3.8 shows details of the construction of a typical molded case circuit breaker.

The breaker contains two separate methods of detecting current flow. The first method uses a bimetallic strip that increases in temperature as the current flow exceeds a specific design value. This increased temperature tends to deform the bimetallic element, which in turn releases the circuit breaker tripping mechanism. This thermal detection scheme is useful primarily in detecting overloads that represent relatively modest levels of current above the device's rating. The second method of detecting current uses a coil of wire around which a magnetic field is established by current flow. Under short-circuit conditions, the current flow can reach several thousand times the normal level. The tremendous, sudden increase in magnetic field rapidly releases the circuit breaker tripping mechanism by means of solenoid action. A graph of both the thermal and the magnetic characteristics are shown in Figure 3.9.

Circuit breaker characteristics normally are plotted on a Log-Log axis system with current plotted along the X axis and trip time on the Y axis. The thermal range of operation is considered to extend from full rated current to about ten times full rated current. Also note the inverse time current nature of the curves. An overload 1.35 times rated current results in tripping within about 1,800 seconds, whereas a current flow of five times rated value produces tripping in about 10 seconds. For current flow above ten times rated value, there is no intentional delay in tripping action. The only actual delay results from the time required for the breaker to release the trip mechanism and the contacts to be forced open. This minimum operating time establishes the horizontal portion of the magnetic or instantaneous trip mode shown in Figure 3.9(b). The combination of both thermal and magnetic action is shown in Figure 3.9(c). In most cases, a thermal-magnetic breaker will carry about 2.5 times its rated value for 60 seconds. The right hand limit, shown in Figure 3.9(c), is determined by the maximum fault current level the breaker can safely withstand without damage to itself. This current level is called the maximum *ampere interrupting capacity (AIC)* rating of the breaker. A circuit breaker then has two discrete current ratings, a *continuous current rating* and an *AIC*

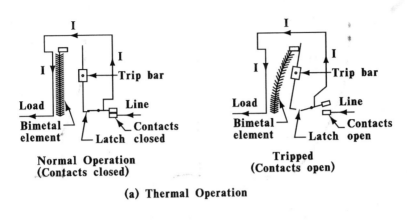

(a) Thermal Operation

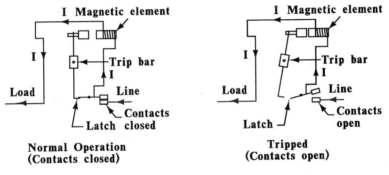

(b) Magnetic Operation

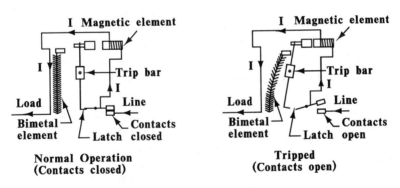

(c) Thermal-Magnetic Operation

Figure 3.8 Thermal-magnetic circuit breaker details (a) thermal operation, (b) magnetic operation, (c) thermal-magnetic operation.

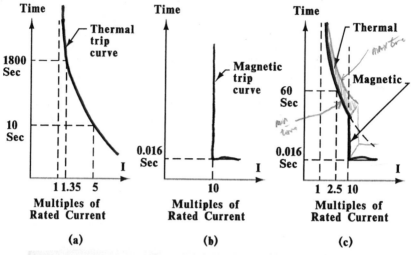

Figure 3.9 Thermal-magnetic circuit breaker trip curve characteristics.

rating. The interrupting rating is discussed further in the chapter on fault current analysis.

Circuit breakers are available in a wide variety of standard trip ratings and in one- two- or three-pole configurations. In two- and three-pole breakers, identical current sensing elements are located in each phase, with a common trip mechanism that opens all phases at the same time, even if the fault occurs on only one phase. This common trip feature prevents damaging single-phase operation of three-phase motors.

Circuit breakers also are available in several physical sizes, called *frame sizes*. The correct frame size is determined largely by the nature of the equipment into which the circuit breaker is to be installed. For example, a lighting panel may be capable of accepting only breakers of a certain frame size, while switchboards may require larger frame sizes to accommodate their specific mounting arrangements.

Circuit breakers are attached to the equipment in which they are installed in one of two methods, *plug-in mounting* or *bolt-on mounting*. As its name implies, the plug-in breaker attaches by means of spring loaded contacts which maintain tight mechanical and electrical contact. Plug-in breakers are easy to install and remove, and are generally found in residential and light commercial applications. Bolt-on circuit breakers are actually attached by screws or bolts to the equipment in which they are

Table 3.5 Typical Molded Case Circuit Breaker Frame Sizes, Trip Settings and Interrupting Ratings.

FRAME SIZE	TRIP SETTINGS
50 AMP	15 20 25 30 35 40 45 50
100 AMP	15 20 25 30 35 40 45 50 60 70 80 90 100
250 AMP	70 80 90 100 110 125 150 175 200 225
400 AMP and 600 AMP	125 150 175 200 225 250 300 350 400 450 500 600
800 AMP and 1200 AMP	250 300 350 400 450 500 600 700 800 1000 1200
1600 AMP	400 450 500 600 700 800 1000 1200 1600
3000 AMP	2000 2500 3000
4000 AMP	4000
5000 AMP	5000
6000 AMP	6000

TYPICAL INTERRUPTING RATINGS (RMS SYM. AMPERES)

240 Volt	480 Volt
10 kA	14 kA
18 kA	
22 kA	25 kA
	30 kA
	35 kA
42 kA	50 kA
65 kA	65 kA
100 kA	
200 kA	150 kA

installed. While somewhat more difficult and time consuming to install, bolt-on breakers offer increased contact pressure and are favored by most industrial users, many of whom consider this method of attachment more reliable. Table 3.5 shows the standard frame sizes, trip ratings, and interrupting capacities of molded case circuit breakers.

Molded Case Circuit Breaker Trip Tolerance

The exact tripping time of a given circuit breaker depends on several variables, ranging from the metallurgical characteristics of the bimetallic strip to required latch release pressure. Slight variations in these and other characteristics result in an operating band rather than a single line, as shown in Figure 3.9. Figure 3.10 shows a typical time current trip curve for a molded case circuit breaker. The lower curve represents the minimum clearing time of a representative sample of tested breakers, and the upper curve represents the maximum clearing time. Note that the horizontal axis is in terms of multiples of the circuit breaker trip rating and the vertical axis is in seconds.

Example 3.8: What is the trip time for the circuit breaker shown in Figure 3.10 if it is subjected to a fault of six times its full load rating? Eight times full load rating?

From Figure 3.10, the breaker will trip in about 3.5 to 11 seconds for a current of six times the device rating. For a current of eight times the device rating, the breaker will trip in about 1.6 to 6.5 seconds.

Solid State Trip Circuit Breakers

With recent advances in digital technology, it is now possible to design circuit breakers with the capability of field adjustments for several key operating parameters. These solid state trip units are fairly expensive presently, so their use generally is restricted to larger circuit breakers.

As shown in Figure 3.11, the solid state trip breaker is equipped with a field adjustable front panel with the control features specified by the engineer during the building design process. A given application may not

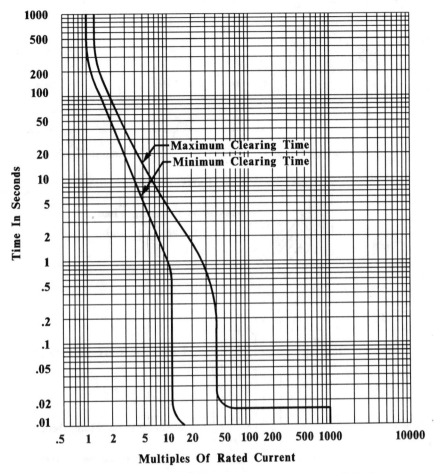

Figure 3.10 Typical thermal-magnetic molded case circuit breaker trip curve.

require all of these features.

Briefly, the available adjustments are as follows:
- The long-time current pickup and time delay allow the adjustment of the current level at which the breaker trip unit begins timing toward tripping action for current levels just above the trip rating.
- The long-time delay allows the adjustment of the time interval between detection and tripping.

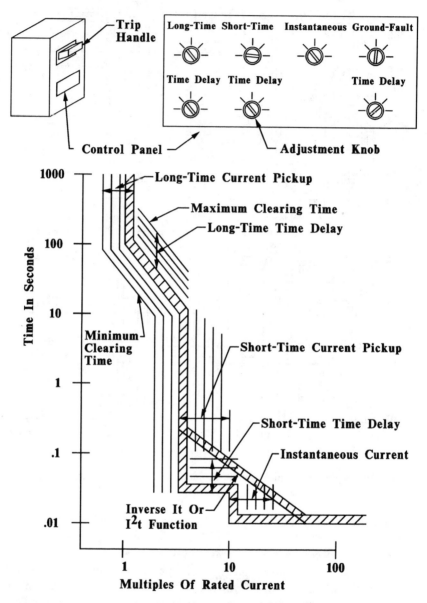

Figure 3.11 Typical trip adjustments for solid-state trip circuit breaker. Not all functions are necessarily required for all applications.

- The short-time current pickup and time delay allow adjustment of the current level at which the breaker detects currents significantly above the trip rating and begins timing toward trip.
- The instantaneous adjustment allows adjustment of the current level at which the breaker immediately initiates tripping action with no intentional time delay.

The ground fault setting controls the current flow to ground at which tripping action is initiated. Ground faults are essentially short-circuits involving ground instead of, or in addition to, phase and neutral conductors. Due to the impedance of the ground path, such faults often reach current levels of only a fraction of bolted three-phase fault values. This low current flow is particularly dangerous because it might not be detected by overcurrent devices as quickly as desired. A ground fault detection system essentially measures the current flow through the phase conductors and that returning through the neutral. In either balanced or unbalanced loads the vector sum of the phase currents equals the neutral current. If both phase and neutral conductors run through a coil, as shown in Figure 3.12, the net induced voltage in the coil is zero. This is due to the canceling effect of the magnetic fields created by the phase and neutral conductors. If a ground fault occurs, some of the outgoing current is shunted to ground and does not return through the neutral path. In this case, the net induced voltage in the coil is not equal to zero. The ground fault system detects the induced voltage, and depending on its trip setting and time delay, trips the device. We will discuss ground faults further in the fault current analysis chapter. The ground fault current pickup, as shown in Figure 3.12, can be adjusted from the full rated current of the device, to only a fraction, perhaps 10%, of the device rating.

In addition to the above adjustments, some solid state trip breakers offer adjustments for the continuous current rating. This adjustment is usually variable from 50% to 100% of the device rating. Finally, some solid state trip breakers feature the ability to shape the lower portion of the trip characteristic curve for both overcurrents and ground faults. This feature shapes the curves as an inverse function of the product of current and time, It, or ampere-squared seconds, I^2t, which is related to the energy associated with the current flow. This feature, shown in both Figure 3.11 and Figure 3.12 as a sloping line, allows the device to better coordinate with the characteristics of downstream thermal-magnetic breakers or fuses.

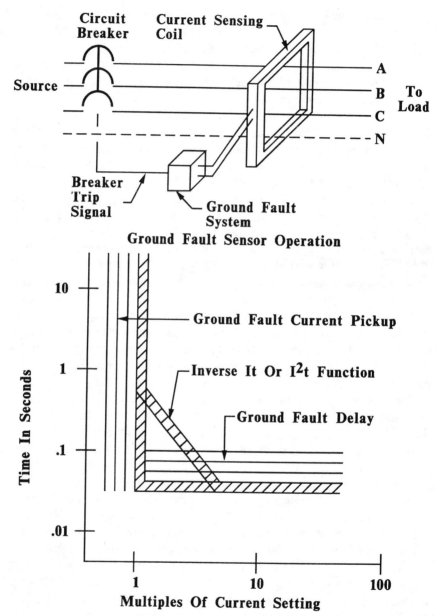

Figure 3.12 Typical ground fault system details and ground fault adjustments for solid-state trip circuit breaker.

Other Considerations in the Application
of Circuit Breakers

Temperature and Continuous Current Rating

Molded case circuit breakers are tested and calibrated to carry 100% of their rated current in open air at a specified temperature, normally 25°C. When mounted in an enclosure, the ambient temperature may be higher. As a result, most molded case circuit breakers are permitted to continuously carry only 80% of their current rating.

Power circuit breakers, such as the solid state trip breakers and a few molded case circuit breakers, are specifically designed to be operated at 100% of their current rating in a 40°C ambient. The manufacturer's technical information should be carefully reviewed to determine the exact capabilities of a specific breaker.

Current Limiting Circuit Breakers

During the past few years, special circuit breakers have been introduced which offer current limiting capabilities. In order to limit current, a device must fully open before the fault current waveform reaches its first cycle peak value. Circuit breakers are electro-mechanical devices, and, as such, are somewhat limited in their speed of operation. Such devices normally require several cycles to open. Current limiting breakers are designed to limit the I^2t energy let-through to less than the I^2t of a half-cycle wave of symmetrical fault current. We will discuss fault current wave symmetry in a later chapter.

Ground Fault Circuit Breakers

A special type of molded case circuit breaker has been developed to help reduce the potential shock hazard associated with electrical equipment.
Research shows that currents through the human body of over 5 ma can induce heart irregularities, leading to cardiac arrest and death. This micro-shock hazard can be reduced by the use of ground fault circuit interrupters (*GFCI*) devices. The solid state GFCI unit monitors the current in the phase and neutral conductors of single-phase circuits, such as those serving receptacles. An imbalance in this current indicates that an unintended path to ground, perhaps through a person, exists. These devices are designed to trip on a ground fault current of 5 ma. GFCI equipment can be manufactured as part of a molded case circuit breaker or as part of a wiring device such as a receptacle. In either case, protection is provided for

people operating equipment connected to the circuit. The *NEC* now requires GFCI devices in several specific locations, such as residential bathrooms.

FUSES

Like circuit breakers, fuses are overcurrent protective devices intended to provide protection for the branch circuits and feeders they serve. Fuses are actually the oldest type of overcurrent device and were developed by Edison about the same time he invented so many other electrical components. As we will discuss further in the chapter on utility systems, Edison essentially developed the utility industry in the late 1800s as a result of his work with the incandescent lamp. In order for his newly invented light source to have practical application, Edison realized that whole new industries would have to be created. Generation, transmission, and distribution of power would have to occur on a widespread basis, the foundation of the utility industry. Buildings would then require interior wiring systems in order to utilize the incandescent lamp. Edison realized that protecting the circuits serving a building's electrical system would be crucial to minimize hazards associated with the cohabitation of man and electricity. His early patent work centered around a "safety conductor for electric lights." In 1880, the year after he developed the first commercially viable incandescent light source, Edison's patent described the fuse as a piece of small conductor with sufficient current-carrying capacity to serve the requirements of a specific circuit size without overheating. At current levels above the designated value, the fuse element would overheat and melt, thus removing the circuit from the system and thereby avoiding damage. He reasoned that this fuse element must be contained in a protective housing to provide the necessary support and also to contain any molten material ejected from the fuse.

Although this basic description of a fuse and its function holds true today, tremendous advances have been made in understanding fuses, their materials, and construction. Today, fuses are used to provide protection for circuits serving equipment ranging from modern versions of Edison's incandescent lamp to very sensitive solid state devices. Let us now review the primary types of fuses used in building systems today.

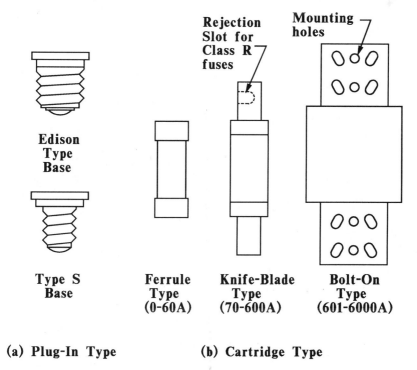

(a) Plug-In Type   **(b) Cartridge Type**

Figure 3.13 Typical fuses (a) plug-in type (b) cartridge type.

Fuse Configurations and Operating Curves

Fuses are available in a wide variety of configurations, ranging from those that bear a striking resemblance to Edison's early fuse to bolt-in units for large switchboards. Figure 3.13(a) shows two variations of plug-in fuses. The first, or Edison type, plug-in is found only in older buildings and almost totally has been replaced by the newer, narrow base configuration, also shown. Adapters are available to facilitate replacement of the older fuse types without the necessity of equipment replacement. Plug-in fuses are found today primarily in only residential applications. Figure 3.13(b) shows fuses of the cartridge type. The ferrule type fuse is available in current ratings of 0 to 60 amperes. Also shown is the knife blade type, available in current ratings of 70 to 600 amperes. The last fuse shown is the bolt-on type, used for large feeders and available in standard ampere ratings from 601 to 6000 amperes.

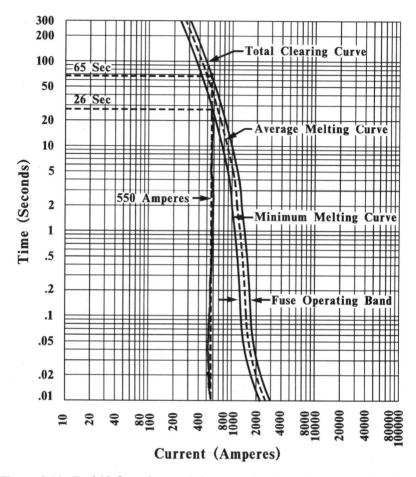

Figure 3.14 Typical fuse characteristic curve showing minimum melting time, average melting time and total clearing time.

Fuses, like circuit breakers, operate on an inverse time current relationship, designed to quickly remove short circuits while permitting short term overloads, such as those associated with motor starting. Figure 3.14 shows a typical fuse characteristic curve, that, like those of circuit breakers, is drawn on a Log-Log axis system. Also like circuit breakers, fuses have an operating band rather than a single operating curve. In the case of fuses, however, the lower curve is referred to as the minimum melting time curve, and the upper curve as the maximum clearing time curve. The total clearing time is the sum of the time required for the fuse element to melt plus the subsequent arcing time. The minimum melting time and total clearing time can vary ±10% from the average current value, shown as a dotted line.

Example 3.9: What is the minimum melting and total clearing time of the fuse shown in Figure 3.14 if subjected to a current level of 550 amperes?

From Figure 3.14, the minimum melting time the fuse for a current level of 550 amperes is about 26 seconds, and the total clearing time is approximately 65 seconds.

The lower axis scale often includes a specified multiplier, such as 10, 100, or 1000, in order to calculate the actual current. Care should be taken in reading these curves to make sure that any such multiplier is understood.

Current Limitation and Dual Element Fuses

One of the greatest advantages of fuses is that certain types have the ability to limit the flow of fault current to a level well below the theoretical peak value. This greatly reduces the stress on the system from both the thermal and magnetic forces associated with large-scale faults. In order to act as a current-limiting device, the fuse must effectively clear the fault before the current wave reaches its first peak. Figure 3.15(a) shows a cross section of a typical current-limiting fuse. The fuse is housed in a sturdy case to safely contain the melting and arcing which occurs during fuse operation. The case is filled with a very high grade silicon quartz, and is hermetically sealed to prevent loss of this filler. The fuse element itself is often made of pure silver and has intentional "weak" links to encourage fuse melting at these locations. The blades provide both support for fuse mounting as well as electrical contact for current flow through the fuse. When a short circuit occurs on the circuit served by the fuse, the silver element heats and melts very rapidly. The melting action fuses the silica filler in the immediate area of the melted link, as shown in Figure 3.15(a). The heat absorbed by the silica reduces the arc heat and helps extinguish the arc itself. The non-conducting nature of the fused silica prevents the restriking of the arc. Larger fuses often contain several current-carrying elements which are parallel, as shown in Figure 3.15(b).

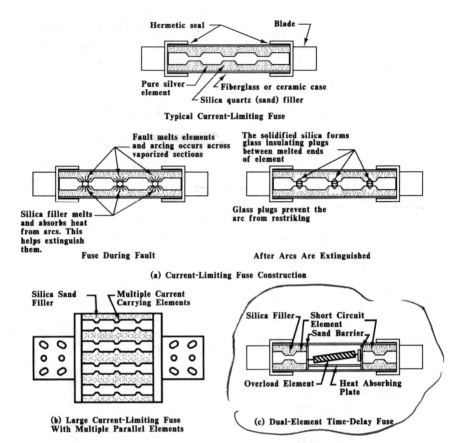

Figure 3.15 Typical fuse details (a) current-limiting fuse operation (b) dual element time-delay fuse construction (c) large current-limiting fuse construction.

The current waveform during the fault is shown in Figure 3.16. The fault current rapidly escalates toward the potential instantaneous peak shown, but prior to reaching this value the fuse element melts and arcing begins to occur. The current is then forced to zero before the first ½ cycle point on the current waveform. The current never reaches the highest potential instantaneous peak, but rather a lower value, called the peak let-through current. The total clearing time of the fuse is the sum of the melting time and the arcing time. The area under the curve represents the energy let-through of the fuse, referred to as its I^2t let-through value.

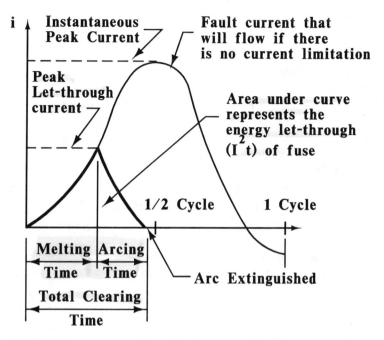

Figure 3.16 Fault current waveform showing current-limiting fuse operation.

Unfortunately, the fast response time required for current limitation also makes the fuse very sensitive to short time overloads, such as those encountered in normal motor starting. The ideal fusible element for maximum current limitation possesses low conductivity, small mass for reduced heating time, and a very high, but sharply defined, melting point. On the other hand, the ideal fusible element for short term overloads might have somewhat lower conductivity, higher mass, and a lower melting point. In order to achieve both current limitation and acceptable overload delay, a dual element design has been created that uses two elements in series as shown in Figure 3.15(c). The overload element responds to currents up to 500% of the fuse rating. This element is not embedded in silica quartz, but is contained in a separate compartment in the center of the fuse. The short circuit elements respond to currents above 500% of the fuse rating. These are located in end compartments which are filled with silica quartz, as described previously. This series combination provides both the desired current-limiting and time delay characteristics. Today, essentially all current-limiting fuses are of the dual-element type.

Now we turn our attention to the various standard fuse classifications.

Fuse Classifications

Figure 3.17 shows the family of fuse classifications as recognized by Underwriters Laboratories (UL). The majority of building fuse applications fall under the current-limiting and non-current-limiting classifications.

Non-Current-Limiting Fuses

This family of fuses consists of the plug fuses shown in Figure 3.13(a), and early cartridge fuses, shown in Figure 3.13(b) called Class H fuses. Class H fuses are capable of safely interrupting fault current levels of only 10,000 amperes, and are not current-limiting. Today, many systems have prospective fault current levels far exceeding the capabilities of this fuse. Class H fuses are rated for 250 or 600 volts and are available in ratings up to 600 amperes. Some Class H fuses are renewable, as the fuse element can be replaced without the necessity of replacing of the entire fuse assembly. Class H fuses are often called "*NEC* fuses." Today, such fuses are available primarily for replacement use in older installations.

Current-Limiting Fuses

We will now examine the more commonly used current-limiting fuses, Classes J, K, L, R, and T.

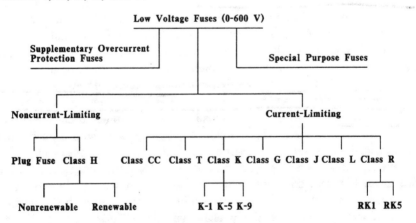

Figure 3.17 Low voltage fuse classifications.

Class J

Class J fuses are current-limiting and are rated for 600 volts or less. They are capable of interrupting faults of up to 200,000 amperes. Their dimensions are not compatible with other fuse classes. Some Class J fuses are labeled as "time delay" which means they must be capable of carrying a current level of five times their rating for at least 10 seconds. These fuses are available in current ratings up to 600 amperes.

Class K

Class K fuses are current-limiting fuses with three discrete ratings: K-1, K-5 and K-9. Each rating has specific limits on peak current let-through and I^2t values. The interrupting rating for the K-1, K-5 and K-9 types are 50,000, 100,000, and 200,000 amperes, respectively. Class K fuses are dimensionally similar to Class H fuses, so the unfortunate possibility does exist for misuse of Class H fuses in Class K applications. Class K fuses may be marked "time delay" if they are capable of carrying five times their current rating for at least 10 seconds.

Class L

Class L fuses are widely used, and are available in ampere ratings of 601 amperes to 6000 amperes and for voltages of 600 volts or less. They are capable of interrupting fault current levels of up to 200,000 amperes, and are designed to be bolted in place. Despite the fact that no specific criteria has been developed for rating the time delay features of Class L fuses, many do, in fact, exhibit significant time delay characteristics. These fuses are excellent candidates for use as primary overcurrent devices for major feeders or sub-feeders.

Class R

Class R fuses are available in 250 or 600 volt ratings and two sub-classifications based on the level of peak let-through current and thermal let-through (I^2t). These two sub-classifications are RK-1 and RK-5, with RK-1 having the most restrictive limits on both peak let-through and I^2t. Class R fuses are available in ratings up to 600 amperes, and are equipped with a slot in the blade designed to match a rejection pin feature in equipment designed for this type of fuse. This pin prevents the insertion

Table 3.6 Standard Ampere Ratings for Low Voltage Fuses.

0-600 Amperes		
15	70	225
20	80	250
25	90	300
30	100	350
35	110	400
40	125	450
45	150	500
50	175	600
60	200	
601-6000 Amperes		
601	1350	3000
650	1500	3500
700	1600	4000
800	1800	4500
1000	2000	5000
1200	2500	6000

of other fuses in equipment intended for Class R fuses, but does allow Class R fuses to be retrofitted in applications where older Class H fuses were originally used. Many engineers feel that the design of larger feeders using Class L fuses (601-6000 A), combined with Class R fuses (up to 600 A) for smaller feeders, provides extremely effective system protection. Class R fuses are time-delay fuses rated to carry 500% of their rated current for at least 10 seconds.

Class T

Class T fuses are relative newcomers to the fuse family and are specifically designed for use in compact installations. They are current-limiting with interrupting capabilities up to 200,000 amperes. They are available in current ratings up to 600 amperes and for 250 or 600 volt operation. These fuses are very fast acting and have both lower peak let-through current and thermal let-through (I^2t) values than other fuse types. They are not interchangeable with other fuses due to their more compact size. They do not have time delay characteristics. Table 3.6 lists standard fuse ratings for

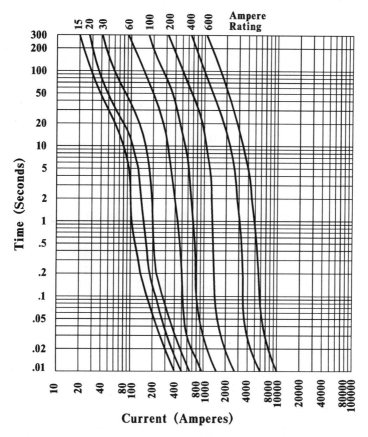

Figure 3.18 Typical characteristic curves for type RK-1 current-limiting fuses, 15-600 amperes.

fuses from 20 amperes to 600 amperes and from 601 amperes to 6000 amperes.

Typical Fuse Curves

Figures 3.18 and 3.19 show typical average melting time curves for a representative group of Class RK-1 and Class L fuses. Remember that the melting of a specific fuse might occur at a current level 10% above or 10% below the average value shown.

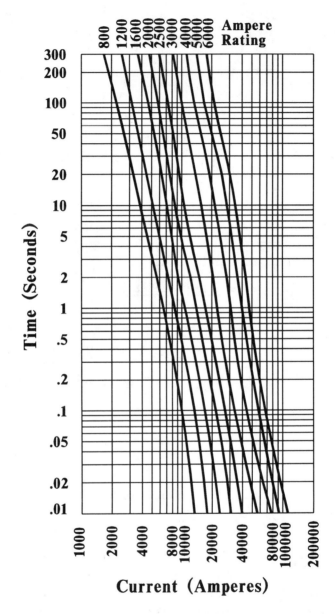

Figure 3.19 Typical characteristic curves for type L current-limiting fuses, 800–6000 amperes.

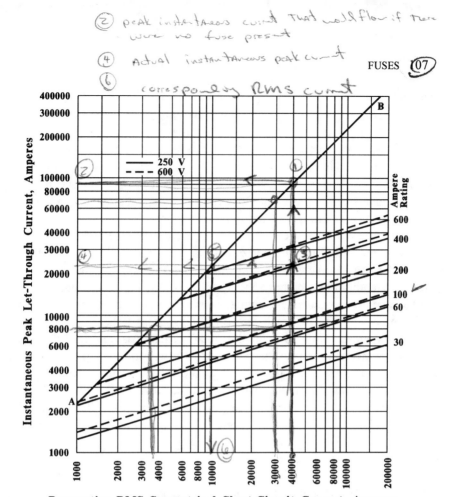

(handwritten annotations on figure:)
(2) peak instantaneous current that would flow if there were no fuse present
(4) Actual instantaneous peak current
(6) corresponding RMS current
7000

Figure 3.20 Typical fuse let-through curves for type RK-1 current-limiting fuses, 30-600 amperes.

Peak Current Let-Through Curves

Figure 3.20 shows peak let-through curves for several typical RK-1 fuse ratings. The reading and interpretation of these curves requires a bit of explanation.

The horizontal axis represents the *rms* prospective short-circuit current. This current value is determined during a fault current study using techniques presented in a later chapter. The vertical axis represents the instantaneous peak current that would flow if no fuse were present. The slope of the line AB is determined from the ratio of the instantaneous peak fault current to the *rms* symmetrical fault current for a fault with a 15% power factor. It should be understood that this power factor is that of the circuit during fault conditions, and not the power factor of the circuit during normal operation. In most circuits, the inductive reactance component predominates under fault conditions and, hence, the low fault power factor. Note that the individual fuse curves stop at the AB line. At current levels below this point, no current-limiting action occurs, above this point the fuse enters its current-limiting region.

After determining the prospective fault current value and fuse size, the graph is entered along the horizontal axis. A line is extended vertically from the prospective short-circuit current to the point it crosses the AB line. Extending the line horizontally to the vertical axis, we read the peak instantaneous current that would flow if no fuse were present. We next determine the point at which the vertical line crosses the appropriate fuse curve, extend this line horizontally to the vertical axis, and read the actual instantaneous peak current permitted to flow with the fuse present. If we stop the horizontal line just drawn at the AB line and extend the line vertically downward to the horizontal axis, we read the maximum *rms* symmetrical fault current permitted to flow by the fuse. This *rms* value can be used to determine the appropriate interrupting rating for devices located downstream, such as circuit breakers.

Let's look at an example.

Example 3.10: A fault current analysis reveals that the prospective fault current level at a specific point in the building's electrical system is 40,000 amperes. The fuse contemplated at this location is a 400A RK-1 fuse rated at 250 V, as shown in Figure 3.20. What is the instantaneous peak current if no fuse were present, the instantaneous peak let-through current of the 400A RK-1 fuse, and the actual *rms* current let-through?

Entering Figure 3.20 at 40,000 amperes, we extend a line vertically to the AB line and horizontally to the vertical axis, and read 93,000 amperes. This is the peak instantaneous current that would flow if no fuse were present.

If we stop our vertical line at its intersection with the 400 A (250 V) fuse line and extend it horizontally to the vertical axis, we read an actual instantaneous peak current of 23,000 amperes. If we stop our horizontal line at the AB axis and read vertically downward, we find the corresponding *rms* current to be about 9800 amperes.

In summary, a potential fault current level of 40,000 amperes *rms* has been reduced to a *rms* value of 9800 amperes by the current-limiting action of the 400A RK-1 fuse.

TRANSFORMERS

One of the greatest advantages of *ac* systems is the ability to change voltage levels by means of transformers. Utilities routinely use step-up transformers at generation sites to raise the generator voltage, often to 500 to 1000 kV, in order to lower the current level and, hence, lower the I^2R losses associated with power transmission over great distances. Within cities, the voltage is often lowered to distribution levels of 25 kV or less for distribution to individual communities or industrial areas. At these locations, other transformers step the voltage down for actual connection to the customer. Within the customer's facility, it is often necessary to once again change voltage levels in order to accommodate specific types or classes of equipment. In this section, we will discuss some of the characteristics of this important component.

The Ideal Transformer

A good place to begin our discussion of transformers is with the ideal transformer, shown in Figure 3.21. A transformer is essentially two windings on a core of magnetic steel. One set of windings is called the *primary* and is connected to the source of power. The other winding, the *secondary*, is connected to the load being served. The primary and secondary are linked by the magnetic flux produced by the primary, which in turn induces a voltage in the secondary winding. The steel core, normally made of a number of layers or laminations of magnetic steel, serves as a path for the magnetic flux. As shown in Figure 3.21, the primary voltage is v_p, the secondary voltage, v_s, the number of primary turns, N_p, and the number of secondary turns, N_s.

When energized, current, I_p, will flow in the primary which will give rise to a flux, φ. This flux will link the windings of the secondary and induce a voltage. If the secondary is connected to a load, a current, I_s, will flow. The primary and secondary voltages involved are given by

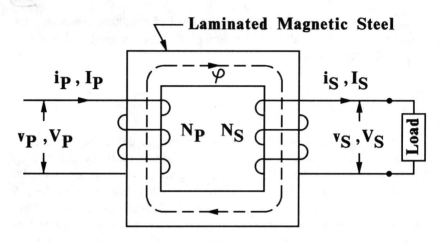

Ideal Transformer

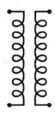

Transformer Symbol

Figure 3.21 The ideal transformer.

$$v_p = N_p \frac{d\varphi}{dt} \tag{3.9}$$

$$v_s = N_s \frac{d\varphi}{dt}. \tag{3.10}$$

From Equations 3.9 and 3.10, we see that

$$\frac{V_p}{V_s} = \frac{N_p}{N_s} \tag{3.11}$$

where V_p and V_s are now the *rms* values of the primary and secondary voltages.

We define the ratio of the transformer's primary turns to secondary turns as

$$a = \frac{N_p}{N_s}.$$ (3.12)

In the ideal transformer, the windings have no resistance and all flux generated by the primary is assumed to link the secondary. As a result, there is no loss in the power transformation and the input power will equal the output power given by

$$V_p I_p = V_s I_s \quad VA.$$ (3.13)

We are now in a position to develop several important relationships for the ideal transformer. From Equation 3.13 we see that

$$I_p = \frac{V_s I_s}{V_p} \quad Amperes$$ (3.14)

where V_p and I_p represent the primary voltage and current. Similarly, the secondary current will be given by

$$I_s = \frac{V_p I_p}{V_s} \quad Amperes.$$ (3.15)

Since $V_p I_p = V_s I_s$ = load VA, we can rewrite Equations 3.14 and 3.15 in terms of kVA and kV as simply

$$I_p = \frac{kVA}{kV_p}$$
$$I_s = \frac{kVA}{kV_s}$$ (3.16)

From Equation 3.16, we see that if the transformer's load kVA is known, along with the primary and secondary voltage, it is possible to compute the primary and secondary current.

If we connect the ideal transformer to a load impedance, Z_L, the secondary current will be

$$I_s = \frac{V_s}{Z_L} \quad Amperes \qquad (3.17)$$

and the secondary kVA will be

$$kVA_s = kV_s \cdot I_s \quad kilovolt-Amperes. \qquad (3.18)$$

We may often find it necessary to determine the value of Z_L, as seen from the primary. To do this we begin with Equation 3.11. Rearranging Equation 3.11

$$V_p = \frac{N_p}{N_s} V_s = a V_s \quad Volts \qquad (3.19)$$

and from Equation 3.13

$$I_P = \frac{V_s}{V_p} I_s = \frac{1}{a} I_s \quad Amperes. \qquad (3.20)$$

Dividing Equation 3.19 by 3.20

$$\frac{V_p}{I_p} = a^2 \frac{V_s}{I_s}. \qquad (3.21)$$

Realizing that

$$\frac{V_p}{I_p} = Z_p \quad Ohms$$

and

$$\frac{V_s}{I_s} = Z_s \quad Ohms$$

we can now rewrite Equation 3.21 as

$$Z_p = a^2 Z_s \quad Ohms \tag{3.22}$$

which gives us an easy way to reflect impedances from the secondary to the primary, or vice versa.   We will make use of this technique in our discussion of short circuit analysis in a later chapter.

Let's look at an example to illustrate our discussion so far.

Example 3.11:   Given an ideal transformer as shown in Figure 3.21 with $V_p = 480$ V, $V_s = 240$ V and $Z_L = 12$ Ω, what is the primary current, I_p, the secondary current, I_s, and the total power delivered by the transformer?

Using Equation 3.17

$$I_s = \frac{V_s}{Z_L} = \frac{240}{12} = 20 \quad Amperes.$$

Using Equation 3.22

$$Z_p = a^2 Z_s = \left[\frac{480}{240} \right]^2 = 4 \cdot 12 = 48 \ \Omega.$$

So that

$$I_p = \frac{V_p}{Z_p} = \frac{480}{48} = 10 \quad Amperes.$$

The total power can be computed using either secondary or primary values as

$$
\begin{aligned}
VA &= V_s I_s = 240 \cdot 20 = 4800 \quad Volt-Amperes \\
&= V_p I_p = 480 \cdot 10 = 4800 \quad Volt-Amperes.
\end{aligned}
$$

As we will see in later chapters, most load calculations will be expressed in terms of kVA and therefore, the primary and secondary currents will be computed using Equation 3.16.   In short-circuit studies, where the load impedance is neglected, we will be concerned with the impedance of the system conductors and will rearrange Equation 3.22 to reflect impedances from primary to secondary.

To summarize, there are two important points regarding the ideal transformer:

a. There is no power loss in the ideal transformer. This is because the windings have no resistance and there are no hysteresis or eddy current losses. In other words, the power consumed by the load is equal to the power flowing into the transformer primary. Also, since there are no losses, the ideal transformer can deliver any amount of power required without concern for overheating.

b. Under no-load conditions, the primary current will be equal to zero. This is because the magnetic steel is assumed to have infinite permeability, and all flux is confined to the core.

The Ideal versus the Real Transformer

Unfortunately, the ideal transformer does not actually exist. The windings of real transformers have resistance and inductive reactance. The flow of current will produce some degree of I^2R heating within the windings. Even under no-load conditions, some small level of primary current flows in order to maintain the losses associated with the transformer's magnetic steel core. As load current flows, these losses increase. A more accurate transformer model is shown in Figure 3.22. The winding resistance and inductive reactance are represented by R_s and X_s respectively.

We now see that under no-load conditions, a small magnetizing current, I_m, will flow through R_m and X_m. If a load is connected to the secondary, the primary current becomes the sum of the magnetizing current, I_m, and the load current, as seen from the primary, I_s/a. The transformer losses will then increase as I_s increases. This model more accurately reflects an actual transformer.

In actuality, the value of R_m and X_m are very great compared with those of R_s and X_s. The total losses in modern transformers are quite small and many have efficiencies in the range of 97% or even higher.

As a practical matter, then, we normally compute load currents and kVA's, for both primary and secondary, using the ideal transformer model. On the other hand, we cannot use the ideal transformer model for fault current studies because the ideal model has no impedance whatsoever. As

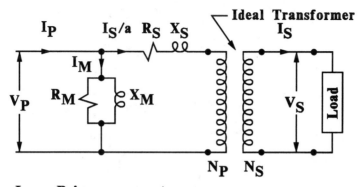

Ip Primary current
IM Magnetizing current
IS Secondary current
RM Magnetizing resistance
XM Magnetizing reactance
RS Series resistance (primary and secondary)
XS Leakage reactance

Figure 3.22 The non-ideal transformer.

we will see, even the small impedance of real transformers significantly limits the fault current allowed to flow through the primary to the secondary.

Transformer Connections

A single transformer can be connected, as shown in Figure 3.23(a), to provide two secondary voltages by simply grounding the center of the secondary winding. In this case, the line-to-line voltage will be equal to twice the line-to-neutral voltage. A common example of this is a transformer with a primary voltage of 480 volts and a secondary voltage of 120/240 volts.

It is also possible to combine three single-phase transformers to provide a three-phase transformer bank, as shown in Figure 3.23(b). The primaries of these single-phase transformers are connected in a delta configuration,

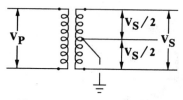

(a) Single-Phase Transformer

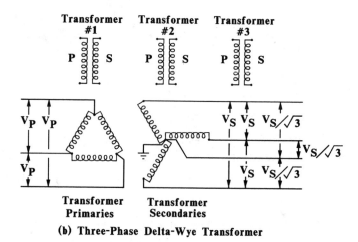

(b) Three-Phase Delta-Wye Transformer

Figure 3.23 Typical transformer connections (a) single-phase (b) three phase.

and the secondaries are connected in a wye configuration, with the center point of the wye grounded. As shown, most transformer primaries are delta connected. If the nature of the secondary load does not require a phase-to-neutral connection, the secondary could be delta connected. Most transformer secondaries are wye connected.

100%Z would allow 100% of rated I to flow when 100% rated voltage is applied.

Transformer Short-Circuit Test and %Z

The basic method of determining the transformer's ability to deliver fault current is shown in Figure 3.24. A transformer is connected to a variable source of primary test voltage, V_T, and an ammeter is placed in the primary as shown. The secondary of the transformer is then short circuited.

The test begins by very slowly increasing the value of V_T and carefully noting the level of primary current, I_p, and the value of V_T. When the

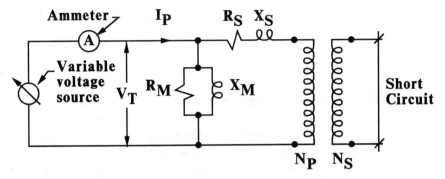

Figure 3.24 Transformer short circuit test connections.

current reaches the transformer's nameplate full load primary current rating, I_p, the value of V_T is recorded. The percent impedance, %Z, of the transformer is then given by

$$\%Z = \frac{V_T}{V_p} \cdot 100 \quad Percent \tag{3.23}$$

%Z = % rated Voltage applied during short circuit test.

where V_p is the transformer's rated primary voltage.

The test can be interpreted as follows: If a voltage of V_T causes full-rated primary current, I_p, to circulate in the primary windings, and, hence, full-rated current to circulate in the secondary as well, then 100% of the rated primary voltage will result in a current of V_p/V_T times the rated primary.

This assumes that the primary voltage can be maintained even under fault conditions, and that the transformer core will be capable of carrying this level of current. In other words, this is a worst case situation. Stated another way, the percent impedance of a transformer is the percentage of the rated primary voltage that will cause rated primary current to circulate in the transformer primary and secondary when the secondary is shorted.

Knowing the transformer's rated secondary current, we can calculate its worst case fault current let-through as

$$I_{sc} = I_{Rated} \cdot \frac{100}{\%Z} \quad Amperes$$

or in terms of the rated kVA

$$I_{fc} = \frac{I_{sec}}{transform\ impedance\ (3\%)}$$

$$I_{sc} = \frac{kVA}{kV} \cdot \frac{100}{\%Z} \quad Amperes \quad Single-Phase \ Transformer$$

$$I_{sc} = \frac{kVA}{kV\sqrt{3}} \cdot \frac{100}{\%Z} \quad Amperes \quad Three-Phase \ Transformer.$$

(3.24)

Example 3.12: A fault current test is run on a 50 kVA single-phase transformer. The rated primary voltage is 480 volts, and the rated secondary voltage is 120/240 volts. During the test, a primary voltage of 21.6 volts is found to circulate full-rated primary current. What is the percent impedance of the transformer and also the maximum secondary fault current capability of the transformer?

Here, $V_p = 480$ V and $V_T = 21.6$ V. Using Equation 3.23

$$\%Z = \frac{21.6}{480} \cdot 100 = 4.5\%.$$

Now, using Equation 3.24, on the secondary side of the transformer

$$I_{sc} = \frac{50}{.240} \cdot \frac{100}{4.5} = 4630 \ Amperes.$$

Standard Transformer kVA Ratings

In reality, the kVA rating of a transformer is partly a function of its ability to dissipate the internal heat generated during normal operation. It is also related to the nature and amount of magnetic steel used in the tramsformer's construction, as well as the thermal limits of the transformer's winding insulation. Transformers are available in a number of standard kVA ratings in both single-phase and three-phase configurations. They are also available with several classes of insulation, permitting operation under a range of ambient temperatures and loading conditions. Available insulation classes are 105, 120, 150, 185, and 220°C.

Table 3.7 lists data for a representative group of single- and three-phase transformers. Take a few moments to review this table. Note that representative dimensions and weights are listed in addition to X/R ratios and percent impedance, %Z. We will need both X/R and %Z in our later discussion of short-circuit analysis.

Table 3.7 Typical Transformer Data.

GENERAL PURPOSE TRANSFORMERS STANDARD THREE-PHASE kVA RATINGS						
kVA	Height (in.)	Width (in.)	Depth (in.)	Weight (lbs)	Avg. (%Z)	Avg. X/R
15.0	23.0	22.25	15.0	230	3.6	1.94
30.0	23.0	22.25	15.0	285	6.4	0.92
45.0	26.0	24.00	15.0	369	6.6	1.13
75.0	30.0	30.00	20.0	590	5.7	1.38
112.5	37.0	30.00	20.0	690	6.1	1.51
150.0	42.0	36.00	24.0	1050	5.5	1.53
225.0	42.0	36.00	24.0	1350	6.6	2.00
300.0	48.0	48.00	29.5	2000	3.6	1.81
500.0	58.0	48.00	29.5	2700	5.0	2.89
750.0	90.0	72.00	54.0	5200	5.0	1.98
1000.0	90.0	72.00	54.0	6000	5.8	2.38

GENERAL PURPOSE TRANSFORMERS STANDARD SINGLE-PHASE kVA RATINGS						
kVA	Height (in.)	Width (in.)	Depth (in.)	Weight (lbs)	Avg. (%Z)	Avg. X/R
3.0	14.25	7.50	7.75	48	-	3.53
5.0	15.75	9.38	9.00	75	-	2.64
7.5	16.00	12.00	10.63	102	-	2.54
10.0	19.00	12.00	10.63	128	-	2.18
15.0	23.00	16.00	15.00	190	5.2	0.60
25.0	23.00	16.00	15.00	230	6.4	1.06
37.5	30.00	20.00	20.00	345	5.6	1.07
50.0	30.00	20.00	20.00	430	5.2	1.21
75.0	37.00	20.00	20.00	450	6.1	1.64
100.0	42.00	24.00	24.00	665	6.5	1.65
167.0	48.00	32.00	29.50	1050	7.0	2.27

Remember that Table 3.7 represents data for representative dry-type indoor transformers normally used in building electrical systems, not outdoor, high voltage transformers used by utilities. Finally, it should be understood that the rated primary and secondary currents are based on the transformer's kVA rating.

Example 3.13: What is the rated primary and secondary current for a 50 kVA, 480 V-120/240 V single-phase transformer? A 75 kVA, 480 V-208Y/120 V, three-phase transformer?

For the 50 kVA single-phase transformer

$$I_p = \frac{kVA}{kV} = \frac{50}{.480} = 104.2 \ \ Amperes$$

$$I_s = \frac{kVA}{kV} = \frac{50}{.240} = 208.3 \ \ Amperes$$

For the 75 kVA, three-phase transformer

$$I_p = \frac{kVA}{kV\sqrt{3}} = \frac{75}{.480 \cdot \sqrt{3}} = 90.2 \ \ Amperes$$

$$I_s = \frac{kVA}{kV\sqrt{3}} = \frac{75}{.208 \cdot \sqrt{3}} = 208.2 \ \ Amperes.$$

Transformer Mounting

Step-down transformers should be located as closely as possible to the load being served in order to minimize the cost of the larger secondary conductors. In Figure 3.1 note that step-down transformers are located near the panels being served. Transformers can be mounted in a number of ways, depending on the available space and the size and weight of the transformer involved. For smaller transformers, it is often possible to arrange mounting on brackets above the panel being served, as shown in Figure 3.25(a). If wall brackets are not feasible, a trapeze support can be designed to provide the necessary support from the ceiling. In either case, the location should be coordinated with the structural engineer in order to assure sufficient wall or ceiling strength. The obvious advantage to wall or ceiling mounting is the reduction in floor space required for electrical

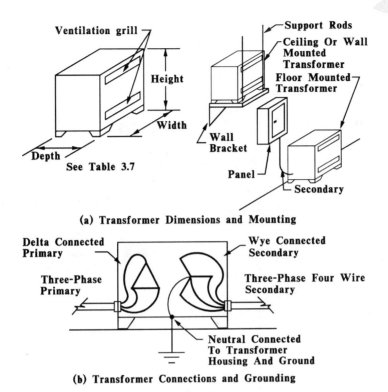

(a) Transformer Dimensions and Mounting

(b) Transformer Connections and Grounding

Figure 3.25 Typical transformer details (a) typical indoor transformer dimensions (see Table 3.7) and mounting arrangements (b) typical delta-wye transformer wiring.

equipment. With floor space at a high premium, this small reduction can be important. When wall or trapeze support is not practical, transformers can be floor-mounted near the panel or equipment they serve.

A word of caution regarding heat build-up is in order. The author once visited a large office building to perform some requested electrical tests. These tests required the installation of power monitoring equipment in the electrical equipment room. The room contained not only the main switchboard, but a 112.5 kVA step-down transformer. The room was found to be so hot that it was not possible to install the monitoring equipment for fear of possible heat damage. The building's designers had forgotten that the 112.5 kVA transformer, even lightly loaded, would generate several kW of heat. No provisions had been made to remove the heat by ventilation or by the air conditioning of the room. Eventually, a ventilation system was installed. The moral of the story? Let the architect and mechanical engineer know if you plan to install equipment which will generate significant amounts of heat.

PANELBOARDS AND SWITCHBOARDS

Panelboards and switchboards serve as locations for overcurrent devices which serve the branch circuits, feeders and sub-feeders of the building's electrical system. They are located as close to the loads they serve as possible, and provide a convenient location for the control of the various circuits connected to the individual overcurrent devices located within. A *panelboard* consists of a metal enclosure containing bus bars to which circuit breakers or fused switches are attached. The interior space of the housing provides sufficient physical space for safe installation of the circuit conductors to their respective overcurrent devices. *Switchboards*, on the other hand, normally serve as locations for larger overcurrent devices, or as main distribution panels for an entire building. Switchboards are physically much larger than panelboards, due to the size of the overcurrent devices involved, and are designed to provide the necessary space for installation of larger cables. Switchboard housings are of metal construction, and often consist of several individual cubicles, or sections, which are fabricated by the manufacturer and shipped to the job site for field assembly.

Lighting Panelboards and Power Panelboards

Lighting panels are, actually, more accurately defined as lighting and appliance branch-circuit panelboards. As the name implies, these panels are intended to serve both lighting branch circuits as well as circuits for receptacles and other equipment. According to *NEC*, Section 384.14, a lighting panelboard must have at least 10% of its overcurrent devices rated 30 amperes or less, for which neutral connections are provided. There is also a limit on the number of individual overcurrent devices located in a single lighting panel enclosure. A single-pole breaker counts as one overcurrent device because it connects to the panel on only one phase. A two-pole breaker counts as two overcurrent devices, and a three-pole breaker as three overcurrent devices. The total number of poles permitted by *NEC*, Article 384.15 for a single panelboard enclosure is 42. Should more overcurrent devices be required, additional panelboard sections can be added. Figure 3.26 shows a typical lighting panelboard with branch circuits for both single-phase and three-phase loads. Lighting panels can be mounted either on the surface or actually recessed in a wall. Surface-mounted panels are normally used in unfinished spaces or in locations where appearance is not critical. The raceways for branch circuits can be

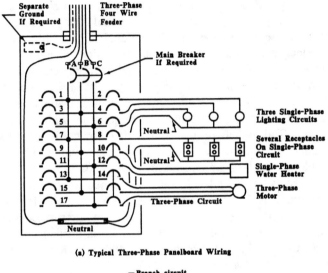

(a) Typical Three-Phase Panelboard Wiring

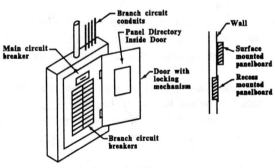

(b) Panelboard Elevation and Mounting

Figure 3.26 Typical panelboard details (a) typical panelboard construction and wiring (b) typical panelboard front details and mounting arrangements.

run through the sides, top or bottom of the panel, depending on installation requirements. In the case of surface mounted panels, both the feeder or sub-feeder serving the panel as well as branch circuits leaving the panel will be exposed. In the case of recessed panels, both panel feeders and branch circuits can be concealed in the wall.

Lighting panels can have either a main circuit breaker, as shown in Figure 3.26, or just termination lugs for cable connection. A panel without a main breaker is referred to as a *main lugs only* or *MLO* panel. The decision of whether or not to provide a main breaker depends on several factors. If the panelboard feeder is protected by an overcurrent device not rated more highly than the rating of the panel, the presence of a main overcurrent device at the panel may be optional. The design engineer may, however, elect to provide a main breaker in order to facilitate shut-down of the panel. In other instances, the panel may be served from a large feeder that serves one or more additional panels. In such cases, a main circuit breaker is required by the *NEC* as well as plain common sense, since the breaker permits isolation of any one panel without the necessity of switching off all panels on the feeder. If the panel is fed directly from the secondary of a transformer, a main breaker is required. Lighting panels are available in standard ampere ratings from 100, 225, and 400 amperes.

Power panels are intended to serve equipment other than lighting and small appliances as described above. Power panels are also available in surface or recessed configurations, although many are surface mounted due to the fact that they are frequently located in unfinished spaces. Power panels are available in standard ratings of 400, 600, 800, 1000, and 1200 amperes.

Both lighting panels and power panels are available in a number of voltage ranges, including 120/240 volts, single-phase, three-wire, 208Y/120 volts, three-phase, four-wire and 480Y/277 volts, three-phase, four-wire systems.

Both lighting panels and power panels, as well as their overcurrent devices, are required to be marked with their maximum *rms* fault current interrupting rating. The lowest available standard rating for 208Y/120 volt systems is 10,000 amperes, and for 480Y/277 volt systems, 14,000 amperes.

Lighting and power panel circuits are normally numbered in a standardized way, as shown in Figure 3.26. The circuits on the left side of the panel begin with Circuit 1, followed by 3, 5, 7, etc. Those on the right are numbered 2, 4, 6, 8, etc. The important point to note is that circuits 1, 3, 5 are each on separate phases.

The advantage here is that a single neutral conductor can be used to serve all three, because, even under unbalanced conditions, the current can never exceed the value of the current in the heaviest loaded circuit. In actuality, under balanced conditions, the neutral will carry no current at all because balanced three-phase currents are completely cancelled in the neutral. A notable exception to the cancellation of current in the neutral is in the case of non-linear loads. In such loads, a significant portion of the current is often in the third harmonic (180 Hz). Third, sixth, ninth, etc., harmonic currents do not cancel in the neutral, but are in fact, additive. We will discuss this and related issues in the chapter on power quality. In the case of such non-linear loads, it is necessary to increase the size of the neutral conductors for three-phase circuits or to install a separate neutral for each circuit.

Switchboards

Switchboards act as the primary distribution points for the building. Unlike panelboards, switchboards are of such size that they are normally free-standing instead of surface- or recess-mounted. Some switchboards are designed to be mounted directly against a wall and, as a result, all wiring connections and maintenance are accomplished from the front or side only. Other switchboards are designed to be mounted several feet from a wall, and have removable access panels at the rear for installation of cable and for maintenance. The switchboard MDP shown in Figure 3.1 acts as the primary distribution point for the system shown. In larger buildings, other switchboards are often served from the main distribution panel, thus forming a system with several layers or levels.

Switchboards are normally 90 inches in height and are manufactured in sections called cubicles, which may vary in width from about 30 inches to 48 inches, and in depth from about 24 inches to 48 inches. The various cubicles which make up the entire switchboard are shipped to the job site, and are then bolted together. Shipment and handling of a fully assembled switchboard is simply not practical.

A switchboard can contain a number of different types of overcurrent devices as well as metering and instrumentation options. Each switchboard is in fact a custom assembly, thus affording the engineer a great deal of design flexibility. Let's discuss a few of these design options.

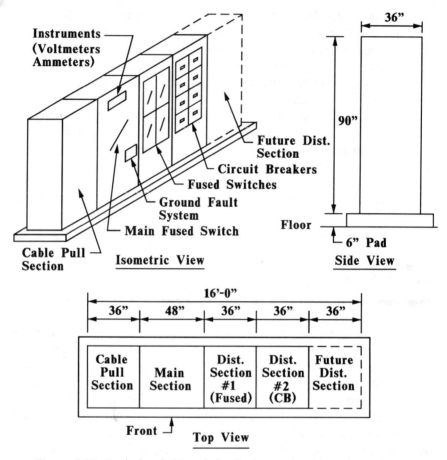

Figure 3.27 Typical switchboard details, isometric, side and top views.

Figure 3.27 shows an isometric view as well as a front, top and side view of a typical switchboard. Figure 3.28 shows a typical internal bussing and connection detail for this same switchboard.

Cable Pull-Sections

The conductors providing power to the switchboard can be either in the form of conduit and cables or busway. In the case of conduit and cable, the point of entry is frequently from the bottom, as shown in Figure 3.1. There are often as many as 10 or more individual conduits. As you can imagine, simply installing and terminating 40 or more large (500 kCMIL

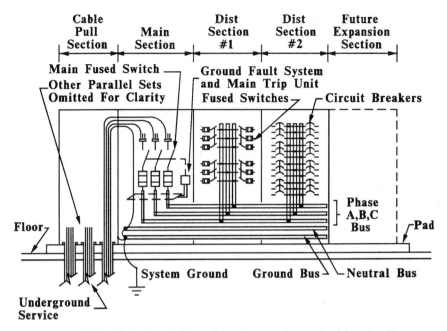

Figure 3.28 Typical switchboard interior bussing and wiring details.

or more) conductors can be quite a challenge. Depending on whether the switchboard is front or rear accessible, and taking into account the number of conduits and the cable size entering the switchboard, a cable pull section might be specified. This area is specifically intended to provide the physical space to set up cable-pulling equipment and to install, and bend, the large number of conductors involved. Of course, this cubicle requires additional floor space, which is always at a premium, but its use is often well worth the effort.

Main Overcurrent Device
The switchboard may or may not require a main overcurrent device, depending on its application. If used as a main switchboard serving the entire facility, it will normally have a main overcurrent device. This device can be either a fused switch, a molded-case or a power circuit breaker. In 480Y/277 volt systems, the main overcurrent device is normally equipped with a ground fault detection system to trip the main device in case of a ground fault.

Branch Overcurrent Device

Overcurrent devices for equipment connected to the switchboard are normally either fused switches or molded-case circuit breakers. In industrial applications, power circuit breakers might be used. Depending on the nature of the system, some of these overcurrent devices may also be equipped with ground fault protection.

Special Metering and Instrumentation Equipment

The switchboard can also contain a wide variety of special metering and instrumentation options tailored to the specific needs of the project. For example, some utilities permit the installation of current as well as potential transformers for utility metering purposes. They often require special security arrangements for such metering in order to avoid unauthorized tampering. Many engineers prefer to include voltmeters and ammeters with front-mounted displays and switches to permit measurement of the various voltages and currents. More sophisticated instrumentation might allow complete monitoring of energy usage, power factors and other important operational parameters. In very modern facilities, this data can be monitored directly by the building's electrical/energy management system, and integrated with mechanical system data to provide a comprehensive assessment of the total electrical and mechanical system performance. Such systems have become very important as the overall cost of energy has increased.

Bus Material and Bus Bracing

Switchboard busses can be made of either copper or aluminum. These bus bars are mounted on insulators, and are mechanically braced to withstand the tremendous level of magnetically induced forces associated with short circuits. The level of bracing required is determined from the results of a fault current study as well as by examining the characteristics of upstream overcurrent devices. Today, many switchboards utilize aluminum bus bars.

Ampacity and Voltage Ratings

Switchboards are available in voltages including 120/240, 208Y/120 and 480Y/277 volts, in ampacities of 1200 to 4000 amperes. In actual practice, applications requiring over 3000 amperes are often sub-divided into two or more smaller switchboards. Common ampacities are 1200, 1600, 2000, 2500, 3000, and 4000 amperes.

Special Requirements

Switchboards are available which can meet a number of environmental requirements, ranging from normal interior mounting to totally enclosed, rain-tight construction for outside mounting. In some cases, step-down transformers mounted inside a switchboard section are available which can be bolted directly between the section housing the transformer's primary overcurrent device and the section served by the transformer.

MOTORS

Motors play an important role in building electrical systems due to their use in so many applications. In mechanical systems, motors are used to power air conditioning compressors, for fans to move air in the building, or for pumps to circulate hot water. In the area of industrial and commercial processes, motors are also widely used. In some industrial facilities, motors are used to power pumps which transfer process liquids, move production line conveyers, run overhead cranes, and stir process vats and mixers. In many industries, powdered materials are moved by large motor-powered blowers. In food stores, refrigerated case systems include a wide variety of small motors ranging from circulation fans to compressors. Motors are also used to power compactors, automatic door openers, and check-out conveyor belts.

Despite the variety of available motors, both *ac* and *dc*, over 90% of all motors in use today are *ac induction motors*. There are a number of reasons for this fact, which include reliability, simplicity of construction, and low cost. Although induction motors are available for single-phase systems, most induction motors are of the three-phase type. As we will see, this is due, in large measure, to the three-phase motor's quiet operation and smooth torque characteristics.

In the early part of this section, we will focus on the characteristics of the three-phase induction motor. In the latter part of this section, we will discuss the theory and operation of single-phase induction motors, and finally, briefly review several techniques for the speed control of induction motors.

Three-Phase Induction Motors

The key to the successful operation of the three-phase induction motor is the rotation of a magnetic field created by the stationary windings of the motor's armature. With properly located and designed windings, the magnetic field created by all three-phases combines to produce a single rotating flux wave of constant magnitude and speed of rotation. The rotor of the induction motor consists of a number of individual conductors, or bars, embedded in slots which run the length of the rotor. The rotating magnetic field induces voltages and currents in these rotor bars. Figure 3.29(a) shows the basic configuration of this kind of motor. The interaction of the magnetic field with the rotor currents creates a force on the rotor that comes close to accelerating to the speed of the rotating magnetic field. The rotor can never reach the speed of the rotating field

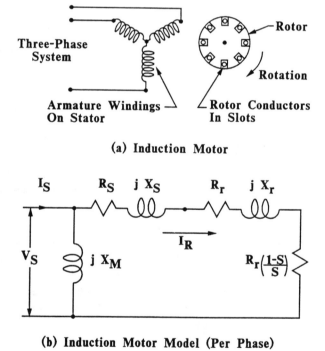

(a) Induction Motor

(b) Induction Motor Model (Per Phase)

Figure 3.29 Three-phase induction motor (a) stator windings and rotor detail (b) model of one phase of a three-phase induction motor.

because the relative motion between the rotor and field which creates the induced currents, and hence, the torque. Even if there is no load connected to the motor, there will be friction and windage losses, so the motor accelerates to a speed just below the speed of rotation of the magnetic field. The speed of rotation of the magnetic field depends on the number of poles and the frequency of the applied three-phase voltage, and is given by

$$\omega_s = 2\frac{\omega}{P} \quad Radians/Sec. \tag{3.25}$$

where ω_s is the synchronous speed of the rotating magnetic field, and ω the supply voltage frequency, both in radians per second. P represents the number of motor poles. Since ω is determined by the utility system frequency, f, the value of ω_s is constant once the number of poles is established.

The speed of rotation of the magnetic field in *RPM* will be given by

$$RPM_s = \frac{120 \cdot f}{P} \quad RPM. \tag{3.26}$$

As mentioned earlier, the induction motor never reaches synchronous speed, but rather settles down to a speed just below ω_s. The rotor then operates at a relative slip with the synchronous speed. This slip, s, is given by

$$s = \frac{\omega_s - \omega_r}{\omega_s} \qquad n_r = (1-s)n_r \tag{3.27}$$

where ω_s and ω_r are the angular speeds of the rotating magnetic field and rotor, respectively. When the motor is first energized, the rotor is at a standstill, so $\omega_r = 0$. In this case, $s = 1$. Should the rotor reach the speed of the revolving magnetic field, $\omega_s = \omega_r$ and $s = (\omega_s - \omega_r)/\omega_s = 0$.

Example 3.14: A four-pole induction motor operates at a slip of 0.05. What is the angular frequency of the magnetic field, ω_s, the speed of the rotor, ω_r?

From Equations 3.25 and 3.26

$$\omega_s = 2\frac{\omega}{P} = \frac{2 \cdot 2\pi \cdot 60}{4} = 60\pi = 188.5 \ \textit{Rad}/\textit{Sec}.$$

and

$$RPM_s = \frac{120f}{P} = \frac{120 \cdot 60}{4} = 1800 \ RPM.$$

Solving Equation 3.27 for ω_r and using s = 0.05 yields

$$\omega_r = (1 - s) \ \omega_s = 0.95 \ \omega_s$$

so that

$$RPM_r = 0.95 \cdot 1800 = 1710 \ RPM.$$

There are a number of common models for the three-phase induction motor, and all represent some degree of compromise. Figure 3.29(b) shows one commonly used model that is sufficiently accurate for our purposes. Take a few moments to examine the model that represents one phase of a three-phase motor. V_s is the per phase stator voltage. I_s is the motor input current or stator current. R_s and X_s represent the stator resistance and inductive reactance, respectively. Note that since R_s is in series with I_s, the I^2R losses in the stator increase as I_s increases. This seems reasonable. The value of X_m represents the motor's magnetizing reactance. R_r and X_r represent the resistance and inductive reactance of the rotor itself, and I_r, the rotor current. It is customary to divide R_r into two parts as shown. The energy dissipated in the $R_r(1 - s)/s$ component represents the actual output power of the motor. Note that when the motor is at a standstill and $s = 1$, this component has a value of zero, indicating that no useful work is being done. In this case, all of the rotor energy is liberated as heat in the R_r component that models the motor under starting or locked rotor conditions. As the motor approaches synchronous speed and the slip approaches its rated value, a larger proportion of energy will be expended in useful work.

The model shown in Figure 3.29 can be used to develop an expression for the torque developed by the complete three-phase induction motor. If the values of V_s, R_s, X_s, R_r and X_r are known and are fixed, the output torque is a function of only the slip, s. The expression for the output torque of a three-phase induction motor is given, without proof, by Equation 3.28.

$$T_d = \frac{3 R_r V_s^2}{s \omega_s \left[\left(R_s + \dfrac{R_r}{s}\right)^2 + \left(X_s + X_r\right)^2\right]} \qquad Newton\ Meters.$$

(3.28)

Let's look at an example of a typical three-phase motor and see how its output torque varies with the motor's speed.

Example 3.15: A 460 volt, three-phase, four-pole induction motor has the following characteristics:

$$R_s = 1.01\Omega \qquad R_r = 0.69\Omega$$
$$X_s = 1.30\Omega \qquad X_r = 1.94\Omega$$

Determine the output torque for values of slip between 1.0 (O *RPM*) and 0.0 (1800 *RPM*) and plot the results.

Figure 3.30 shows the torque/speed curve for this motor using Equation 3.28, and the above values for the various motor parameters. Note that the motor has a significant level of torque at startup in order to begin acceleration of the motor's load. The torque/speed characteristics of a representative load are also shown in Figure 3.30. Since the torque output of the motor exceeds the torque requirements of the load, the motor will, in fact, accelerate the load. As the motor accelerates toward synchronous speed, it first reaches the peak value shown at a slip value of about 0.20 (1440 *RPM*) called the breakdown torque. As the motor accelerates even further, the relative motion between the magnetic field and the rotor decreases to the point that the output torque begins to decline. At the point where the output torque exactly equals the load's torque requirements, a point of equilibrium is reached and stable operation begins.

Further study of the operating characteristics of induction motors reveals several important characteristics. First, induction motors experience a very high current inrush at startup that can reach levels of up to six times the normal full load current. Secondly, the maximum overall motor efficiency, as well as the highest operating power factor levels, are attained at slips of about 0.04 or 4%. From a design viewpoint, the first point

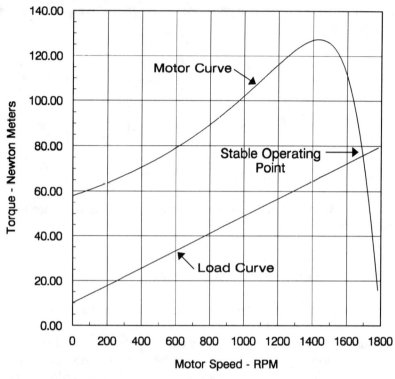

Figure 3.30 Typical torque speed curves for three-phase induction motor and load.

means that we must be very careful in the design of circuits serving induction motors in order to avoid nuisance tripping of overcurrent devices due to the high momentary inrush currents at motor startup. At the same time, the motor is best operated at loads very close to its maximum output rating for maximum overall efficiency and highest power factor. As we will see later in this section, it is often desirable to vary the speed of the motor in order to match varying torque requirements.

Motor Classes, Frame Sizes, Full Load Currents

Induction motors are classified according to their torque characteristics at *startup* (*locked rotor*), at *maximum output* (*breakdown*), *starting current*, and *slip at full load*. Most modern motors are classified as B, C, D, or F, depending on their characteristics. The appropriate standards for these motors are found in the *National Electrical Manufacturers Association (NEMA)* Standard MG-1.

Class B

Class B motors are the standard motor design most commonly found in today's building systems, and are referred to as general purpose motors. For this classification, the starting torque for 1800 *RPM* motors ranges from 80 to 275% of the motor's full load torque rating. For smaller motors, 1/2 to 3 HP, the value is somewhat over 200%, and for large motors, over 100 HP, the value is 100% or less. The breakdown torque ranges from 300 to 175% of the full load torque (300% for 1 HP, 175% for 500 HP). These machines are rated at a maximum full load slip of 5% (s = .05), and have an inrush current level of up to 600% of their full load current rating.

Class C

Class C motors are equipped with two sets of rotor bars, each with different resistance values. One set of bars has a somewhat higher resistance value and is used to produce increased starting torque. As the motor accelerates, the first set of rotor bars is replaced by the second, lower resistance set, for normal operation. The additional costs associated with constructing this type of motor can be significant. NEMA standards indicate that motor starting torque varies between 250% (7½ HP) and 200% (200 HP) of its full load torque rating. In this type of motor, the maximum torque output (breakdown) varies between 200% (5 HP) and 190% (200 HP) of its full load torque rating. These motors are rated for a maximum full load slip value of 5% (s = .05). The maximum inrush current during startup can be 600% of the motor's full load current rating. These motors are generally referred to as *high torque* motors.

Class D

Class D motors are designed with rotor bars of higher-than-normal resistance and actually achieve their maximum torque value at startup. NEMA standards reveal that these motors have startup torques of 275% of the motor's full load torque rating. These motors operate at full load slips of up to 15% (s = .15), and have maximum inrush currents of 600% of the motor's full load rating. Class D motors are referred to as *high torque, high slip* motors.

Class F

Class F motors are rather special motors, and are designed for low starting current. They are available only in ratings over 30 HP. These motors have a starting torque of 125% of the full load torque rating and a maximum torque of 160% of the full load torque rating. These motors have a maximum full load slip of 5% ($s = .05$) and a maximum current at startup of 400% of the rated full load value. Class F motors are referred to as *low starting current, low torque* motors.

Frame Size

It became important, as early as 1928, to standardize certain features of the mounting arrangements for motors. This standardization was important for motor manufacturers and equipment manufacturers, as well as for ultimate users. Today, standard mounting arrangements can be determined by reference to the appropriate NEMA requirements for the motor's frame size. Frame size is indicated by three numbers followed by a letter of the alphabet. These letters, *U* or *T*, refer to changes in shaft diameter standards due to the creation of more compact, high horsepower motors. The *U* designation refers to changes which occurred in 1953 and the *T* to changes in 1964. For example, a motor frame size might be indicated as 365T. Figure 3.31 shows that, according to NEMA requirements, the shaft centerline is nine inches above the mounting plate. The distance between

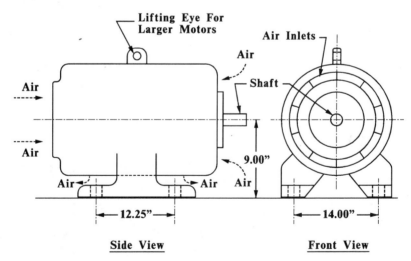

Figure 3.31 Side and front view of typical induction motor.

Table 3.8   Full-Load Current in Amperes - Single-Phase Alternating Current Motors.

The following values of full-load currents are for motors running at usual speeds and motors with normal torque characteristics. Motors built for especially low speeds or high torques may have higher full-load currents, and multispeed motors will have full-load current varying with speed, in which case the nameplate current ratings shall be used.

The voltages listed are rated motor voltages. The currents listed shall be permitted for system voltage ranges of 110 - 120 and 220 - 440.

HP	115 V	200 V	208 V	230 V
1/6	4.4	2.5	2.4	2.2
¼	5.8	3.3	3.2	2.9
1/3	7.2	4.1	4.0	3.6
½	9.8	5.6	5.4	4.9
¾	13.8	7.9	7.6	6.9
1	16	9.2	8.8	8
1½	20	11.5	11	10
2	24	13.8	13.2	12
3	34	19.6	18.7	17
5	56	32.2	30.8	28
7½	80	46	44	40
10	100	57.5	55	50

the mounting holes is 14 inches across the motor, and 12.25 inches along the length of the motor.

Full Load Currents

In order to standardize the design process and eliminate potential confusion, the *NEC* has established standard full load current values for various horsepower ratings and voltages. For single- and three-phase motors, this data is summarized in *NEC* Tables 430-148 and 430-150, respectively. Portions of these tables are reprinted as Table 3.8 for Single-Phase Motors, and Table 3.9 for Three-Phase Motors. We will make use of this data in the process of designing circuits to serve motors.

Table 3.9 Full Load Current* - Three-Phase Alternating Current Motors.

HP	115 V	200 V	208 V	230 V	460 V
INDUCTION TYPE SQUIRREL-CAGE AND WOUND-ROTOR AMPERES					
½	4	2.3	2.2	2	1
¾	5.6	3.2	3.1	2.8	1.4
1	7.2	4.1	4.0	3.6	1.8
1½	10.4	6.0	5.7	5.2	2.6
2	13.6	7.8	7.5	6.8	3.4
3		11.0	10.6	9.6	4.8
5		17.5	16.7	15.2	7.6
7½		25.3	24.2	22	11
10		32.2	30.8	28	14
15		48.3	46.2	42	21
20		62.1	59.4	54	27
25		78.2	74.8	68	34
30		92	88	80	40
40		119.6	114.4	104	52
50		149.5	143.0	130	65
60		177.1	169.4	154	77
75		220.8	211.2	192	96
100		285.2	272.8	248	124
125		358.8	343.2	312	156
150		414	396.0	360	180
200		552	528.0	480	240

* These values of full-load current are for motors running at speeds usual for belted motors and motors with normal torque characteristics. Motors built for especially low speeds or high torques may require more running current, and multispeed motors will have full-load current varying with speed, in which case the nameplate current rating shall be used.

The voltages listed are rated motor voltages. The currents listed shall be permitted for system voltage ranges of 110 - 120, 220 - 240, 440 - 480, and 550 - 600.

Motor Power Factor and Efficiency

As mentioned previously, induction motors operate at their highest power factor and best overall efficiency when operated close to full load. Below full load, both power factor and efficiency decline substantially. The efficiency of a motor is defined as:

$$\% \; Efficiency \;\; = \;\; \frac{Mechanical \;\; Output \;\; Power \; \cdot \; 100}{Electrical \;\; Input \;\; Power}. \qquad (3.29)$$

Since motor output is commonly specified in terms of horsepower, we can convert this to kW by multiplying the horsepower rating by .746 kW/horsepower.

As an example of the effect of motor load on efficiency and power factors, let us consider the performance of a representative three-phase induction motor rated 10 HP. At full load, the efficiency is 85%, while at one-half load, the efficiency declines to 83%. For this same motor, a typical value for power factors might be 88% at full load and only 77% at half load.

In addition to the effects of motor loading on efficiency and power factors, the horsepower rating itself plays a rather important role. Figure 3.32 shows a representative curve for motor efficiency and for power factors as a function of motor size. Note that both curves are for fully loaded motors, and that both overall efficiency and power factors are higher for larger motor sizes.

Motor Enclosures, Service Factors and Insulation

Motor Enclosures

Motor enclosures are available which can meet a wide range of applications, ranging from open motor enclosures for normal interior applications to totally enclosed motors for use in hazardous locations where flammable liquids or gasses are present. Open motors are available which are designed to withstand any liquids dripped or splashed on the housing. Open motors have air intakes at the ends and internal fan and exhaust ports on the bottom. Totally enclosed motors can be either non-ventilated or fan-cooled, depending on the requirements of the application.

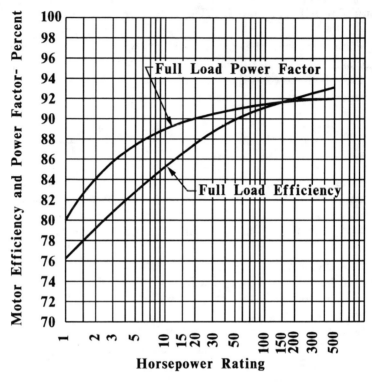

Figure 3.32 Graph of three-phase induction motor full load efficiency and power factor as a function of various standard horsepower ratings.

Service Factors

Some motors are designed to operate at a slightly higher level than the full load rating. This overload ability is expressed in terms of the motor service factor. For example, a service factor of 1.15 means that a 20 horsepower motor can be loaded continuously to 23 horsepower without damage.

Motor Insulation

Motors are available with several types of insulation ratings based on the maximum permissible temperature difference between the interior of the motor and its surrounding ambient. The ambient is assumed to be 40°C. The standard insulation classes are A, B, F, and H. For a motor with a service factor of 1.0, Class A, B, F, and H insulations are rated for 60, 80,

105, and 125°C, respectively. For a motor with a service factor of 1.15, A, B, and F insulations are rated for 70, 90, and 115°C, respectively.

Single-Phase Induction Motors

In many applications, it is desirable to use induction motors even though there is no three-phase system available. Applications include residential air conditioning systems, washers, dryers and fans. Small commercial and industrial facilities might include the above motor types, plus other small process motors. Unfortunately, as noted in an early part of this section, induction motor operation depends on the establishment of a rotating magnetic field. In the case of single-phase systems, this is not possible. The stator coils would simply produce a pulsating magnetic field. It is interesting to note that if one manually spins a single-phase motor, it will continue to spin, and will accelerate to operating speed. The motor is carried through the null torque points of the magnetic field by its momentum. The problem is, simply, how to initiate rotation. There are two common approaches to the solution.

Resistance Starting

In the resistance starting method, a second winding of higher resistance and lower reactance is added to the motor. This winding is connected through a centrifugal switch which opens once the motor reaches about 75% of its rated speed. During startup, the current in the main winding, which has lower resistance and higher inductive reactance, will lag current in the starting winding by about 25°. This creates a starting torque of about 150% of the full rated value. Such motors are called *resistance split-phase motors* for obvious reasons. Regrettably, single-phase motors are much noisier than three-phase motors due to the vibration induced by the pulsating magnetic field. A second drawback can be the potential failure to open of the centrifugal switch, that can easily result in the overheating and destruction of the motor. The basic design of a resistance start motor is shown in Figure 3.33(a).

Capacitance Starting

In the case of a capacitance start motor, the main winding still has low resistance and high inductive reactance. The start winding has a moderate level of both resistance and inductive reactance. In this case, however, a

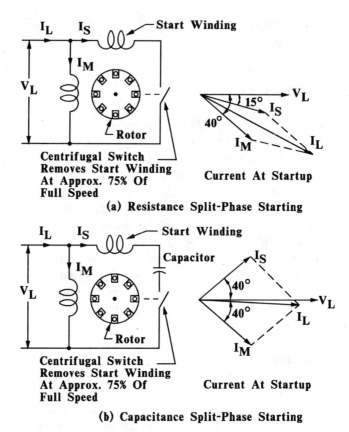

(a) Resistance Split-Phase Starting

(b) Capacitance Split-Phase Starting

Figure 3.33 Single-phase induction motor starting techniques (a) resistance split-phase starting (b) capacitance split-phase starting.

capacitor is added in series with the start winding and centrifugal switch. The value of the capacitor is chosen so as to produce a leading power factor for the start winding. The current in the start winding in this instance leads the current in the inductive main winding by almost 90°. This can produce starting torque of up to 450% of the full load value. Again, however, the motor is much noisier than a three-phase motor, and is also at risk due to the potential failure of the centrifugal switch. These motors are often referred to as *capacitance split-phase motors*. The basic design of a capacitance start motor is shown in Figure 3.33(b).

Speed Control of Induction Motors

The immense popularity of the induction motor, especially the three-phase variety, stemmed directly from its very simple construction, as well as its robust and reliable operating characteristics. Unfortunately, speed control was a serious problem, and in applications where such control is necessary, induction motors lost some of their usefulness. During the past few years, even more applications for variable speed motors have evolved, making this a very important topic. A number of plans for varying the speed of induction motors have been developed, and we will briefly review several of these.

Stator Voltage Control

A review of Equation 3.28 suggests one approach to induction motor speed control, and that is by varying the input voltage, V_s. As V_s is reduced, the speed will be reduced somewhat and the slip will increase. Overall, however, the speed is not reduced as much as we would like, but the input current will be increased substantially. In fact, if V_s is reduced beyond a certain point, overheating and motor damage can occur. As a result, speed control by varying V_s is effective for only very slight speed reduction.

Wound Rotor Motors

In the wound rotor motor, the windings of the rotor are made accessible by means of slip rings on the motor's shaft. The wound rotor motor achieves speed control by varying the value of the rotor resistance and, hence, the shape of the torque speed curve. This approach can be used to achieve high starting torque by adjusting the variable resistor to a higher value during starting, and then adjusting it downward during normal operation. The higher value of resistance also lowers the starting current. If the resistance is then increased slightly during normal operation at constant load, the slip will increase and the motor will slow down. Unfortunately, however, motor efficiency decreases drastically. In addition, the extra equipment required adds both cost and potential maintenance headaches.

Electronically Controlled Drives

Recent advances in electronic power conversion technology have finally made effective speed control of induction motors a reality. Speed control is achieved by varying the frequency of the applied voltage, and often the applied voltage as well. One of the most popular techniques is called *volts-*

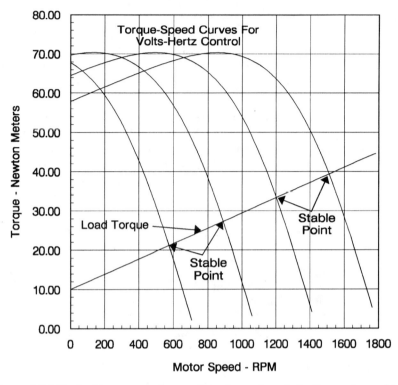

Figure 3.34 Typical torque speed curve for a three-phase induction motor and load for volts-Hertz speed control.

hertz control. In this control plan, *ac* voltage at line frequency, 60 Hz, is connected to the variable speed drive system, which transforms the *ac* voltage to *dc*, and then inverts the *dc* to a variable frequency and variable voltage output. A typical family of torque speed curves for a motor controlled by a volts-hertz system is shown in Figure 3.34.

Such systems are beginning to find very widespread use in many induction motor applications. Unfortunately, due to the nature of the power electronic devices, very significant current waveform distortions are produced at the input to the converter which can result in high levels of harmonic current in the building's distribution system. The problems associated with the increased use of solid state power conversion technology will be discussed further in the chapter on power quality.

References

Beeman, Donald. 1955. *Industrial Power Systems Handbook*. New York: McGraw-Hill Book Company, Inc.

Croft, Terrell, and Wilford Summers. 1987. *American Electricians' Handbook*. New York: McGraw Hill Book Company, Inc.

Elgerd, Olle I. 1977. *Basic Electric Power Engineering*. Massachusetts: Addison-Wesley Publishing Company.

Fink, Donald G., and John M. Carroll. 1969. *Standard Handbook for Electrical Engineers*. New York: McGraw Hill Book Company, Inc.

Hughes, S. David. 1988. *Electrical Systems in Buildings*. Boston: PWS-Kent Publishing Company.

Institute of Electrical and Electronic Engineers, Inc. 1986. *IEEE Recommended Practice for Electric Power Distribution for Industrial Plants. ANSI/IEEE Std. 141-1986*. New York: Institute of Electrical and Electronic Engineers, Inc.

Institute of Electrical and Electronic Engineers, Inc. 1990. *IEEE Recommended Practice for Electric Power Systems in Commercial Buildings. IEEE Std. 241-1990*. New York: Institute of Electrical and Electronic Engineers, Inc.

National Electrical Manufacturers Association. 1987. *NEMA Standards Publication No. MG-1 Motors and Generators*. Washington, D.C: National Electrical Manufacturers Association.

National Fire Protection Association. 1989. *National Electrical Code - 1990*. Massachusetts: National Fire Protection Association.

The Okonite Company. 1984. *Engineering Data. Copper and Aluminum Conductor Electrical Cables*. New Jersey: The Okonite Company.

Rashid, Muhammad Harunur. 1988. *Power Electronics - Circuits Devices and Applications*. New Jersey: Prentice Hall, Inc.

Walsh, Edward M. 1967. *Energy Conversion*. New York: The Ronald Press Company.

4

Introduction to System Design

In this chapter we introduce the basic building electrical system design methodology, beginning with the typical organization of plans and specifications. We then discuss the role of construction codes, such as the *National Electrical Code, NEC*. We will review some of the more important factors in the system planning process, such as architectural coordination, mechanical and plumbing systems, owner/tenant criteria and utility coordination. In the following sections, we discuss the building power system design process, beginning with a load study, distribution system planning, branch circuit and feeder design and panelboard and switchboard calculations. We also discuss system grounding, voltage drop analysis and protective device coordination.

Many of these concepts will be demonstrated in Chapter 5, which contains the design of a typical small industrial building.

BUILDING PLANS AND SPECIFICATIONS

The final product of the building design process is a set of plans and specifications. Together, they represent the design documents which convey all information necessary for the contractors to bid and finally construct the complete building. Each member of the design team, architect, electrical engineer, mechanical/plumbing engineer, civil engineer, structural engineer and landscape architect prepares their respective portion of plans and specifications. The drawings prepared by each team member are sequentially numbered and normally carry a prefix indicative of the nature of the drawing content. For example, architectural plans might begin with sheet A-1, electrical plans with E-1, mechanical with M-1, plumbing with P-1, civil with C-1, structural with S-1 and landscape with LS-1. The entire package is normally bound together and can consist of fifty to well over one hundred sheets, even for a medium sized building. In addition, the building owner or tenant might assemble a set of criteria drawings that detail the nature of specific requirements related to actual equipment to be installed in the building. This information is used in the engineers' planning process to make sure that all necessary electrical and mechanical provisions are made a part of the basic building design. Such owner requirements may appear on numbered sheets, beginning with T-1. Such plans are normally not a part of the construction documents, but are incorporated by the design team in their respective plans and specifications.

The specifications cover items such as workmanship, approved materials, method of assembly, sequence of construction, special conditions, handling of drawing conflicts, construction permits and shop drawings. Each engineering discipline prepares their own portion of the specifications that are then assembled and bound. As in the case with the building plans, each discipline is normally assigned a specification section number. Once the general specification format is established, by the architect, each discipline can prepare their own sections. In some cases, a specification consultant is retained to coordinate preparation and assembly of project specifications.

Although there is no universal format, an average specification might be divided into the following divisions.

Division: 0   ***Bidding Requirements, contract forms and conditions of the contract***

            1   ***General Requirements***
Summary of the work, testing and laboratory services, material and equipment. Schedule of finishes

            2   ***Site Work***
Earthwork, termite control, drilled piers

            3   ***Concrete***
Concrete framework, concrete reinforcement, cast-in-place concrete, tilt-up concrete construction

4 *Masonry*
Brickwork, limestone

5 *Metals*
Structural steel, steel joists, metal arching, metal fabrication, specialty fabrication

6 *Wood and Plaster*
Rough carpentry, millwork and related items

7 *Thermal and Moisture Protection*
Building insulations, preformed roofing and siding, soffit cladding system, membrane roofing, flashing and sheet metal, roof accessories, sealants

8 *Doors and Windows*
Metal doors and frames, aluminum doors and frames, special doors, access panels, hardware, glazing

9 *Finishes*
Gypsum wallboard system, ceramic tile, acoustical ceilings, resilient flooring, carpeting, painting, wall coverings

10 *Specialties*
Tackboards and pegboards, toilet partitions, fire extinguisher, wire mesh partitions, toilet and bath accessories

11 *Equipment*
Loading dock equipment

12 *Furnishings*
Horizontal blinds, entrance mats

13 *Special Construction*
Interior space frames

14 *Conveying Systems*
Elevators, escalators, conveyors

15 *Mechanical Systems*
General mechanical requirements, plumbing, sprinkler system, heating, ventilating and air conditioning

16 *Electrical Systems*
Electrical requirements, general provisions, raceways, conductors, outlet boxes, wiring devices, service switchboard, panelboards, overcurrent protective devices, grounding, dry-type transformers lighting, fire alarm systems

Not all buildings require all of these items.

CONSTRUCTION CODES

Every building project is subject to compliance with national and local construction codes. One common goal these codes have is to assure the safe construction of buildings and their electrical and mechanical systems. Since fire safety is of major concern, many codes result from the extensive research and standards development activities of the National Fire Protection Association. This organization, located in Quincy, Massachusetts, publishes a highly regarded set of standards covering almost every phase of construction activity. These are gathered in a multi-volume set of standards. One of these codes, NFPA-70, is called the *National Electrical Code*, or simply the *NEC*. Essentially all cities and towns include NFPA-70 as part of their own local code requirements. The NFPA has acted as sponsor of the *NEC* since 1911. The original code document was developed in 1897 as the result of efforts by those concerned with building construction, such as insurance companies, architects and engineers. The *NEC* is subject to periodic revision and updating by a number of technical committees and is revised and republished every three years. Proposed amendments to the *NEC* can be suggested by any concerned individual or group. The appropriate technical committee receives proposed changes at the annual meeting of the organization.

Organization of the *National Electrical Code*

The 1990 Edition of the *National Electrical Code* is organized into nine chapters covering the range of electrical construction activities.

Chapter 1   includes definitions of electrical terms, as well as general requirements for electrical installations.

Chapter 2   covers wiring and protection of electrical systems, including required outlets, branch circuit and feeder calculations, electric service requirements, overcurrent protection and system grounding.

Chapter 3   covers wiring methods and materials as well as topics such as temporary services, electrical raceways, fittings, busways, flat cable systems, outlet boxes, cabinets, wireways and switchboards and panelboards.

Chapter 4   deals with general electrical equipment including lighting fixtures, space heating equipment, motors, motor controllers, air conditioning and refrigeration equipment, generators, transformers and storage batteries.

Chapter 5 addresses the requirements of special occupancies, including buildings where hazardous material or substances might exist, such as petrochemical facilities. This chapter classifies various hazardous areas according to the nature of the hazardous material present. Chapter 5 also deals with occupancies such as health care facilities, places of assembly, theaters, agricultural buildings, mobile homes, recreational vehicles, floating buildings, marinas and boat yards.

Chapter 6 deals with special electrical equipment such as electrical signs, elevators, dumbwaiters, escalators, moving walks, welders, x-ray equipment, swimming pools and fountains.

Chapter 7 covers special systems such as emergency lighting, control circuits and fiber optic cables.

Chapter 8 deals with communications systems. This chapter includes requirements for system grounding and for cables associated with radio and television equipment.

Chapter 9 the final chapter, is devoted to Tables and Examples. Included are tables of raceway capacity data, dimensions of various cable types, and examples of service calculations for dwelling-type occupancies.

An additional reference tool is the *National Electrical Code Handbook*, also published by the National Fire Protection Association. This publication, presently in its fifth edition, contains the complete text of the *NEC*, along with explanations of many of the more difficult code requirements. The handbook also includes a number of illustrations covering various aspects of the *NEC*.

A few cautionary words about the *NEC* and *NEC Handbook* may be in order. When using the *NEC*, the engineer must remember that the provisions contained within the codes are *minimum* requirements. The code is not a design guide. The engineer must exercise good judgment during the design process. Although never allowing a design to fall below code requirements, the engineer may find it prudent to exceed code requirements in order to meet special design conditions. After all, the engineer is being paid for his professional knowledge and experience. On very rare occasions the engineer may find reason to request a slight alteration of the *NEC* or other code in order to meet special conditions associated with a given project. The local code authorities should be consulted before a decision involving such a code alteration is finalized.

SYSTEM PLANNING

During the design process, numerous factors having impact on the electrical system must be carefully considered. Each project is different and poses unique challenges, however, some concerns are fairly universal. A few of these are as follows.

Architectural Considerations

During the early design phase, the architect prepares preliminary floor plans and cross sections of the building. During this process, the electrical engineer must clearly coordinate with the architect for necessary electrical equipment space. Failure to allow sufficient space for switchboards, panelboards and transformers may cause untold grief, expensive field changes and outright unworkable equipment locations. The nature of the architect's proposed ceiling system should also be determined. This is important in the luminaire selection process. If the architect has special decorative or accent lighting requirements or plans items such as lighted fountains, the electrical engineer will have to make provisions for the necessary circuits.

Mechanical/Plumbing Coordination

Nowhere does poor design team coordination show up more quickly than in the exchange of information between the mechanical/plumbing and electrical engineers. During the very early design phase, the mechanical/plumbing engineer must provide preliminary information along with all relevant electrical data. The electrical engineer uses this data in preparing a preliminary load estimate, which, in turn, is used to make general decisions on the layout of the building power distribution system. At the same time, the electrical engineer must prepare preliminary lighting plans showing the location of all luminaires. The mechanical engineer uses this information in estimating the heat generated by the lighting system, as well as in selecting locations for ceiling mounted air supplies and air returns. The mechanical and electrical engineers must have a clear understanding of whose plans and specifications will specify control system requirements as well as motor control devices and disconnect switches. It is also important for the electrical engineer to appraise the mechanical engineer of the available voltages since this can quite possibly alter the model of mechanical and plumbing equipment. Both engineers must keep each other informed when changes to previously transmitted data occur. In all cases, the transfer of information should be in written form. Verbal communications are virtually useless months or even years later when a problem occurs and a reconstruction of events becomes necessary. Effective written documentation in the project file can save everyone a great deal of trouble.

Owner/Tenant Requirements

On some projects, the owner/tenant has very few, if any, specific electrical requirements beyond normal electrical system design. The engineer is free to locate equipment, lighting, receptacles, etc., as he deems appropriate. In other cases, the owner/tenant may have very well developed ideas on many aspects of the electrical system layout. This information is often conveyed to the electrical engineer in the form of tenant requirement plans. These plans might, for example, provide equipment locations and electrical load data on building process equipment or locations for electrical office equipment such as copiers, computers, etc. In some cases, the engineer must actively seek information on these requirements by meeting with the owner/tenant's representative.

In either case, the engineer can then combine this information along with architectural requirements, mechanical/plumbing requirements and his own design concept to develop the most effective electrical design.

As in the case of mechanical/plumbing coordination data, owner/tenant coordination should be maintained in written form. In many cases, such information should be routed to the engineer through the architect, because more than one engineering discipline might be involved. An example might be an electrical process oven that requires electrical power as well as mechanical ventilation and special floor construction for support of the added weight.

Electric Utility and Telephone Company Coordination

One often overlooked area of coordination is between the electrical engineer and the utility and telephone companies. In actuality, the engineer should contact the utility company very early in the design phase. The utility company may have to make utility system modifications such as increased substation size, extension of overhead or underground lines, or even ordering special transformers. By giving the utility early warning of the project, delays resulting from these factors are often avoided. The utility company will require an early load estimate from the engineer in order to make appropriate engineering decisions, while the engineer will need information on the available voltages for both his own planning and for coordination with the mechanical/plumbing engineer. The telephone company may need to make provisions for additional lines to the area, a time consuming project.

As the design progresses, the electrical engineer will require information on the utility's prospective fault current level in order to properly select protective device interrupting ratings. Utilities are normally very cooperative in supplying the necessary information and willing to make every effort to meet the project's requirements. Here again, effective, timely communication is the key ingredient to success.

SYSTEM DESIGN

In this section we introduce the basic concepts associated with system planning and design. We begin by discussing some factors to be considered in planning the distribution system, including panelboard and switchboard locations. Following this, we discuss basic design methodology by which circuits are designed for various items of building electrical equipment such as lighting fixtures, receptacles and mechanical equipment. We will then see how panelboards and switchboards are selected to serve such loads. Next, we learn how voltage drop analysis helps us assure the wiring system is capable of delivering voltages within the operating range of the various items of equipment. Finally, we will see that an overcurrent coordination study is necessary to assure the proper operating sequence for the system's circuit breakers and fuses.

Distribution System Planning

The design of a building electrical system begins with the assembly of load data. This load study includes all known items of electrical equipment, such as heating, ventilating and air conditioning equipment, electric water heaters, owner and tenant equipment and finally, the building lighting and receptacle system.   In many cases, the mechanical loads, or at least an estimate of them, is available early in the project.   The owner or tenant can normally identify major items of owner furnished electrical equipment early in the design phase.   The lighting and receptacle systems are under the direct control of the engineer, so an estimate of these is fairly straight forward.   The assembled load data can be used to prepare a preliminary sketch of the building electrical system and identify major loads as well as potential locations for panelboards and switchboards.   The preliminary load estimate may also be utilized by the utility to begin preliminary planning for equipment.

The planning process begins with a review of architectural plans along with the assembled load data.   One of the key items is the location of panelboards and switchboards as close as possible to the loads they serve. This close proximity reduces the length of branch circuits and feeders as well as wiring cost.   The location of panelboards near the loads they serve also provides a readily accessible location for turning off any circuit in case of emergency.   A review of the architectural plans and the location of major equipment or equipment groups will normally suggest general locations of the necessary panelboards.   The goal is to utilize the minimum number of panelboards while maintaining as much flexibility as possible. In today's buildings, floor space is always at a premium.   Owners and architects often regard electrical and mechanical equipment as a necessary evil to be given only the space absolutely necessary.   This view is understandable since the space taken up by these systems cannot be leased or used for revenue producing purposes.   It is essentially nonproductive space.   By the same token, however, these systems must have adequate room for the safe installation and maintenance of the components involved. The initial architectural plan review should also suggest the optimal location for the building's main electrical panel.   For small buildings, this might only be a small area of a wall, while for large buildings, a separate electrical equipment room might be in order.   The time to identify space requirements for electrical rooms is during the very early design development phase while the architect is still engaged in overall space planning. During this phase, the building design is fluid, and setting aside the necessary space is much easier than later after every square foot of space has been allocated for other purposes. Knowledgeable architects recognize the importance of this task and are generally agreeable to arranging the necessary space requirements very early. A word of caution may be in order. The engineer should avoid agreeing to a space allocation so restrictive that only small design alterations will force expansion of the required space during the construction phase. Such changes are embarrassing, expensive and very difficult to explain.

Some years ago it was quite common to locate mechanical, plumbing, electrical and telephone equipment within the same space. Ruptured water lines often inundated electrical equipment causing damage and even fires. Today, this practice is not quite as common, however, should the situation arise, it is crucial that the electrical equipment be clearly located out of harm's way and the assigned space be clearly understood by all concerned. It is astonishing how often, a small square shown on the mechanical plans may grow to a ten foot by ten foot monster in the field, consuming all space intended for electrical equipment. This situation is often made worse since the mechanical equipment is frequently already in place when the electrical contractor begins panelboard installation.

Panelboards are often recessed in corridor walls with locked doors to prevent tampering. The only possible drawback is the difficulty of adding additional circuits after the walls are constructed. Empty conduit stubbed above the ceiling for future use frequently relieves the problem, if allowed for on the plans. The battle for electrical equipment space has gone on since the first electrical system was installed and will probably continue indefinitely.

Once potential space for electrical equipment has been identified, it is necessary to evaluate the anticipated size of distribution panels, panelboards, fire alarm system and other equipment which must occupy the space. A sketch of the space, drawn to scale and based on the actual dimensions of the equipment, is very useful. The *National Electrical Code* clearly specifies the safe working clearances around electrical equipment and these minimum dimensions must be carefully followed. Local construction codes often contain additional clearance criteria as well as equipment room exit path requirements.

Finally, the location of the main electrical distribution panel should be as close as possible to the location of the utility service transformer. Often the electrical equipment room is located on an exterior wall with the utility transformer placed immediately outside.

Design Methodology

Before proceeding with the specific techniques for designing branch circuits and feeders, let us summarize the general approach.

System design begins with the location and identification of all equipment to be connected to the power system. Panelboards are then located and circuits designed to serve the various loads. The aggregate of all loads connected to a given panel or switchboard determines the appropriate panel ampere rating. The loads of the various panels and switchboards are reflected back to the main distribution panel. These panel loads, along with items of equipment connected directly to the main distribution panels, determine its appropriate rating. We must also consider that some items of equipment do not operate concurrently. Also, not all circuits, such as those for receptacles or cooking equipment, are fully loaded all the time. Although it is possible to design a system to accommodate the simultaneous operation of all equipment, it is not realistic from a cost standpoint. The goal of the design process is to provide the necessary capacity to meet the building needs under its maximum actual demand load. It should be clear that system design is essentially a "roots upward" process and is driven by the individual system loads. Figure 4.1 summarizes the above design process in flow chart form.

Let us now see how circuits are designed to serve the individual loads. We will then study how the loads are combined to select the correct overcurrent devices and ampere ratings for panelboards and switchboards.

Branch Circuits

In this section we discuss the design of branch circuits such as those serving individual pieces of equipment such as motors, resistance heaters or receptacle circuits. We will divide these circuits into two broad categories, non-motor and motor circuits. The reason for this division lies in the high starting current characteristics of motors as discussed in Chapter 3.

Before beginning our discussion, it may be useful to review the various ways circuits can be connected to single-phase and three-phase systems. Figure 4.2 shows a summary of these various circuit configurations. The circuits are connected to panelboards by means of circuit breakers, as shown in Figure 3.26.

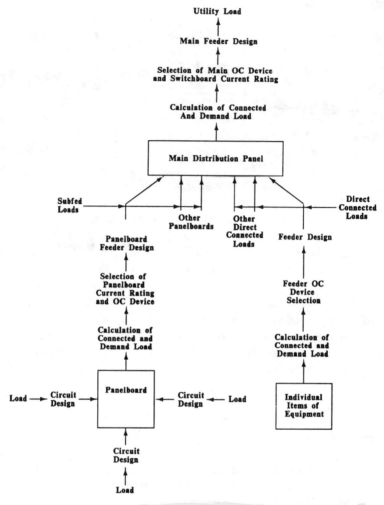

Figure 4.1 Typical circuit design and load calculation sequence.

Figure 4.2(a) shows the ways single-phase circuits can be derived from single-phase or three-phase systems. Single-phase circuits can be connected to single-phase systems in two ways. The first is by means of a single pole circuit breaker serving a load connected between phase and neutral. This is typical of most lighting and receptacle circuits. The second method is by means of a two pole circuit breaker connected

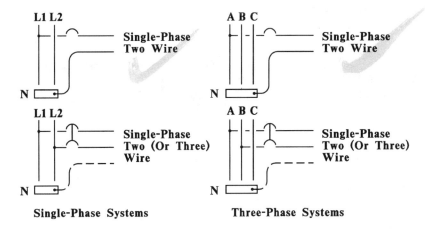

Single-Phase Systems   Three-Phase Systems

(a) Single-Phase Circuits

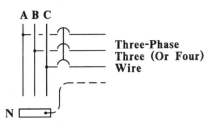

(b) Three-Phase Circuits

Figure 4.2 Single-phase and three-phase circuits.

between Line 1 and Line 2. This circuit may or may not include a neutral depending on the nature of the load being served. An example would be an electric range that typically utilizes the line-to-line voltage for the oven heating element and connects the range top units between phase and neutral. Single-phase circuits can also be connected to three-phase systems. The most common connection is by means of a single pole circuit breaker serving a load connected between phase and neutral. This is again typical of many receptacle and lighting circuits. The second method is by means of a two pole circuit breaker connected between Phase A and B, B and C, or C and A. Here again, a neutral might or might not be required. The connection for three-phase circuits is shown in Figure 4.2(b). In this case, a three pole circuit breaker is required. The neutral may or may not be required, depending on the nature of the load itself.

Non-Motor Circuits

For circuits serving non-motor loads, first determine the load's current requirements and then select the appropriate circuit breaker or fuse. Remember that molded case circuit breakers are not allowed to carry 100% of their rated current on a continuous basis, so we must limit the circuit current to 80% of the breaker rating. The overcurrent device, oc, is then selected as follows:

$$OC = 1.25 \cdot \text{Load Amps} \qquad \text{(Select next standard circuit} \qquad (4.1)$$
$$\text{breaker or fuse)}$$

The proper wire size is then selected based on the overcurrent device selected.

NOTE: The following examples assume THW copper conductors. Refer to Table 3.2 for current-carrying capacities, Table 3.4 for conduit capacities and Tables 3.5 and 3.6 for standard circuit breaker and fuse ratings.

Example 4.1: A 120 volt circuit carries a load of 1550 VA. What are the correct molded case circuit breaker and branch circuit conductor sizes?

$$I = \frac{kVA}{kV} = \frac{1.55}{.12} = 12.92 \quad A$$

$$OC = 1.25 \cdot 12.92 = 16.15 \text{ A} \quad (20 \text{ A/1P Circuit Breaker})$$

For a 20 A overcurrent device the correct wire size would be 2#12 conductors in a 1/2" conduit.

Example 4.2: A 240 V single-phase circuit serves a 4.5 kW water heater (no neutral required). What are the correct molded case circuit breaker and branch circuit conductor sizes?

$$I = \frac{kVA}{kV} = \frac{4.5}{.240} = 18.75 \quad A$$

$$OC = 1.25 \cdot 18.75 = 23.44 \text{ A} \quad (25 \text{ A/2P Circuit Breaker})$$

The correct wire size is 2#10 in a 1/2" conduit.

Example 4.3: A 208 V, three-phase load requires 37.0 amperes. What are the correct circuit breaker and wire sizes?

$$OC = 1.25 \cdot 37.0 = 46.25 \text{ A} \quad (50 \text{ A/3P Circuit Breaker})$$

The correct wire size is 3#8 conductors in a 3/4" conduit.

Motor Circuits

For motor circuits, we take into consideration the motor's inrush current during startup. We must also recognize the differences between molded case circuit breakers and fuses. Time delay fuses, such as the RK-1, are able to withstand the momentary starting current of motors somewhat better than molded case circuit breakers. As a result, we select the overcurrent device size differently, depending on whether the device is a circuit breaker or fuse. Also, some circuits serve more than one motor and require special consideration. See Figure 4.3 for single- and three-phase motor circuits. Actual circuits involve starters and other equipment not shown.

First, let us treat the case of a single motor served by a circuit breaker or fuse. Here, FLA refers to the motor's full load amp rating.

For a molded case circuit breaker:

$$OC = 1.75 \cdot FLA \quad \text{(Select Next Standard Circuit Breaker)} \quad (4.2)$$
$$Wire = 1.25 \cdot FLA \quad \text{(Select Next Standard Wire Size)} \quad (4.3)$$

For a time delay fuse:

$$OC = 1.25 \cdot FLA \quad \text{(Select Next Standard Fuse Rating)} \quad (4.4)$$
$$Wire = 1.25 \cdot FLA \quad \text{(Select Next Standard Wire Size)} \quad (4.5)$$

Note that the motor's full load current is determined from Table 3.8 for single-phase motors and Table 3.9 for three-phase motors.

NOTE: The following examples assume the use of THW copper conductors. Refer to Table 3.2 for current-carrying capacities, Table 3.4 for conduit capacities, and Tables 3.5 and 3.6 for standard circuit breaker and fuse ratings.

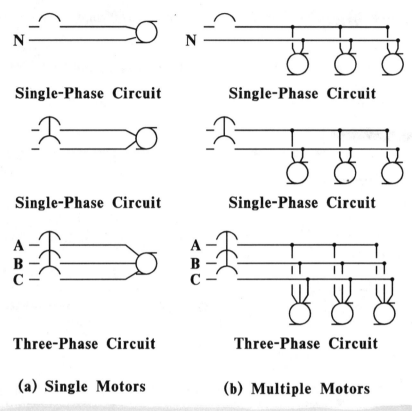

Single-Phase Circuit   Single-Phase Circuit

Single-Phase Circuit   Single-Phase Circuit

Three-Phase Circuit   Three-Phase Circuit

(a) Single Motors   **(b) Multiple Motors**

Figure 4.3 Single and multiple motor circuits for single-phase and three-phase systems.

Example 4.4: Design a circuit to serve a 10 HP, 208 V, three-phase motor. The overcurrent device is a molded case circuit breaker.

From Table 3.9, the full load current for a 10 HP motor is 30.8 A.

OC = 1.75 · 30.8 = 53.9 A (60 A/3P Circuit Breaker)
Wire = 1.25 · 30.8 = 38.5 A   (3#8-3/4" Conduit)

If the overcurrent device is a fuse, then the correct ampere rating would be

OC = 1.25 · 30.8 = 38.5 A   (40 A Fuse)
The wire size would remain unchanged.

Example 4.5: A 240 volt single-phase motor is rated at 3 HP. What is the correct fuse overcurrent device size and wire size?

From Table 3.8, we see that the column labeled 230 volts can be used for voltage ranges of 220 to 240 volts. The full load current for a 3 HP motor is 17.0 A.

$$OC = 1.25 \cdot 17.0 = 21.25 \text{ A} \quad (25 \text{ A Time Delay Fuse})$$
The correct wire size is 2#10-1/2" Conduit

Circuits Serving Several Motors

If several motors are connected to the same circuit, such as the compressor motor and associated fans in a rooftop air conditioning unit, we select the circuit overcurrent device and wire size as follows:

For a Molded Case Circuit Breaker

$$OC = 1.75 \cdot FLA \text{ (Largest Motor)}$$
$$+ \text{ Sum of FLA for Smaller Motors}$$
(Select Next Standard Circuit Breaker) $\qquad$ (4.6)

$$Wire = 1.25 \cdot FLA \text{ (Largest Motor)}$$
$$+ \text{ Sum of FLA for Smaller Motors}$$
(Select Next Standard Wire Size) $\qquad$ (4.7)

For a Time Delay Fuse

$$OC = 1.25 \cdot FLA \text{ (Largest Motor)}$$
$$+ \text{ Sum of FLA for Smaller Motors}$$
(Select Next Standard Fuse Rating) $\qquad$ (4.8)

$$Wire = 1.25 \cdot FLA \text{ (Largest Motor)}$$
$$+ \text{ Sum of FLA for Smaller Motors}$$
(Select Next Standard Wire Size) $\qquad$ (4.9)

Example 4.6: A rooftop air conditioning unit has two motors, a 7½ HP compressor and a 1.0 HP condensing fan. Both motors are rated at 208 V, three-phase.

Design a circuit for this unit using a molded case circuit breaker as the overcurrent device.

From Table 3.9, the full load current (FLA) for the 7½ HP compressor is 24.2 A and for the 1 HP fan, 4.0 A.

OC = 1.75 · 24.2 + 4.0 = 46.35 A
(50 A/3P Circuit Breaker)

Wire = 1.25 · 24.2 + 4.0 = 34.25 A
(3#8-3/4" Conduit)

In actual practice the air conditioning unit may have specific overcurrent device requirements listed on its nameplate. In such cases this data will determine circuit design.

Example 4.7: Three 480 V, three-phase, 3 HP motors are to be served from a single circuit. Design the circuit based on the use of time delay fuses.

From Table 3.9, the full load current of the 3 HP motors is 4.8 A each.
OC = 1.25 · 4.8 + 4.8 + 4.8 = 15.6 A   (20 A Time Delay Fuses)
Wire = 1.25 · 4.8 + 4.8 + 4.8 = 15.6 A   (3#12-1/2"C)

Panelboards and Switchboards

Panelboards and switchboards are the nerve centers of the electrical distribution system and house the overcurrent devices which protect the components from overload or short circuit damage. Figures 3.26, 3.27 and 3.28 show details of panelboard and switchboard construction. In this section, we will learn how to select the correct panelboard or switchboard ampere rating.

Before proceeding, we must understand the concept of connected and demand loads. A list of connected loads represents the sum of all of the loads connected to the panel or switchboard and does not take into account whether the loads are seasonal or likely to be utilized to their full capacity. The demand load, on the other hand, considers the seasonal nature of the various loads as well as the likely maximum demand for items such as receptacles and cooking equipment. The demand load is used in the selection of panelboard and switchboard ratings because it represents the maximum carried by the equipment.

Connected loads are frequently grouped together in several general classes as follows.

Connected Loads
1. Lighting
2. Receptacles
3. Resistance Heat (Seasonal)
4. Heat Motors (Seasonal)
5. Air Conditioning Motors (Seasonal)
6. Motors
7. Other Loads
8. Water Heating
9. Kitchen
10. Spare Capacity

These totals represent total loads for each category, including loads served directly from the panel or switchboard as well as subfed loads in downstream panels. The last item, spare capacity, represents the engineer's best estimate of the additional electrical equipment which might be added in the future. This spare capacity can often be further divided into two categories. The first is future equipment which the owner can clearly identify at the time of design and for which provisions must be made. The second category represents normal additions to the building load which must be estimated by the engineer based upon his experience with the particular type of building involved. The latter category is obviously the more difficult to quantify.

Based on the connected loads, a demand load can be prepared as follows:

Computed Load	**Demand Factor**
1. Lighting (Continuous)	1.25 (See Note 1)
2. Receptacles - 1st 10 kVA @ 100%	1.00 (See Note 2)
- Remainder @ 50%	0.50 (See Note 2)
3. Resistance Heat (Seasonal)	1.0 or 0.0 (See Note 3)
4. Heat Motor (Seasonal)	1.0 or 0.0 (See Note 3)

5. Air Conditioning Motors
    (Seasonal)   1.0 or 0.0 (See Note 3)
6. Motors (Non Seasonal)   1.0
7. Other Loads   1.0 or 1.25 if continuous
8. Water Heating   1.0
9. Kitchen   0.65 to 1.0 (See Note 4)
10. Spare Capacity   1.0
11. Largest Motor   0.25 (See Note 5)

NOTES:

1. Lighting can be classified as a continuous load and circuits serving such loads cannot be loaded to more than 80% of their rating.
2. The actual equipment connected to general purpose receptacles is generally unknown, so an estimate must be made. Many engineers allow 180 VA (.18 kVA) for each duplex receptacle. All receptacles will not be required to provide this load simultaneously, so a demand factor is assigned to the total connected load to arrive at the actual demand load. Many engineers consider the first 10 kVA of receptacle load to have a demand factor of 100% and the remainder a demand factor of 50%. If knowledge of the actual load and demand factor of the receptacle circuit is available, this should obviously supersede the above approach, which provides only a minimum design standard.
3. In most buildings the heating and air conditioning systems do not operate simultaneously. The panel or switchboard must be capable of providing power to the heaviest of the two. After a load comparison is made, a demand factor of 1.0 is assigned to the larger load and a demand factor of 0 is assigned to the smaller. In many cases, the same air handling (blower) motors function both during the heating and the air conditioning season.
4. Kitchen loads are recognized as having a demand factor of 1.0 or less, depending on the number of individual items of kitchen equipment present. Table 220-20 of the *NEC* lists a demand factor of 1.0 for kitchens with one or two items of equipment and ranging down to a demand factor of 0.65 for kitchens with over six items of equipment.
5. It is necessary to include an additional 25% of the largest motor to correctly select the total panelboard or switchboard ampacity. This is analogous to multiplying the largest motor FLA by 1.25 when selecting the correct feeder size.

As a practical matter, the kVA rating of step-down transformers can be selected using the same methodology.

Let's look at an example to see how this works.

Example 4.8: A 208Y/120 V panelboard serves the following loads.

Compute the total connected load and the calculated demand load.

1. Lighting   15.0 kVA @ PF = 0.95
2. Receptacles   22.0 kVA @ PF = 0.90
3. Resistance Heat   9.0 kW
4. Air Conditioning 12.0 kVA @ PF = 0.88
5. Motors   6.0 kVA @ PF = 0.88
6. Water Heating   4.5 kW
7. Spare Capacity   10.0 kVA @ PF = 0.90

The largest motor is rated 5 HP (16.7 FLA @ 208 V, 3-phase, PF = 0.88)

Solution
The total connected load is:

1. Lighting   14.25 + j4.68
2. Receptacles   19.80 + j9.59
3. Resistance Heat   9.00 + j0.00
4. Air Conditioning 10.56 + j5.70
5. Motors   5.28 + j2.85
6. Water Heating   4.50 + j0.00
7. Spare   9.00 + j4.36
                     ─────────────
                     72.39 +j27.18
                     (77.32 kVA)

$$\theta = \cos^{-1}(PF)$$
$$Q = |S|\sin(\theta)$$
$$P = |S|\cos(\theta)$$

The computed demand load will be:

1. Lighting   (14.25 + j4.68) 1.25 = 17.81 + j5.85
2. Receptacles   (9.00 + j4.36) 1.00 = 9.00 + j4.36
                 (10.80 + j5.23) 0.50 = 5.40 + j2.62

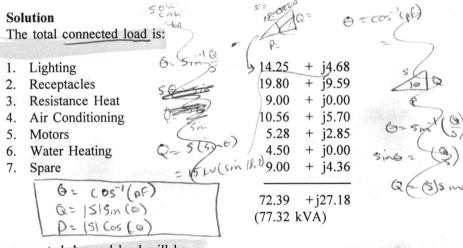

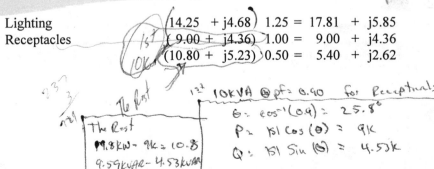

3. Resistance Heat   – o because A.l. is large   0.00 =   -
4. Air Conditioning   (10.56 + j5.70) 1.00 = 10.56 + j5.70
5. Motors   (5.28 + j2.85) 1.00 = 5.28 + j2.85
6. Water Heating   (4.50 + j0.00) 1.00 = 4.50 + j0.00
7. Spare   (9.00 + j4.36) 1.00 = 9.00 + j4.36
8. Largest Motor 5 HP   (5.30 + j2.86) 0.25 = 1.33 + j0.72
       └ An Additional 25%.)

                                          62.88 +j26.46
                                          (68.22 kVA)

The total demand load of 68.22 kVA can then be converted to current as follows:

$$I = \frac{kVA}{kV \sqrt{3}} = \frac{68.22}{.208 \sqrt{3}} = 189.36 \; A.$$

The next standard circuit breaker rating is 200 amperes, so the circuit serving the panelboard will be protected by a 200 A/3P circuit breaker. This in turn requires a circuit consisting of 4#3/0-THW copper conductors in a 2" conduit. As a practical matter, panelboards are generally available in ratings of 100, 225 and 400 amperes. As a result, the panel itself will actually have a rating of 225 amperes.

Example 4.9: The load of the previous example is to be served by a 480-208Y/120 V step-down transformer. What is the correct transformer kVA rating?

The total computed demand load is 68.22 kVA, so the next standard transformer rating would be 75.0 kVA. The transformer kVA rating is then used to select the correct primary and secondary overcurrent rating and feeder size. This is addressed in the next section.

Transformers

The actual kVA rating of a transformer is determined based on its load demand as discussed in the previous section. Once the kVA rating has been selected, the process of circuit design can begin.

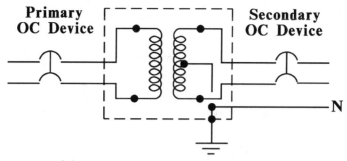

(a) Single-Phase Transformer

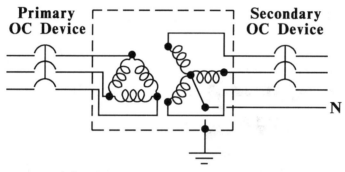

(b) Three-Phase Transformer

Figure 4.4 Transformer primary and secondary overcurrent devices, single-phase and three-phase systems.

Transformers must have overcurrent device protection on both the primary and secondary side, as shown in Figure 4.4.

These overcurrent devices are generally selected as follows:

OC(PRI) = 1.25 · FLA(PRI)   (Select Next Standard
                                                Device Rating)   (4.10)

OC(SEC) = 1.25 · FLA(SEC)   (Not to exceed 125% of
                                                Secondary Rated Current)   (4.11)

Here the FLA for the primary and secondary is determined from the transformer's rated kVA and voltage, not the load current. The primary and secondary conductors are selected based on the primary and secondary overcurrent device ratings.

The above overcurrent device selection process applies to molded case circuit breakers or fuses. Under some conditions, the primary overcurrent device can be selected as high as 250% of the transformer's rated primary current in order to avoid nuisance tripping due to inrush current. The secondary overcurrent device cannot exceed 125% of the rated secondary current.

Some examples will help clarify the process.

Example 4.10: A 45 kVA step-down transformer with a 480 V three-phase primary and 208Y/120 V secondary is selected to serve a lighting panel. Select the correct overcurrent device and circuit size for this transformer.

For a 45 kVA transformer

$$I_{PRI} = \frac{kVA}{kV\sqrt{3}} = \frac{45}{.480\sqrt{3}} = 54.13 \ A$$

$$I_{SEC} = \frac{kVA}{kV\sqrt{3}} = \frac{45}{.208\sqrt{3}} = 124.91 \ A$$

So the overcurrent devices are

$$OC_{PRI} = 1.25 \cdot 54.13 = 67.66 \ A \quad (70 \ A/3P \ \text{Circuit Breaker})$$

$$OC_{SEC} = 1.25 \cdot 124.91 = 156.14 \ A \quad (150 \ A/3P \ \text{Circuit Breaker})$$

The correct primary and secondary feeders are

Primary = 3#4-1" Conduit
Secondary = 4#1/0-2" Conduit

Example 4.11: A 50 kVA step-down transformer with a 480 V single-phase primary and 120/240 V secondary is to serve a panelboard for an industrial process. Select the correct overcurrent devices and wire sizes.

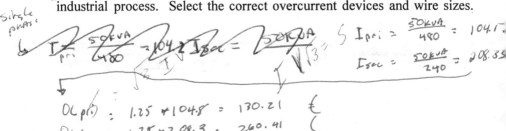

$$I_{PRI} = \frac{50}{.480} = 104.17 \ A$$

$$I_{SEC} = \frac{50}{.240} = 208.33 \ A$$

these values to determine wire size.

The overcurrent devices are

$OC_{PRI} = 1.25 \cdot 104.17 = 130.21$ A   (150 A/2P Circuit Breaker)

$OC_{SEC} = 1.25 \cdot 208.33 = 260.41$ A   (250 A/2P Circuit Breaker)

The associated wire sizes are

Primary = 2#1/0-1¼" Conduit
Secondary = 3#250 kCMIL-2½" Conduit

Note that in both of the above examples, the secondary overcurrent device can be a main breaker located in the panelboard connected to the transformer secondary.

System Grounding

There are two very important reasons for grounding all electrical system components. First, ground to reduce any voltage differential between adjacent pieces of equipment. Such a voltage differential could result in a shock hazard to operating personnel. Secondly, by effectively grounding all electrical system components, we provide a low impedance path for fault currents. This low impedance path facilitates rapid overcurrent device operation and hence, quick removal of the faulted piece of equipment from the electrical system.

The ground path for many electrical components such as lighting fixtures and receptacles is the metallic raceway system. In the case of non-metallic raceways, a separate ground conductor is installed. Transformers are grounded at their secondaries, as shown in Figures 3.25 and 4.4. Remember the building's electrical service neutral and ground are connected at only one location, namely at the main distribution panel. This is shown in Figure 3.28. The neutral and ground must never be interchanged because the neutral normally carries current, while the ground conductor only carries current under abnormal conditions, such as short circuits between conductors and grounded equipment housings. A severe safety hazard results if the raceway system ground path is connected as the neutral conductor because current would flow through all raceways, panelboards and switchboard housings, as well as other equipment housings.

There are several methods of providing the connection to ground at the main service location, shown in Figure 3.28. One of the oldest methods is by means of connection to a metallic cold water pipe. Although this technique provides an effective ground path, the increased use of plastic water pipe often makes this approach impossible. Another approach is by connection to driven ground rods. A third approach is connection to the reinforcing steel inside the building foundations. System grounding also depends on the resistivity of the soil itself. Soil resistivity varies widely from one part of the country to another. For example, in coastal areas where salt marshes are common, soil resistivity might vary between 590 and 7000 ohm-centimeters. In locations where clay, shale, gumbo or loam are prevalent, the value may vary from 340 to 16,300 ohm-centimeters. If sand or gravel is added, the same soil might have a resistivity of 1020 to 135,000 ohm-centimeters. In areas where gravel, sand or stones are prevalent, the value could range from 59,000 to 458,000 ohm-centimeters. Cured concrete has an average resistivity of 2000 ohm-centimeters or about the value of average soil, which frequently makes connection to building footings a very effective approach, especially when the interconnection between footings by means of the steel building structure is considered.

The variation of soil resistivity with both moisture content and with temperature is another complicating factor. Soil with a high moisture content has a lower level of resistivity than does dryer soil. Soil at higher temperatures will have a lower level of resistivity than soil with a lower temperature.

It should be clear that predicting the value of soil resistivity at a given site prior to construction is very difficult. Some electrical specifications simply specify that a ground resistance of less than 25 ohms be obtained by connection to a water pipe, driven ground rod, building footing steel or combination of the above. In many cases, a ground resistance value of less than three ohms is attainable. The contractor is often required to measure the ground resistance and add additional ground rods, etc., until the measured value falls below the maximum specified. Figure 4.5 shows some of the common system grounding techniques.

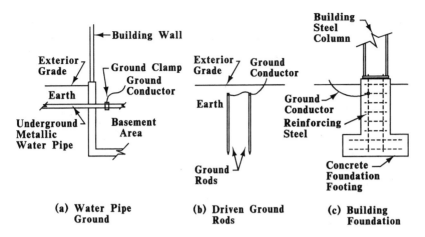

Figure 4.5 Typical electrical system ground connections.

Voltage Drop Analysis

Chapter 2 discussed the various commonly used system voltages and also the standard tolerance limits for various levels of the power system. Recall that Figure 2.14, ANSI Standard C84.1-1989 establishes the maximum total percentage voltage drop within the building system as 5%. The question then becomes how to allocate this allowable voltage drop between the various system conductors? It is customary to allow a maximum total of 3% voltage drop for the various feeders and 2% for the branch circuits.

Figure 4.6 shows a phasor diagram of the voltage relations for voltage drop analysis. Note that e_s represents the sending end voltage, the voltage at the beginning of the circuit. e_r represents the receiving end voltage, the voltage at the load itself. The angle φ represents the power factor angle. An inspection of the figure reveals that an approximate value of the voltage drop is given by

$$V_{LN} = I (R \cos \varphi + X \sin \varphi) \quad Volts. \qquad (4.12)$$

This is the line-to-neutral voltage drop, the voltage drop in one conductor, one way. The line-to-line voltage drop, V_{LL}, can be computed by multiplying the results of Equation 4.12 by one of the following constants:

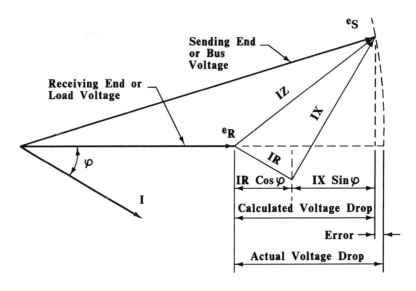

V	= Voltage drop in circuit, line to neutral.
I	= Current flowing in conductor.
R	= Line resistance for one conductor, in ohms.
X	= Line reactance for one conductor, in ohms.
φ	= Angle whose cosine is the load power factor.
Cos φ	= Load power factor, in decimals.
Sin φ	= Load reactive factor, in decimals.

Figure 4.6 Phasor diagram of circuit voltages for voltage drop analysis.

For Single-Phase systems, multiply by 2.000
For Three-Phase systems,   multiply by 1.732.

The exact expression for the voltage drop is given by

$$V_{LN} = e_s + IR \cos \varphi + IX \sin \varphi$$
$$- \sqrt{e_s^2 - (IX \cos \varphi - IR \sin \varphi)^2}. \tag{4.13}$$

Either Equation 4.12 or 4.13 yields results which are sufficiently accurate for normal use. For our purposes, we will use Equation 4.12. We can rewrite this equation as

$$V_{LN} \sim I(R\cos\phi + X\sin\phi)$$

$$V_{LN} = I Z_{eq} \quad Volts. \tag{4.14}$$

Where Z_{eq}, the equivalent impedance, is given by

$$Z_{eq} = R\cos\varphi + X\sin\varphi. \tag{4.15}$$

Remember that φ represents the power factor of the circuit's load.

The values for R and X are normally taken from *NEC* Table 9, which is reprinted in Table 3.1. Note that the values in this table are for conductors at 75°C. This temperature represents fully loaded conductors and the resulting voltage drop will be higher than would be the case for cooler, unloaded conductors. This more closely represents the worst case voltage drop.

The line-to-line voltage drop is normally converted to a percentage by dividing by the circuit voltage and multiplying the result by 100.

$$V\% = \frac{V_{LL}}{V} \cdot 100 \quad \% \tag{4.16}$$

Let's look at a few examples.

Example 4.12: A 120 volt branch circuit carries a load of 10.0 A, PF = 0.90. The conductors are #12, THW, copper in rigid steel conduit and the circuit length is 100 ft.

What is the voltage drop?

First we refer to Table 3.1. For #12 THW Copper in Rigid Steel Conduit, we see that

$$X = 0.068\Omega/1000 \text{ ft.}$$
$$R = 2.000\Omega/1000 \text{ ft.}$$

For 100 ft.

$$X = 0.068 \cdot \frac{100}{1000} = 0.0068 \quad \Omega$$
$$R = 2.000 \cdot \frac{100}{1000} = 0.200 \quad \Omega.$$

or use *—lecture & pots*
F- tables

Using Equation 4.15

φ = cos θ

fro-table F # = 0.3 Z_{eq} = $R \cos \varphi + X \sin \varphi$

$V_{drop} = \dfrac{F \cdot A \cdot ps \cdot Dist}{Line\ U}$

= $0.200 \cdot .90 + 0.0068 \cdot \sin(A \cos(0.90))$
= $0.180 + 0.003$ = 0.183 Ω.

Using Equations 4.14 and 4.16 *— Lecture & notes → F values*

$= \dfrac{0.3 \times 10A \times 100ft}{120v}$

(1.155)

$= 2.9\%$

V_{LN} = $10.00 \cdot 0.183$ = 1.83 V
V_{LL} = $2 \cdot 1.83$ = 3.66 V
$V\%$ = $\dfrac{3.66}{120} \cdot 100$ = 3.05 %

— use A #10 wire.

This far exceeds the 2% allowance for branch circuits.

Example 4.13: What conductor size should be used to limit the voltage drop in the previous example to 2%?

Let's try #10.

F# = 0.19

Again, from Table 3.1, for #10, THW, Copper in Steel Conduit

$V_{drop} = \dfrac{0.19 \times 10A \times 100 \times 1.115}{120v}$

X = 0.063 Ω/1000
R = 1.200 Ω/1000

= 1.837

$$X = 0.063 \cdot \frac{100}{1000} = 0.0063 \ \Omega.$$

So

$$R = 1.20 \cdot \frac{100}{1000} = 0.120 \ \Omega$$

and

Z_{eq} = $0.120 \cdot .90 + 0.0063 \cdot .436$ = 0.111 Ω
V_{LN} = $10.0 \cdot 0.111$ = 1.11 V
V_{LL} = $2.0 \cdot 1.11$ = 2.22 V
$V\%$ = $\dfrac{2.22}{120} \cdot 100$ = 1.85 %.

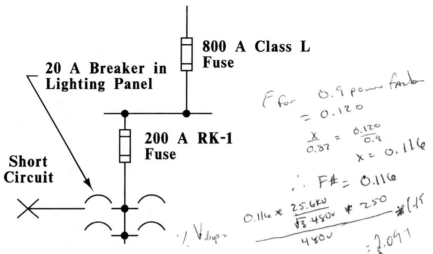

Figure 4.7 Typical system overcurrent protection devices for coordination study. 800 A class L fuse, 200 A RK-1 fuse and 20 A molded case circuit breaker.

which is just under the maximum allowable value of 2%.

Example 4.14: A 480 V, three-phase feeder is 250 ft. in length and carries a load of 25.6 kVA @ PF = 0.87. The conductor size is #8 THW Copper in steel conduit. What is the voltage drop?

Referring to Table 3.1, for #8 THW copper conductors

$$X = 0.065 \cdot \frac{250}{1000} = 0.016 \ \Omega$$

$$R = 0.78 \cdot \frac{250}{1000} = 0.195 \ \Omega$$

$$Z_{eq} = 0.195 \cdot .87 + 0.016 \cdot 0.493 = 0.178 \ \Omega.$$

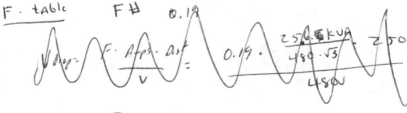

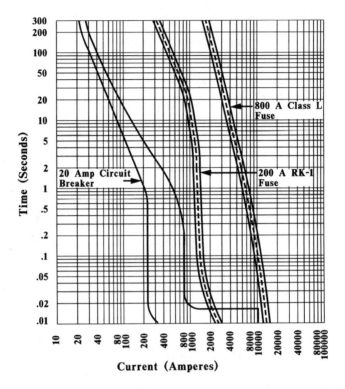

Figure 4.8 Coordination study of 800 A Class L fuse, 200 A RK-1 fuse and 20 A molded case circuit breaker.

$$I = \frac{kVA}{kV \sqrt{3}} = \frac{25.6}{.480 \sqrt{3}} = 30.79 \ A$$

$$V_{LN} = 30.79 \cdot 0.178 = 5.481 \ V$$

$$V_{LL} = 1.732 \cdot 5.481 = 9.493 \ V$$

$$V\% = \frac{9.493}{480} \cdot 100 = 1.98 \ \%$$

Overcurrent Device Coordination

Another important factor to consider in the design of a building electrical system is overcurrent device coordination. Proper coordination requires that the overcurrent device nearest the fault trip without tripping other overcurrent devices upstream. This isolates the faulted circuit while minimizing the interruption of service to equipment on other circuits. For example, if a short circuit occurs inside a residential appliance, we expect the circuit breaker serving the appliance to trip without also tripping the main breaker in the panelboard. Should the main breaker trip, the entire residence would be without power. This case would probably represent only an inconvenience, but in other applications, such as in hospitals, proper overcurrent device coordination can become more critical.

One way to verify the proper overcurrent device coordination is to plot the characteristic trip curve of the various overcurrent devices on a common time-current graph, such as those shown in Sections 3.5 and 3.6. An example will illustrate how this is accomplished. Figure 4.7 shows a main distribution panel, a lighting panel feeder and lighting branch circuit breaker. Let us consider a short circuit on the lighting branch circuit protected by a 20 ampere single pole circuit breaker. The panelboard feeder is protected by a 200 ampere RK-1 fuse and the main distribution panel overcurrent device is a 800 A Class L fuse.

Figure 4.8 shows a plot of all three overcurrent devices on the same time-current graph. Note that the 20 ampere breaker and the 800 ampere fuse curve do not overlap. The 20 ampere breaker and the 200 ampere fuse overlap in the range between 2000 and 3000 amperes indicating that for faults of this magnitude or greater, either the circuit breaker or fuse might operate.

References

Atkinson, Robert S., P.E. 1987. *ECALC Software Manual*. Nashville: Atkinson & Associates, PC

Beeman, Donald. 1955. *Industrial Power Systems Handbook*. New York: McGraw-Hill Book Company.

Croft, Terrell, and Wilford Summers. 1987. *American Electricians Handbook*. New York: McGraw Hill Book Company.

Fink, Donald G., and John M. Carroll. 1968. *Standard Handbook for Electrical Engineers*. New York: McGraw Hill Book Company.

Institute of Electrical and Electronic Engineers, Inc. 1990. *IEEE Recommended Practice for Electric Power Systems in Commercial Buildings. IEEE Std. 241-1990*. New York: Institute of Electrical and Electronic Engineers, Inc.

Institute of Electrical and Electronic Engineers, Inc. 1986. *IEEE Recommended Practice for Electric Power Distribution for Industrial Plants. ANSI/IEEE Std.*

141-1986. New York: Institute of Electrical and Electronic Engineers, Inc.

Institute of Electrical and Electronic Engineers, Inc. 1986. *IEEE Recommended Practice for Protection and Coordination of Industrial and Commercial Power Systems. ANSI/IEEE Std. 242-1986.* New York: Institute of Electrical and Electronic Engineers, Inc.

Institute of Electrical and Electronic Engineers, Inc. 1982. *IEEE Recommended Practice for Grounding of Industrial and Commercial Power Systems. ANSI/IEEE Std. 142-1982.* New York: Institute of Electrical and Electronic Engineers, Inc.

National Fire Protection Association. 1990. *National Electrical Code - 1990.* Quincy, MA: National Fire Protection Association.

National Fire Protection Association. 1990. *National Electrical Code - 1990 Handbook.* Quincy, MA: National Fire Protection Association.

5

System Design Example

In this chapter, we will review the design of a typical small industrial building. The design process will be developed based on the design requirements of the architectural and mechanical plans as well as the owner's own criteria. Based upon this information, we will make a preliminary determination of the location for electrical panelboards and distribution equipment. Based on these locations we will prepare a rough riser diagram of the distribution system. We will also prepare a list of anticipated electrical drawings to help in the planning process. We will then discuss the layout of the other building electrical system components, such as receptacles and lighting fixtures, and their associated circuit design. Finally, based on the electrical system loads, we will select the correct overcurrent devices, panelboard feeders and transformer ratings for the system.

DESIGN CRITERIA

The project involves the design of a new manufacturing facility for the Spectrum Plastics Company. Spectrum Plastics has retained the services of Reginald Paisley, AIA, a local architect. Working closely with Spectrum Plastics, the architect developed basic building space requirements which indicated the need for a facility of approximately 30,000 ft^2. Spectrum Plastics subsequently purchased a suitable site and the architect began refining his original conceptual design. Based on this preliminary work and the actual building site, the architect developed a site plan and floor plans. During the development of the building floor plans, Mr. Paisley selected the mechanical/plumbing consultant, Gompers and Associates, and the electrical consultant, John Matthews & Associates. During the preliminary design phase, Spectrum Plastics prepared drawings of the specific electrical requirements for the process equipment.

Architectural Plans

The results of the architect's preliminary work is shown in Figures 5.1, 5.2 and 5.3 and consists of a site plan, office area plan and manufacturing area plan. These are architectural sheets A-1, A-2 and A-3 respectively. A review of Sheet A-1 shows that the building is composed of two parts, an office area and a manufacturing area, with a total of 30,240 ft^2. The site has been designed to allow truck delivery of raw material and shipment of finished products by means of a rear loading dock area. A truck parking area is also provided. An employee parking area is located adjacent to the building. Since Spectrum Plastics operates two full shifts, these exterior areas will require adequate lighting.

Sheet A-2, Figure 5.2, shows the floor plans for the 4,320 ft^2 office area. The drawing is divided into two parts. The office layout shows the interior arrangement of various rooms that make up the area. The ceiling grid layout shows the location of the ceiling tiles. These tiles are supported by a modular ceiling grid system that, in this case, is 2 ft. by 2 ft. Both lighting fixtures, mechanical diffusers and return air grills must fit within this modular pattern. Note that the architect has established a column grid system based on structural column locations. Locations within the building are frequently specified by their column line coordinates, for example, the north exit door is located between column lines 7 and 8 on column line A.

Sheet A-3, Figure 5.3, shows the floor plan of the manufacturing area. Note that the manufacturing area consists of areas for raw material storage, manufacturing, parts storage, assembly, warehousing and shipment. The building column lines are located on a 24 ft. by 30 ft. grid and the ceiling height is 21 ft. The entire manufacturing area is 144 ft. by 180 ft., or 25,920 ft^2.

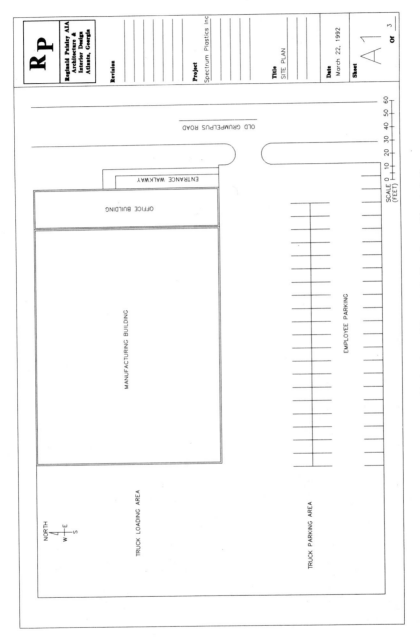

Figure 5.1 Preliminary architectural site plan. Sheet A-1.

183

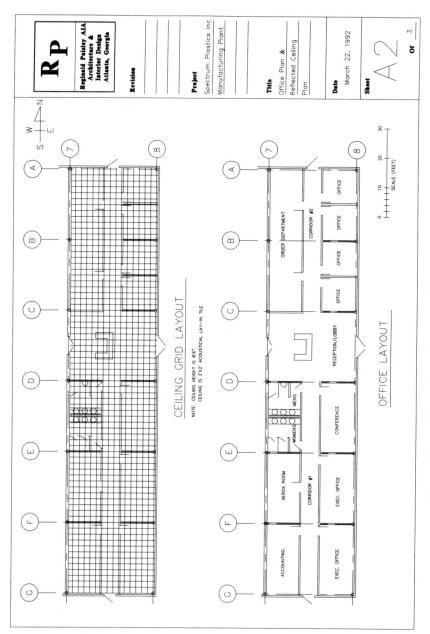

Figure 5.2 Preliminary architectural office area plan. Sheet A-2.

184

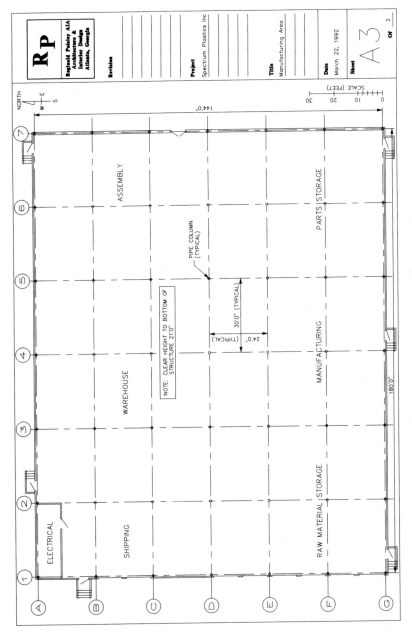

Figure 5.3 Preliminary architectural manufacturing area plan. Sheet A-3.

185

Owner's Criteria

During the preliminary design phase, the owner was asked to furnish electrical data on various items of process equipment planned for the building. The electrical provisions for this equipment must be a part of the electrical design. Figures 5.4 and 5.5 show sheets T-1 and T-2 which were developed by Spectrum Plastics. Sheet T-1 shows a summary of the electrical characteristics of owner furnished equipment for both the office and manufacturing areas. Sheet T-1 shows the location of equipment listed for this area. Sheet T-2 shows the location of equipment in the manufacturing area. From this sheet, we can also gain an overall view of how equipment is located. Note that bulk materials are stored in the storage area in relatively low stacks, about 6 ft. in height. There are storage shelves in the parts storage area and much larger shelves for storage of cartons of finished products in the warehouse. The building lighting design must accommodate these areas as well as the manufacturing and assembly areas. Not all owners are capable of developing their own criteria to the extent shown. In many cases, the electrical engineer must work directly with the owner's representative to determine equipment electrical requirements.

Preliminary Mechanical and Plumbing System Coordination

During the very early part of the design phase, the mechanical and plumbing engineer performs the necessary design analysis to determine the characteristics of air conditioning, heating and plumbing equipment. The electrical requirements for these system components are forwarded to the electrical engineer for distribution system planning purposes.

Figures 5.6 and 5.7, sheets M-1 and M-2, show preliminary mechanical and plumbing system requirements for this project. As noted in earlier chapters, the mechanical and electrical engineers must carefully coordinate the electrical requirements, including voltage, of the various items of mechanical and plumbing equipment.

From Sheet M-1 we see that the office area utilizes two air conditioning systems, AC #1 and AC #2, along with associated Fans, F-1 and F-2. The office area is heated by two electric duct heaters, EDH-1 and EDH-2. We also note that an electric drinking fountain, EDF, is located near the restroom area. Finally, an electric water heater, WH-1, is located just inside the manufacturing area at Column B-7.

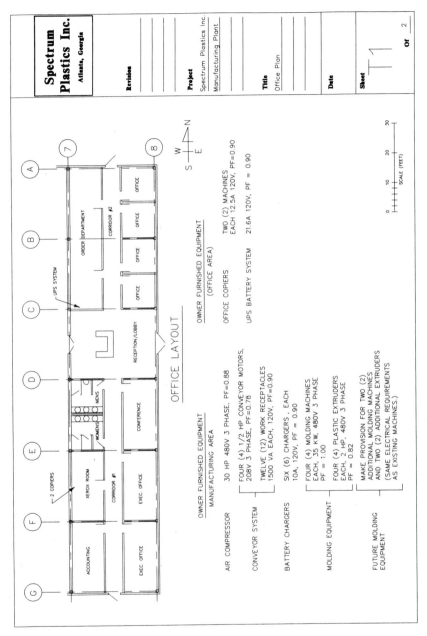

Figure 5.4 Owner electrical criteria - office area. Sheet T-1.

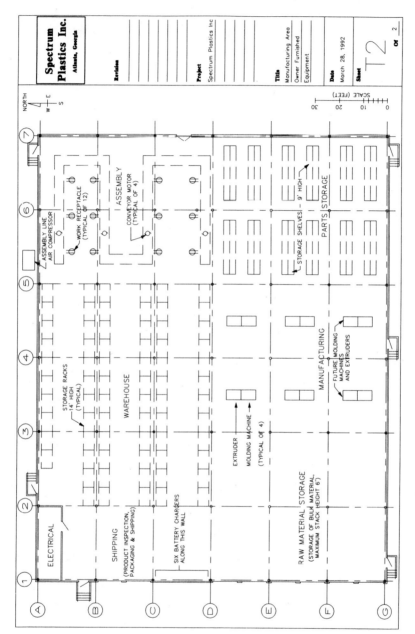

Figure 5.5 Owner electrical criteria - manufacturing area. Sheet T-2.

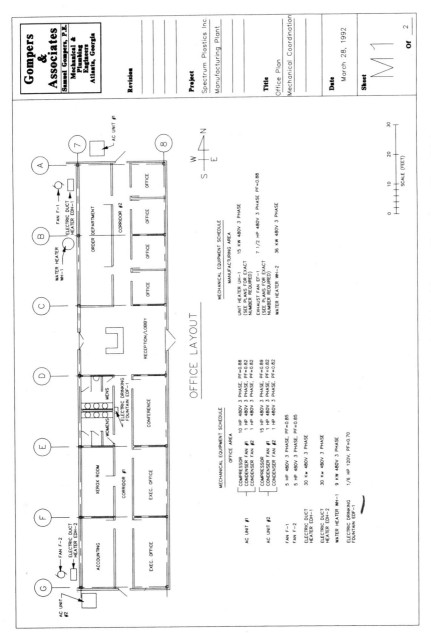

Figure 5.6 Preliminary mechanical/plumbing - office area. Sheet M-1.

189

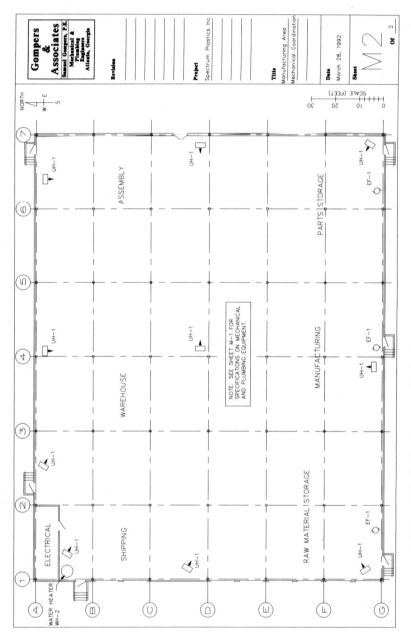

Figure 5.7 Preliminary mechanical/plumbing - manufacturing area. Sheet M-2.

Figure 5.7, Sheet M-2, shows the mechanical and plumbing requirements for the manufacturing area. This area is not air conditioned, but is ventilated by three exhaust fans, EF-1, located along the south wall of the building. Heat is provided by a number of electric unit heaters, UH-1, located throughout the area. An additional water heater, WH-2, is located near the rear of the building.

SYSTEM PLANNING

Based on the information contained in sheets A-1, A-2, A-3, T-1, T-2, M-1 and M-2, the electrical engineer can begin rough preliminary planning. One of the early items in the planning process is the location of major electrical distribution system components.

The location of the main electrical distribution panel, MDP, is frequently determined by the location of nearby utility lines. During the preliminary design the electrical engineer found that the utility company had existing overhead lines near the north side of the site. The architect made available an electrical equipment room at the northwest corner of the building, near Column B-2. The utility's preferred means of service was a pad-mounted transformer, located along the north wall of the building, near the electrical equipment room.

The next step was the selection of the appropriate system voltage. The utility had both 208Y/120 V and 480Y/277 V transformers available. Since much of the owner's equipment was to be 480 V, the choice of a 480Y/277 V service seemed appropriate. With this system, the lighting could be designed based on 277 volt equipment, and the major air conditioning system components on 480 volts, three-phase operation. Step-down transformers could be provided to supply the necessary receptacle and small equipment power requirements.

Panelboard Location Planning

Let us first consider the location of panelboards for the office area. This area utilizes 277 volt fluorescent lighting, 480 V air conditioning, and 120 V receptacles. As a result, we will require both a 480Y/277 V panelboard and a 208Y/120 V panelboard. Since space within the office area is at a premium, we will locate the panelboards, HA (480Y/277 V) and LA

(208Y/120 V), just inside the manufacturing area, at Column C-7. The panelboards will be surface-mounted, and the step-down transformer for Panelboard LA can be mounted on the wall above the panel. In order to reduce the cost of running a separate feeder for HA and LA from the main distribution panel, MDP, the step-down transformer for LA will be connected to Panelboard HA.

The manufacturing area poses several design challenges. From Sheets T-1 and T-2 it is clear that there are two areas of concentrated equipment load. The first is the actual manufacturing area itself. The owner plans four plastic molding machines and extruders, plus the capacity for an additional two molding machines and extruders. This equipment is rated for 480 volts, three-phase operation. The distance from the MDP location would make individual circuits for the mold machine and extruders prohibitively expensive. A panelboard location nearer the equipment would both shorten the length of circuits, and also provide a convenient location for disconnection of power to individual pieces of equipment for maintenance, or in case of emergency. As a result of these considerations, a 480Y/277 V panelboard, HD, is to be located at Column E-4 to serve the process equipment in this area.

A second area of concentrated load is the assembly area. Here workers assemble plastic components as they move along an assembly line. Both compressed air tools and electrical tools are employed. The assembly line conveyor is powered by several small three-phase motors. Since the equipment in this area is rated for 208 V or 120 V, a 208Y/120 V panelboard located nearby seems in order. Panelboard LC is located near column B-7 near the assembly area. The transformer for this panelboard is mounted on the wall above the panel, and is served from the main distribution panel, MDP.

The loads for the remainder of the manufacturing area consist of lighting, mechanical and other small equipment, and general purpose receptacles. The total load exceeds the capacity of a single panelboard, so dividing the load between panelboards for lighting and mechanical system components would be in order. Panelboard HB, 480Y/277 V, is located in the electrical equipment room, and serves the manufacturing area lighting requirements as well as the three building exhaust fans, EF-1. We also connect the building exterior lighting to this panelboard. Panelboard HC, 480Y/277 V, is also located in the electrical equipment room, and serves the electric unit heaters located throughout the building. The receptacles and small equipment circuits are served from the last panelboard, LB. This

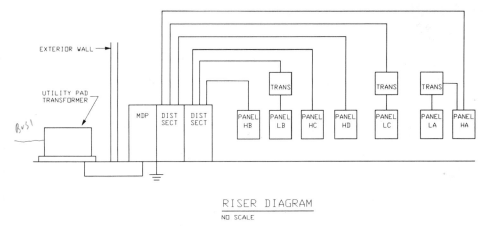

RISER DIAGRAM
NO SCALE

Figure 5.8 Preliminary riser diagram.

panelboard is located with Panels HB and HC in the electrical equipment room.

Based upon this analysis of required panelboard locations, a rough riser diagram for the proposed system can be prepared, as shown in Figure 5.8. At this point, we have not yet selected the correct panelboard current rating or feeder overcurrent devices and wire sizes. These things are determined after the loads connected to each panelboard have been finalized.

Lighting Coordination

Lighting design will be discussed in the second part of this book, beginning with Chapter 9. For the present time, let us assume that the office area will be illuminated by lay-in fluorescent luminaires, and the manufacturing area by industrial High Intensity Discharge luminaires. After an analysis of the lighting equipment for each area of the office and manufacturing area, a lighting plan is prepared, as shown in Figures 5.9 and 5.10. Note that these preliminary sheets show luminaire locations only, and do not contain other electrical circuit information. These plans are forwarded to the architect and mechanical engineer for coordination purposes. The mechanical engineer utilizes the data to confirm the heat contribution of the lighting system to the air conditioning load, and also to coordinate the location of air supply diffusers and returns with the luminaires. The

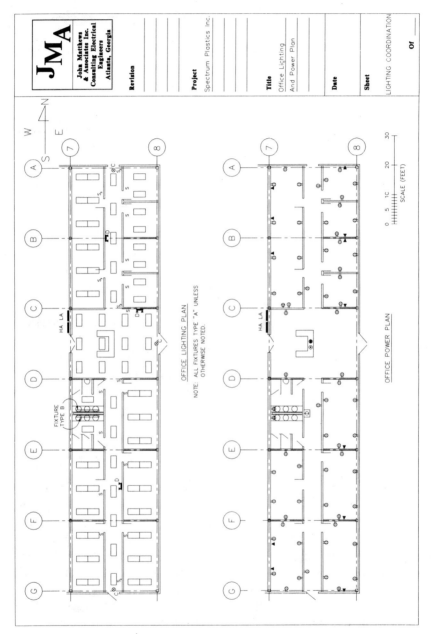

Figure 5.9 Lighting and receptacle layout - office area.

OFFICE LIGHTING PLAN

NOTE: ALL FIXTURES TYPE "A" UNLESS OTHERWISE NOTED.

FIXTURE TYPE B

HA LA

OFFICE POWER PLAN

HA LA

SCALE (FEET)

0 5 1C 20 30

JMA

John Matthews & Associates Inc.
Consulting Electrical Engineers
Atlanta, Georgia

Revision

Project
Spectrum Plastics Inc.

Title
Office Lighting
And Power Plan

Date

Sheet
LIGHTING COORDINATION

or ___

194

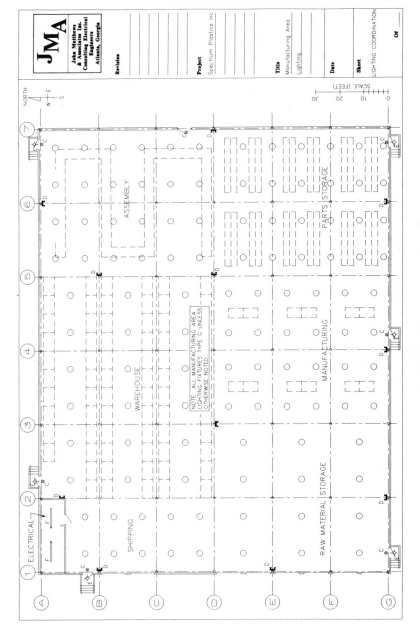

Figure 5.10 Lighting layout - manufacturing area.

195

architect will use this information to make sure the overall symmetry of the ceiling arrangement of luminaires and diffusers meets the design appearance goals. He may also prepare a separate sheet showing ceiling tile locations, luminaire locations and diffuser locations. This sheet is referred to as a reflected ceiling plan. The architect may also review the locations of receptacles and telephone outlets in the office area to assure that these locations work well with possible office furniture locations.

ELECTRICAL SHEET ORGANIZATION

Although there is no universal rule concerning electrical sheet organization, many engineers organize their sheets along the following lines: Sheet E-1 is frequently reserved for the overall site electrical plan. This sheet often shows the utility service location as well as the telephone service arrangements. The telephone service is normally just an empty conduit run underground to the property line location that has been determined by the local telephone company. The telephone company then installs the necessary telephone cable to serve the building.

The electrical sheets following E-1 are often devoted to lighting and power plans for the building. In large, multi-story buildings these sheets can number up to twenty to over one hundred. It is common practice to depict the lighting system and power systems on separate plans in order to reduce clutter.

The final sheets of the electrical set often contain panelboard schedules, notes, lighting fixture schedules, symbols and a riser diagram.

For buildings with sophisticated fire alarm or communications systems, separate floor plans are included with the details of these systems.

The overall objective is to organize the electrical plans in a clear, concise, and easy to read manner, that facilitates the construction process, and conveys all necessary data to the contractor in clear, unambiguous terms.

For our project, a reasonable arrangement of electrical plans might be as follows:

E-1   Electrical Site Plan
E-2   Office Area Lighting and Power Plan
E-3   Manufacturing Area Lighting Plan
E-4   Manufacturing Area Power Plan

BRANCH CIRCUIT DESIGN

During the branch circuit design phase, the various items of electrical equipment are connected to the selected panelboard and assigned circuit numbers and wire and conduit sizes. The loads for each circuit, along with the circuit overcurrent device sizes, are recorded on panelboard schedules. The completed panelboard schedule data is often entered into a computer database that facilitates circuit design, as well as overall panelboard and system load calculations.

Typically the smallest branch circuit conductor size employed is #12 AWG, and the smallest branch circuit breaker is 15 amperes. For receptacles and lighting circuits, 20 ampere circuit breakers are common. As mentioned in Chapter 3, molded case circuit breakers are not allowed to carry more than 80% of their capacity on a continuous basis. For a 20 ampere circuit, this means that a limit of 16 amperes is appropriate. For a 120 V circuit, this represents 1920 VA, and for a 277 V circuit, 4432 VA. These are maximum values, and circuits are not necessarily loaded to this extent.

To illustrate these points, let us consider a small portion of the office area shown in Figure 5.11. We should examine the receptacle layout first. In office areas there are no code requirements governing the number of receptacles, so the engineer must base the design on experience. In small offices, such as those shown, it might be desirable to locate a receptacle on each wall. The *National Electrical Code* does suggest an allowance of 180 VA per receptacle, thus limiting the maximum number of receptacles on a circuit to about 10. Considering the increased density of office equipment, many engineers limit such circuits to five to eight receptacles. In order to minimize wiring costs, the receptacle circuits are frequently installed using a common neutral conductor. Recall that in the case of closely balanced three-phase systems, the neutral current will be very small. It is important to make sure that each of the three circuits originates from a circuit breaker on a separate phase.

In the case shown in Figure 5.11, it was decided to connect six

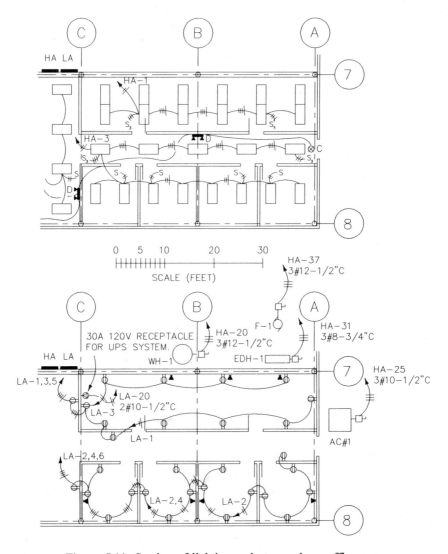

Figure 5.11 Section of lighting and power plan - office area.

In the case shown in Figure 5.11, it was decided to connect six receptacles on each of three circuits. These circuits are assigned to panelboard LA, circuits 2, 4, and 6. Each circuit will be assigned a load of 1080 VA (6 times 180 VA). The homerun indicates that three-phase conductors—short hash marks—and one neutral—the long hash mark—are to be included. Since the smallest conductor size permitted is #12, this homerun is understood to mean 4#12 conductors in a 1/2" conduit. Conductor sizes larger than #12 are normally stipulated on the plans, on the panelboard schedule, or in notes on the drawings. For example, circuit LA-20, also shown in Figure 5.11, requires 2#10-1/2"C because it serves the UPS system (21.6 A, 120 V) shown on the tenant criteria.

A portion of the lighting plan for this same area is shown in Figure 5.11. Each luminaire has an input of 200 VA including lamps and ballast losses. Note that 12 luminaires (2400 VA) are assigned to circuit HA-1. The small offices and the corridor are grouped on HA-3 with a total of 2600 VA. The reader should examine the schedules for Panelboards LA and HA to see how these loads, as well as others, were recorded.

Finally, a review of the symbols list on Sheet E-5 reminds us that the various raceways shown in Figure 5.11 are to be concealed in the wall or ceiling (curved lines). Had the circuits been shown with straight lines, the circuits would be installed exposed, as in the manufacturing area.

The branch circuits for the various items of building equipment are designed in accordance with the requirements discussed in Chapter 4. We will select several examples for the purpose of illustrating the process. Figures 5.12 through 5.20 show Electrical Sheets E-1 through E-9.

Example 5.1: Design the circuit for AC Unit #1. Use THW copper conductors.

From the mechanical criteria, we can see that this unit has one compressor, rated 10 HP, and two condenser fans, rated 1 HP each. All motors are 480 V, three-phase. From Table 3.9 we see that the 10 HP motor has a full load amp rating of 14 amperes. The 1 HP motors have a rating of 1.8 amperes each.

In accordance with our work in Chapter 4 for multiple motor circuits, we compute the overcurrent device size as follows:

$$OC = 1.75 \cdot 14 + 1.8 + 1.8 = 28.1 \text{ A} \qquad \text{(30 A/3P Circuit Breaker)}$$
$$\text{Wire} = 1.25 \cdot 14 + 1.8 + 1.8 = 21.1 \text{ A} \qquad \text{(3#10-1/2"C)}$$

The total unit load is given by

$$kVA = 14 \cdot .480 \sqrt{3} + \left(2 \cdot 1.8 \cdot .480 \sqrt{3} \right) = 14.63 \quad kVA$$

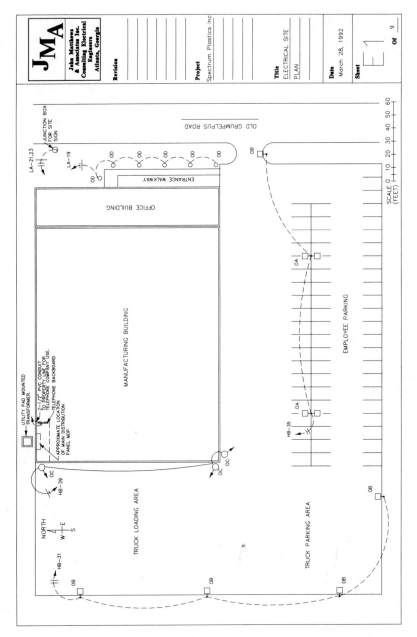

Figure 5.12 Electrical site plan. Sheet E-1.

200

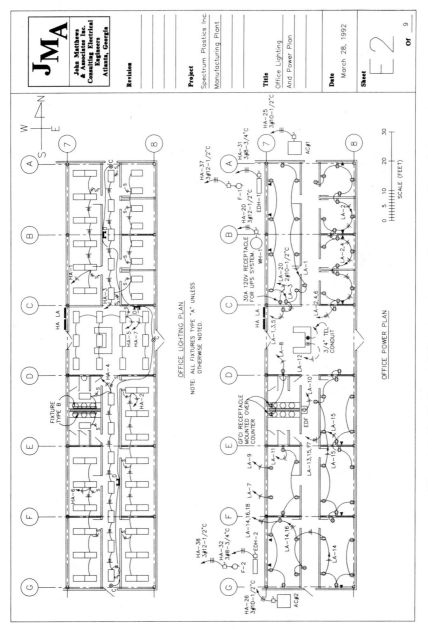

Figure 5.13 Power and lighting plan - office area. Sheet E-2.

201

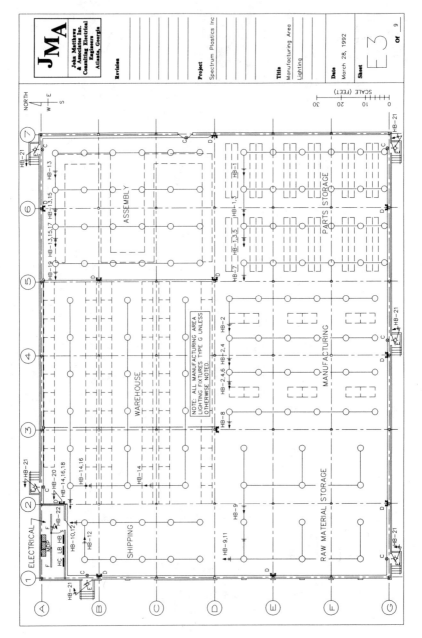

Figure 5.14 Lighting plan - manufacturing area. Sheet E-3.

202

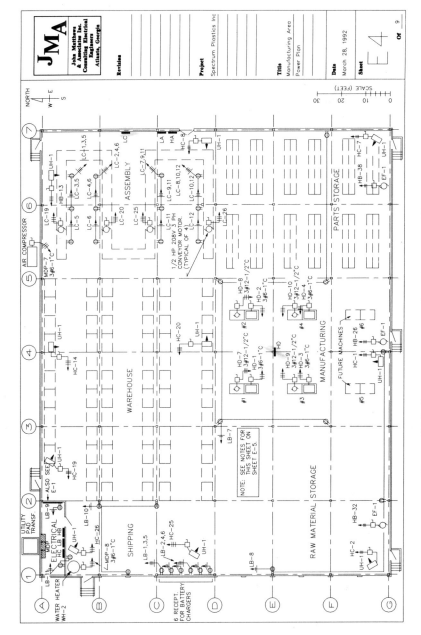

Figure 5.15 Power plan - manufacturing area. Sheet E-4.

203

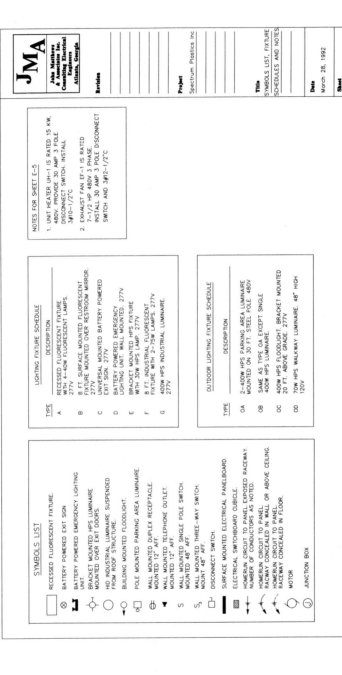

Figure 5.16 Symbols, lighting fixture schedule, notes. Sheet E-5.

SYMBOLS LIST

☐ RECESSED FLUORESCENT FIXTURE.

⊗ BATTERY POWERED EXIT SIGN

☇ BATTERY POWERED EMERGENCY LIGHTING UNIT.

⌁ BRACKET MOUNTED HPS LUMINAIRE MOUNTED OVER EXIT DOORS.

◯ HID INDUSTRIAL LUMINAIRE SUSPENDED FROM ROOF STRUCTURE.

⌐◯ BUILDING MOUNTED FLOODLIGHT.

⊕ POLE MOUNTED PARKING AREA LUMINAIRE.

▼ WALL MOUNTED DUPLEX RECEPTACLE. MOUNTED 12" AFF.

▼ WALL MOUNTED TELEPHONE OUTLET. MOUNTED 12" AFF.

S WALL MOUNTED SINGLE POLE SWITCH. MOUNTED 48" AFF.

S₃ WALL MOUNTED THREE-WAY SWITCH. MOUNT 48" AFF.

☐ DISCONNECT SWITCH.

▬ SURFACE MOUNTED ELECTRICAL PANELBOARD.

▨ ELECTRICAL SWITCHBOARD CUBICLE.

⊣⊢ HOMERUN CIRCUIT TO PANEL. EXPOSED RACEWAY. NUMBER OF CONDUCTORS AS NOTED.

⤙ HOMERUN CIRCUIT TO PANEL. RACWAY CONCEALED IN WALL OR ABOVE CEILING.

⤙ HOMERUN CIRCUIT TO PANEL. RACEWAY CONCEALED IN FLOOR.

◯ MOTOR

⊖ JUNCTION BOX

LIGHTING FIXTURE SCHEDULE

TYPE	DESCRIPTION
A	RECESSED FLUORESCENT FIXTURE WITH 4-40W FLUORESCENT LAMPS. 277V
B	8 FT. SURFACE MOUNTED FLUORESCENT FIXTURE MOUNTED OVER RESTROOM MIRROR. 277V
C	UNIVERSAL MOUNTED BATTERY POWERED EXIT SIGN. 277V
D	BATTERY POWERED EMERGENCY LIGHTING UNIT. WALL MOUNTED. 277V
E	BRACKET MOUNTED HPS FIXTURE WITH 30W HPS LAMP. 277V
F	8 FT. INDUSTRIAL FLUORESCENT FIXTURE WITH 2-75W LAMPS. 277V
G	400W HPS INDUSTRIAL LUMINAIRE. 277V

OUTDOOR LIGHTING FIXTURE SCHEDULE

TYPE	DESCRIPTION
OA	2-400W HPS PARKING AREA LUMINAIRE MOUNTED ON 30 FT. STEEL POLE. 480V
OB	SAME AS TYPE OA EXCEPT SINGLE 400W HPS LUMINAIRE.
OC	400W HPS FLOODLIGHT, BRACKET MOUNTED 20 FT. ABOVE GRADE. 277V
OO	70W HPS WALKWAY LUMINAIRE. 48" HIGH 120V

NOTES FOR SHEET E-5

1. UNIT HEATER UH-1 IS RATED 15 KW, 480V PROVIDE 30 AMP 3 POLE DISCONNECT SWITCH. INSTALL 3#10-1/2"C

2. EXHAUST FAN EF-1 IS RATED 7-1/2 HP 480V 3 PHASE INSTALL 30 AMP 3 POLE D'ISCONNECT SWITCH AND 3#12-1/2"C

JMA
John Matthews
& Associate Inc.
Counting Electrical
Engineers
Athens, Georgia

Revision

_____
_____
_____
_____
_____

Project
Spectrum Plastics Inc

Title
SYMBOLS LIST, FIXTURE SCHEDULES AND NOTES

Date
March 28, 1992

Sheet

E5

9 of ___

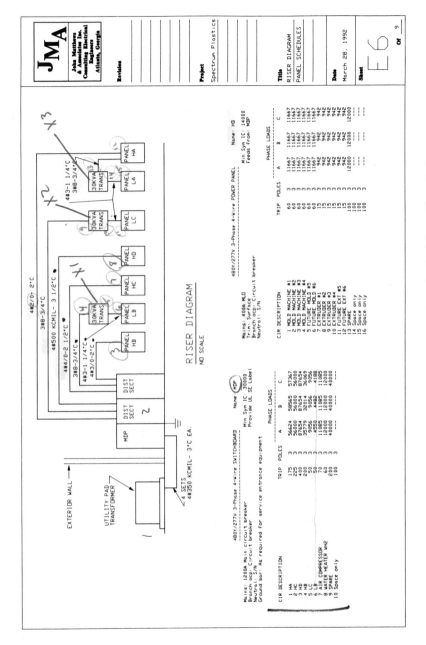

Figure 5.17 Riser diagram, panelboard schedules. Sheet E-6.

480Y/277V 3-Phase 4-Wire LIGHTING PANEL Name: HA

Mains: 225A MLO
Trim: Surface
Neutral: S/N

Min Sym IC: 14000
Feeds From: MDP

CIR	DESCRIPTION	TRIP	POLES	A	B	C	POLES	TRIP	DESCRIPTION	CIR
1	ORDER DEPT LTS	20	1	2400			1	20	CONF - EXEC LTS	2
3	PRI OF-CDR 2 LT	20	1	3600	2600		1	20	RESTRM-CDR #1 LT	4
5	LOBBY LT	20	1	2400		2400	1	20	ACCOUNT-XEROX LT	6
7	AC UNITS			450	2400		1	20		8
9	SPARE	20	1	3000	3000		1	20	SPARE	10
11	SPARE	20	1	3000		3000	1	20	SPARE	12
13	Space only	--	1				--		Space only	14
15	Space only	--	1				--		Space only	16
17	Space only	--	1				--		Space only	18
19	LA	50	3	8266 3000			3	15	WATER HTR WH-1	20
21	-				8657 3000					22
23	-					7659 3000				24
25	AC UNIT #1	30	3	4878			3	40	AC UNIT #2	26
27	-			6818	4878 6818					28
29	-					6818				30
31	ELEC DUCT HTR #1	50	3	10000 10000			3	50	ELEC DUCT HTR #2	32
33	-				10000 10000					34
35	-					10000 10000				36
37	FAN F-1	15	3	2106 2106			3	15	FAN F-2	38
39	-				2106 2106					40
41	-					2106 2106				42

208Y/120V 3-Phase 4-Wire LIGHTING PANEL Name: LA

Mains: 100A Main circuit breaker
Trim: Surface
Neutral: S/N

Min Sym IC: 10000
Feeds From: HA

CIR	DESCRIPTION	TRIP	POLES	A	B	C	POLES	TRIP	DESCRIPTION	CIR
1	ORDER DEPT RECPT	20	1	720			1	20	OFFICE RECEPT	2
3	ORDER DEPT RECPT	20	1	1080	720		1	20	OFFICE RECEPT	4
5	ORDER DEPT RECPT	20	1			720 1080	1	20	OFFICE RECEPT	6
7	XEROX MACHINE	20	1				1	20	RESTRM-LOBBY REC	8
9	XEROX MACHINE	20	1	1500 540	1500		1	20	EDP	10
11	XEROX ROOM RECPT	20	1		750		1	20	RECEPTION RECPT	12
13	CONF-CDR #2 RECP	20	1	900	720		1	20	EXEC OFFICE RECP	14
15	EXEC OFFICE RECP	20	1		720		1	20	ACCT-CDR #1 RECP	16
17	EXEC OFFICE RECP	20	1	900	720		1	20	ACCOUNTING RECPT	18
19	WALKWAY LIGHTS	20	1	700			1	30	UPS SYSTEM	20
21	SITE SIGN	20	1	2592	1000 900		1	20	SPARE	22
23	SITE SIGN	20	1		900	1000	1	20	SPARE	24
25	SPARE	20	1	500		900	1	20	SPARE	26
27	SPARE	20	1		500		--		Space only	28
29	SPARE	20	1			500	--		Space only	30

Figure 5.18 Panelboard schedules. Sheet E-7.

JMA
John Matthews
& Associates Inc.
Consulting Electrical
Engineers
Atlanta, Georgia

Revision

Project
Spectrum Plastics

Title
PANEL SCHEDULES

Date
March 28, 1992

Sheet
E-7 of 9

JMA
John Matthews & Associates Inc.
Consulting Electrical Engineers
Atlanta, Georgia

Revision

Project
Spectrum Plastics

Title
PANEL SCHEDULES

Date
March 28, 1992

Sheet
E8
of 9

Panel: HB — 480Y/277V 3-Phase 4-Wire LIGHTING PANEL
Mains: 225A MLO, Trim: Surface, Neutral: S/N
Min Sym IC: 25000, Feeds From MDP

CIR	DESCRIPTION	TRIP	POLES	A	B	C	POLES	TRIP	DESCRIPTION	CIR
1	PARTS STORAGE LT	20	1	2880			1	20	MANUFACTURING LT	2
3	PARTS STORAGE LT	20	1		2880		1	20	MANUFACTURING LT	4
5	PARTS STORAGE LT	20	1			2880	1	20	MANUFACTURING LT	6
7	PARTS STORAGE LT	20	1	2880			1	20	MANUFACTURING LT	8
9	STORAGE LTG	20	1		2880	2400	1	20	SHIPPING LTG	10
11	STORAGE LTG	20	1			2400	1	20	SHIPPING LTG	12
13	ASSEMBLY LTG	20	1	2880			1	20	WAREHOUSE LTG	14
15	ASSEMBLY LTG	20	1		2880		1	20	WAREHOUSE LTG	16
17	ASSEMBLY LTG	20	1			2880	1	20	WAREHOUSE LTG	18
19	ASSEMBLY LTG	20	1	2880			1	20	EXIT & EMERG LTG	20
21	EXIT DOOR LIGHTS	20	1	1450			1	20	ELECTRICAL RM LT	22
23	SPARE	20	1		3000		1	20	SPARE	24
25	SPARE	20	1			3000	3	20	EF-1	26
27	SPARE	20	1	3000						28
29	SPARE	20	1		3000	3048				30
31	SITE LIGHTING	20	2		900		3	20	EF-1	32
33	—			900		3048				34
35	SITE LIGHTING	20	2		1125	3048				36
37	—			1125			3	20	EF-1	38
39	BLDG FLOOD LTS	20	1	3048	1350					40
41	Space only	—	—			3048			—	42

Panel: HC — 480Y/277V 3-Phase 4-Wire LIGHTING PANEL
Mains: 225A MLO, Trim: Surface, Neutral: S/N
Min Sym IC: 25000, Feeds From MDP

CIR	DESCRIPTION	TRIP	POLES	A	B	C	POLES	TRIP	DESCRIPTION	CIR
1	UH-1	25	3	5000			3	25	UH-1	2
3	—				5000				—	4
5	—					5000			—	6
7	UH-1	25	3	5000			3	25	UH-1	8
9	—				5000				—	10
11	—					5000			—	12
13	UH-1	25	3	5000			3	25	UH-1	14
15	—				5000				—	16
17	—					5000			—	18
19	UH-1	25	3	5000			3	25	UH-1	20
21	—				5000				—	22
23	—					5000			—	24
25	UH-1	25	3	5000			3	25	UH-1	26
27	—				5000				—	28
29	—					5000			—	30
31	SPARE	20	1	3000			1	20	SPARE	32
33	SPARE	20	1		3000		1	20	SPARE	34
35	SPARE	20	1			3000	1	20	SPARE	36
37	Space only	—	—				1	—	Space only	38
39	Space only	—	—				1	—	Space only	40
41	Space only	—	—				1	—	Space only	42

Figure 5.19 Panelboard schedules. Sheet E-8.

Figure 5.20 is a set of three panelboard schedules.

Panel LB — 208Y/120V 3-Phase 4-Wire LIGHTING PANEL — Name: LB — Min Sym IC: 10000 — Feeds from MDP
Mains: 100A Main circuit breaker — Trim: Surface — Neutral: S/N

CIR	DESCRIPTION	TRIP	POLES	A	B	C	POLES	TRIP	DESCRIPTION	CIR
1	BATTERY CHARGER	20	1	1200			1	20	BATTERY CHARGER	2
3	BATTERY CHARGER	20	1	1200	1200		1	20	BATTERY CHARGER	4
5	BATTERY CHARGER	20	1		1200	1200	1	20	BATTERY CHARGER	6
7	MANUFACT. RECPT	20	1	900		1200	1	20	SHIPPING RECEPT	8
9	TELEPHONE RECPT	20	1	540	180		1	20	SHIPPING RECEPT	10
11	ELEC RM RECEPT	20	1	540		360	1	20	SPARE	12
13	SPARE	20	1		500		1	20	SPARE	14
15	SPARE	20	1	500		500	1	20	SPARE	16
17	SPARE	20	1		500	500	1	20	SPARE	18
19	Space only	--	--				--	--	Space only	20
21	Space only	--	--				--	--	Space only	22
23	Space only	--	--				--	--	Space only	24

Panel LC — 208Y/120V 3-Phase 4-Wire LIGHTING PANEL — Name: LC — Min Sym IC: 10000 — Feeds from MDP
Mains: 100A Main circuit breaker — Trim: Surface — Neutral: S/N

CIR	DESCRIPTION	TRIP	POLES	A	B	C	POLES	TRIP	DESCRIPTION	CIR
1	RECEPTACLE	20	1	1500			1	20	RECEPTACLE	2
3	RECEPTACLE	20	1	1500	1500		1	20	RECEPTACLE	4
5	RECEPTACLE	20	1		1500	1500	1	20	RECEPTACLE	6
7	RECEPTACLE	20	1	1500		1500	1	20	RECEPTACLE	8
9	RECEPTACLE	20	1	1500	1500		1	20	RECEPTACLE	10
11	RECEPTACLE	20	1		1500	1500	1	20	RECEPTACLE	12
13	SPARE	20	1	1000			1	20	SPARE	14
15	SPARE	20	1	1000	1000		1	20	SPARE	16
17	SPARE	20	1		1000	1000	1	20	SPARE	18
19	CONVEYOR MOTOR	15	3	264			3	15	CONVEYOR MOTOR	20
21	-	-		264	264		-	-	-	22
23	-	-			264	264	-	-	-	24
25	CONVEYOR MOTOR	15	3	264			3	15	CONVEYOR MOTOR	26
27	-	-		264	264		-	-	-	28
29	-	-			264	264	-	-	-	30

Figure 5.20 Panelboard schedules. Sheet E-9.

JMA
John Matthews & Associates Inc.
Consulting Electrical Engineers
Atlanta, Georgia

Revision

Project
Spectrum Plastics

Title
PANEL SCHEDULES

Date
March 28, 1992

Sheet
E-9
9 of

208

The unit is connected to Circuit HA-25 as shown on Sheet E-2. Compare the above data with the information on the panelboard schedule for HA. It should be remembered that the ac unit nameplate might contain specific information on the maximum overcurrent device rating. This would take precedence over the above calculations.

Example 5.2: Design a circuit for the office UPS system shown on Sheet T-1. The unit requires 21.6 amperes, 120 V. The correct overcurrent device and wire size are:

OC = 1.25 · 21.6 = 27 A   (30 A/1P Circuit Breaker)
Wire Size (Based on 30 A)   (2#10-1/2"C)

The total load of this UPS system is

$$kVA = .120 · 21.6 = 2.592 \ kVA$$

This equipment is served by Circuit LA-20, as shown on Sheet E-2. Compare the above data with that shown on this sheet and on the appropriate panelboard schedule.

Example 5.3: Design a circuit for one of the molding machines and one of the extruder motors.

From Sheet T-1, we see that the molding machines are rated at 35.0 kW, 480 V, three-phase and the extruder motors are 2 HP, 480 V, three-phase. We will use THW copper conductors.
a) Molding Machine
The correct overcurrent device and wire size are as follows.

$$I = \frac{kW}{kV \sqrt{3} \ PF} = \frac{35.0}{.480 \sqrt{3} · 1.0} = 42.1 \ A$$

OC = 1.25 · 42.1 = 52.63 A   (60 A/3P Circuit Breaker)
Wire Size (Based on 60 A)   (3#6-1"C)

b) 2 HP Extruder
From Table 3.9 this motor has a full load current rating of 3.4 A (2827 VA). The correct overcurrent device and wire sizes are as follows.

$$OC = 1.75 \cdot 3.4 = 5.95 \text{ A} \qquad \text{(15 A/3P Circuit Breaker)}$$
$$\text{Wire} = 1.25 \cdot 3.4 = 4.25 \text{ A} \qquad (3\#12\text{-}1/2"\text{C})$$

Molding machine #1 and Extruder #1 are served from circuits HD-1 and HD-7, respectively. Compare the above results with the circuits shown on Sheet E-4 and the appropriate panelboard schedule.

Example 5.4: Design a circuit for the shipping area lighting.

From Sheet E-3 we see that this area contains 10 400 W HID luminaires, Type G. Each luminaire has an input of 480 VA, 277 V. The total load of 4800 VA is too much for a single 277 V circuit, so two circuits will be necessary. If we install five luminaires on each circuit, the total load will be

$$VA = 5 \cdot 480 = 2400 \text{ VA per circuit}$$

The current will be

$$I = \frac{2.4}{.277} = 8.66 \ A$$

which is well under the 16 A capacity of a 20 A breaker. These luminaires are connected to circuits HB-10 and 12. Compare the circuit loads shown on the appropriate panelboard schedules.

PANELBOARD AND SWITCHBOARD DESIGN

Once all circuits served by a particular panelboard or switchboard are identified and designed, it is possible to select the correct overcurrent device and feeder. The process for determining the actual demand load was discussed in Chapter 4, and we will now utilize this technique for several representative panelboards.

Example 5.5: Select the correct panelboard ampere rating, overcurrent device and feeder for Panelboard LA. The panelboard is connected to Panelboard HA by means of a step-down transformer.

From the panelboard schedule, we find the following connected loads. We will use a power factor of 0.95 for lighting and 0.90 for receptacles and spare capacity.

Connected Load

Lighting	2.70 kVA @ PF =0.95	2.57 + j 0.84
Receptacles	15.99 kVA @ PF =0.90	14.39 + j 6.97
UPS System	2.59 kVA @ PF =0.90	2.33 + j 1.13
Spares	3.30 kVA @ PF =0.90	2.97 + j 1.44

$$22.26 + j10.38$$
$$(24.56 \text{ kVA})$$

Calculated Demand

		D.F.	
Lighting	2.57 + j 0.84	1.25	3.21 + j 1.05
Receptacles	9.00 + j 4.36	1.00	9.00 + j 4.36
	5.39 + j 2.61	0.50	2.70 + j 1.31
UPS	2.33 + j 1.13	1.00	2.33 + j 1.13
Spares	2.97 + j 1.44	1.00	2.97 + j 1.44

$$20.21 + j 9.29$$
$$(22.24 \text{ kVA})$$

This load requires a 30.0 kVA transformer.

The correct overcurrent devices for the primary and secondary are as follows:

$$I_{PRI} = \frac{30}{.480 \sqrt{3}} = 36.1 \ A$$

$$I_{SEC} = \frac{30}{.208 \sqrt{3}} = 83.3 \ A.$$

$OC_{PRI} = 1.25 \cdot 36.1 = 45.1$ A   (50 A/3P Circuit Breaker)
$Feeder_{PRI} = 3\#8\text{-}3/4"C$

$OC_{SEC} = 1.25 \cdot 83.3 = 104.1$ A   (100 A/3P Circuit Breaker)
$Feeder_{SEC} = 4\#3\text{-}1\text{-}1/4"C$

Panelboard LA will be rated for 100 amperes, 208Y/120 V with a 100 ampere main circuit breaker.

Example 5.6: Select the correct panelboard ampere rating, overcurrent device and feeder for Panelboard HA.

From the schedule for Panelboard HA we find the following connected loads. We will again use a power factor of 0.95 for lighting and 0.90 for receptacles and spares. We will use the motor power factors from the mechanical drawings. In actuality, these power factors will probably be determined from standard motor reference data.

Connected Load

Lighting	18.95 kVA @ PF =0.95	18.00 + j 5.92
Receptacles	15.99 kVA @ PF =0.90	14.39 + j 6.97
Heat	60.00 kW @ PF =1.00	60.00 + j 0.00
Air Cond.	35.03 kVA @ PF = (See Mechanical Data)	30.68 + j16.91
Other	2.59 kVA @ PF =0.90	2.33 + j 1.13
Motors	12.64 kVA @ PF =0.85	10.74 + j 6.66
Water Heating	9.00 kW @ PF =1.00	9.00 + j 0.00
Spares	18.30 kVA @ PF =0.90	16.47 + j 7.98

$$161.61 + j45.57$$
$$(167.91 \text{ kVA})$$

Computed Demand

			D.F.	
Lighting	18.00 +	j 5.92	1.25	22.50 + j 7.40
Receptacles	9.00 +	j 4.36	1.00	9.00 + j 4.36
	5.39 +	j 2.61	0.50	2.70 + j 1.31
Heat	60.00 +	j 0.00	1.00	60.00 + j 0.00
Air Cond.	30.68 +	j16.91	0.00	-
Other	2.33 +	j 1.13	1.00	2.33 + j 1.13
Motors	10.74 +	j 6.66	1.00	10.74 + j 6.66
Water Heat.	9.00 +	j 0.00	1.00	9.00 + j 0.00
Spares	16.47 +	j 7.98	1.00	16.47 + j 7.98
Largest Motor	5.37 +	j 3.33	0.25	1.34 + j 0.83

$$134.08 + j29.67$$
$$(137.32 \text{ kVA})$$

$$I = \frac{137.32}{.480 \sqrt{3}} = 165.2 \ A \quad (175 \, A/3 \, P \quad \textit{Circuit Breaker})$$

Feeder = 4#2/0-2"C

Panelboard HA will be rated for 225 amperes (next standard rating), 480Y/277 V, and will be served by a 175 A/3P circuit breaker in the main distribution panel, MDP.

Example 5.7: Select the correct panelboard ampere rating, overcurrent device and feeder for Panelboard HD. This is the panelboard serving the molding machines and extruders.

From the panelboard schedule, we find the following connected loads:

Connected Load

Motor	11.31 kVA @ PF = 0.82	9.27 +	j 6.47
Other	140.00 kVA @ PF =1.00	140.00 +	j 0.00
Spares	70.00 kVA @ PF = 1.00	70.00 +	j 0.00
	5.66 kVA @ PF =0.82	4.64 +	j 3.24
	36.00 kVA @ PF = 0.90	32.40 +	j15.69

256.31 + j25.40
(257.57 kVA)

Computed Demand

			D.F.		
Motors	9.27 +	j 6.47	1.00	9.27 +	j 6.47
Other	140.00 +	j 0.00	1.25	175.00 +	j 0.00
Spares	70.00 +	j 0.00	1.00	70.00 +	j 0.00
	4.64 +	j 3.24	1.00	4.64 +	j 3.24
	32.40 +	j15.69	1.00	32.40 +	j15.69
Largest Motor	2.32 +	j 1.62	0.25	0.58 +	j 0.41

291.89 + j25.81
(293.03 kVA)

$$I = \frac{kVA}{kV\sqrt{3}} = \frac{293.03}{.480\sqrt{3}} = 352.5 \ A \quad (400A/3P \ \ Circuit \ Breaker)$$

Feeder = 4#500 kCMIL-3-1/2"C

Panelboard HD will have an ampere rating of 400 amperes, 480Y/277 V and will be served by a 400 A/3P circuit breaker in the main distribution panel, MDP. Note that the molding machines are classified as a continuous load since they are operational 24 hours a day to maintain the plastic at the correct temperature. The extruder motors are non-continuous, since they operate only when the machines are actually in production.

Example 5.8: Select the correct switchboard ampere rating, overcurrent device and main building feeder for Switchboard MDP. The total connected load for MDP is the sum of the various loads for the connected equipment, as follows:

Connected Load

Lighting	80.38 kVA @ PF = 0.95	76.36 + j 25.10
Receptacles	43.71 kVA @ PF = 0.90	39.34 + j 19.05
Heat	210.00 kW @ PF = 1.00	210.00 + j 0.00
Air Cond.	35.03 kVA @ PF = (See Mechanical Data)	30.68 + j 16.91
Motors	87.82 kVA @ PF = (See Equipment List)	75.90 + j 43.94
Mold Machines	140.00 kW @ PF = 1.00	140.00 + j 0.00
UPS System	2.59 kVA @ PF = 0.90	2.33 + j 1.13
Water Heating	45.00 kW @ PF = 1.00	45.00 + j 0.00
Spares	292.45 kVA @ PF = 0.90	263.21 + j127.48

882.82 + j233.61
(913.21 kVA)

Computed Demand

		D.F.	
Lighting	76.36 + j 25.10	1.2	95.45 + j 31.38
Receptacles	9.00 + j 4.36	1.00	9.00 + j 4.36

	30.34 + j 14.69	0.50	15.17 + j 7.35
Heat	210.00 + j 0.00	1.00	210.00 + j 0.00
Air Cond.	30.68 + j 16.91	0.00	-
Motors	75.90 + j 43.94	1.00	75.90 + j 43.94
Mold Machines	140.00 + j 0.00	1.25	175.00 + j 0.00
UPS System	2.33 + j 1.13	1.00	2.33 + j 1.13
Water Heating	45.00 + j 0.00	1.00	45.00 + j 0.00
Spares	263.21 + j127.48	1.00	263.21 + j127.48
Largest Motor	29.27 + j 15.80	0.25	7.32 + j 3.95

$$898.38 + j219.59$$
$$(924.83 \text{ kVA})$$

$$I = \frac{kVA}{kV\sqrt{3}} = \frac{924.83}{.480\sqrt{3}} = 1112.4 \ A \quad (1200A/3P \ Circuit \ Breaker)$$

Feeder = 4 Sets 4#350 kCMIL-3"C

The main overcurrent device will be equipped with a ground fault detection system in accordance with the requirements of the *NEC.*

Before finalizing the design, a voltage drop study should be undertaken to verify that the selected branch circuits and feeders will maintain the load voltage within tolerance levels. Also, a coordination study should be completed to verify proper coordination of the main overcurrent device and the various feeder overcurrent devices.

The above analysis assumes the system will be operated under normal operating conditions. In the next chapter, we will study the effects of short-circuits on the building's electrical system and its components. We will then learn how to perform a fault current study in order to gain the necessary information to properly select the interrupting rating of system components. The proper selection of these interrupting ratings will assure the safe interruption of potentially destructive fault currents.

References

Atkinson, Robert S., P.E. 1987. *ECALC Software Manual.* Nashville: Atkinson & Associates, PC.

Beeman, Donald. 1955. *Industrial Power Systems Handbook.* New York: McGraw-Hill Book Company.

Croft, Terrell, and Wilford Summers. 1987. *American Electricians Handbook.* New York: McGraw Hill Book Company.

Institute of Electrical and Electronic Engineers, Inc. 1990. *IEEE Recommended Practice for Electric Power Systems in Commercial Buildings. IEEE Standard 241-1990.* Institute of Electrical and Electronic Engineers, Inc.

Institute of Electrical and Electronic Engineers, Inc. 1986. *IEEE Recommended Practice for Electric Power Distribution for Industrial Plants. ANSI/IEEE Std. 141-1986.* New York: Institute of Electrical and Electronic Engineers.

Institute of Electrical and Electronic Engineers, Inc. 1986. *IEEE Recommended Practice for Protection and Coordination of Industrial and Commercial Power Systems. ANSI/IEEE Std. 242-1986.* New York: Institute of Electrical and Electronic Engineers, Inc.

6

Fault Current Analysis

A fault current study is one of the most crucial parts of the design process. Fault current analysis allows us to predict the maximum available fault current at various points in the electrical system. This information is used to design and specify components which can withstand the tremendous forces involved in faults without damage, and without injury to nearby operating personnel. Figure 6.1 shows the damage to components of a 3000 A 208Y/120 V switchboard resulting from a fault-related accident. A fork-lift truck inadvertently struck the switchboard, resulting in phase-to-ground and phase-to-phase faults. The copper bus bars inside the switchboard partially melted, and in the process, ejected molten droplets of copper. This molten material, which moves at very high velocities, poses a tremendous burn hazard to nearby personnel. In addition, the explosive nature of large scale faults propels fragments of equipment and enclosures at shrapnel-like speeds, again posing severe risks of injury to personnel. The operator of the fork-lift that precipitated the accident shown in Figure 6.1 was very fortunate that the steel housing of the switchboard largely deflected the blast of molten copper, and that he escaped with only minor burns. Upon disassembly, the interior of the switchboard was littered with shards of resolidified copper, visible in the author's hand.

Accidents like this rarely have such fortunate endings. In addition to personal injury, the damage to the electrical system often forces the shutdown of the entire building or plant, therefore resulting in a loss of production. Let us now discuss the nature of faults.

First, it should be realized that faults are short-circuit conditions in which the normal level of current flow is suddenly increased by a factor of

Figure 6.1 Fault damaged 3000 A switchboard (photo: Mary Matthews).

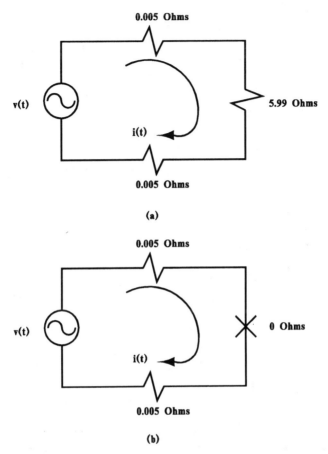

Figure 6.2 Purely resistive *ac* system (a) before fault (b) during fault.

hundreds or even thousands. Figure 6.2(a) shows a 120 V single phase *ac* circuit with a load impedance of 5.99Ω, and a total conductor impedance of 0.01Ω. The total current flow under normal conditions is

$$I_{Normal} = \frac{V}{Z} = \frac{120}{(0.005 + 5.99 + 0.005)} = 20 \ A.$$

Here we have assumed that the load impedance, as well as the conductor impedance, is purely resistive. In actual circuits, both resistive and reactive components will be present.

In Figure 6.2(b) the load impedance is replaced by a short circuit impedance of zero ohms. This condition can result from tools accidentally coming into contact with energized conductors, or through accidents such as the one described previously. In any event, the load's normal impedance, that limits current flow, is replaced by a far lower impedance. The only impedance now limiting current flow is that of the system itself, here represented by the resistance of the conductors. The short circuit current will now be

$$I_{sc} = \frac{V}{Z} = \frac{120}{(0.005 + 0.005)} = 12,000 \ A.$$

An increase in current flow by a factor of 600! The thermal energy released is proportional to the square of the *rms* current and the time involved. The magnetically induced forces are proportional to the square of the peak current.

The thermal energy will elevate the temperature of the conductors, resulting in possible damage to the insulation material. The very high current flow also creates a powerful magnetic field around the conductor and this field interacts with the current flow in nearby conductors, creating large forces on the conductors themselves. These forces can reach very high levels and, in the absence of proper design, severe physical damage can occur. Equipment has been known to literally explode during high level faults. In some cases the forces may deform conductors, or move bus bars into contact with other conductors or grounded surfaces, thus escalating the fault and therefore the subsequent level of damage. Remember, all this can occur in about one cycle of 60 Hz current flow, or about 0.0167 seconds!

TYPES OF FAULTS

From the preceding discussion it should be clear that there are several possible types of faults. Faults can occur between a phase conductor and neutral, or between a phase conductor and ground. Faults can also occur between phase conductors, or between several phase conductors and neutral and/or ground. In the example discussed in the last section the fault impedance was assumed to be zero. This is an example of a *bolted* fault. Such faults can occur due to wiring errors or dropped tools, such as a screwdriver, that might weld itself across conductors to form a very low impedance path.

Faults also often occur under conditions where the impedance at the point of fault is not zero. These higher impedance faults can occur when the point of fault contact is not as solid, as in the case of a bolted fault. These higher impedance faults often involve arcing through air and can result in a tremendous release of heat.

As you can imagine, bolted faults normally produce the highest flow of current. Of the bolted faults, the three-phase fault is normally the worst case and, hence, forms the basis of most fault current analysis. High impedance faults or arcing faults typically produce a lower current flow due to the increased circuit impedance. In order to provide maximum system protection, large scale faults must be removed as rapidly as possible to minimize system damage. The nature of arcing faults poses a challenge in system protection because the current magnitude might well be insufficient to prompt rapid protective device operation. Finally, arcing faults can easily result in very serious and extensive damage due to the erosion of conductor and enclosure material by the arc. We will first address bolted faults and in a later section, take up the matter of higher impedance faults.

The danger associated with faults is clearly recognized by construction code authorities such as the *National Electrical Code (NEC)*.

110-9. Interrupting Rating. Equipment intended to break current at fault levels shall have an interrupting rating sufficient for the system voltage and the current which is available at the line terminals of the equipment.

Equipment intended to break current at other than fault levels shall have an interrupting rating at system voltage sufficient for the current that must be interrupted.[1]

This article requires that devices such as circuit breakers and fuses be capable of safely interrupting the available fault at the point in the system

---

[1] Reprinted with permission from NFPA 70-1990, the *National Electrical Code* ®, Copyright © 1989, National Fire Protection Association, Quincy, MA 02269. This reprinted material is not the complete and official position of the NFPA on the referenced subject which is represented only by the standard in its entirety.

where the device is located. It is also necessary that the physical construction of panelboards and switchboards containing these overcurrent devices be capable of withstanding the magnetically induced forces associated with the maximum fault current flow.

In order to comply with the requirements of Article 110-9, the engineer must perform a fault current study of the building's planned distribution system. Based upon the results of this study he must then specify overcurrent device interrupting ratings, and switchgear and panelboard bracing requirements.

THE NATURE OF FAULT CURRENTS IN *AC* SYSTEMS

Before proceeding with our discussion of fault current analysis techniques, let us pause to consider the nature of fault currents in *ac* systems (Greenwood 1971). Figure 6.3 shows a single-phase *ac* circuit with impedance $Z = R + jX$.

This impedance represents the total system impedance during a fault with the normal load impedance removed. The sinusoidal voltage is given by

$$v = v_{max} \sin(\omega t + \theta) \quad V \qquad (6.1)$$

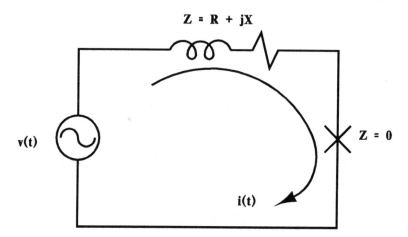

Figure 6.3 *AC* system with both resistive and reactive components during fault.

and

$$\omega = 2\pi f \quad Radians / Sec.$$

Here ω is the angular frequency of the voltage wave in radians per second and f represents the system frequency in cycles per second. θ represents the angular point on the voltage wave at which the fault is initiated.

When the fault occurs, the voltage drop around the loop will be given by

$$RI + L\frac{di}{dt} = v_{max} \sin(\omega t + \theta) \quad V. \qquad (6.2)$$

Using the trigonometric substitution

$$\sin(A + B) = \sin(A)\cos(B) + \cos(A)\sin(B) \qquad (6.3)$$

Equation 6.2 becomes

$$RI + L\frac{di}{dt} = v_{max}[\sin(\omega t)\cos(\theta) + \cos(\omega t)\sin(\theta)]. \qquad (6.4)$$

Remember that $\cos(\theta)$ and $\sin(\theta)$ are now constants. We can now take the Laplace Transform of both sides of Equation 6.4, yielding

$$Ri(s) + Lsi(s) - Li(0) = v_{max}\left[\frac{\omega \cos(\theta)}{(s^2 + \omega^2)} + \frac{s \sin(\theta)}{(s^2 + \omega^2)}\right] \qquad (6.5)$$

Solving for $i(s)$ and remembering $i(o) = O$

$$i(s) = \frac{v_{max}}{L(\frac{R}{L} + s)} \cdot \left[\frac{\omega \cos(\theta)}{(s^2 + \omega^2)} + \frac{s \sin(\theta)}{(s^2 + \omega^2)}\right]. \qquad (6.6)$$

If we let

$$C_1 = \frac{v_{max}}{L}\omega \cos(\theta), \quad C_2 = \frac{v_{max}}{L}\sin(\theta) \quad \text{and} \quad \alpha = \frac{R}{L}$$

Equation 6.6 becomes

$$i(s) = \frac{C_1}{(\alpha + s)(s^2 + \omega^2)} + \frac{C_2 s}{(\alpha + s)(s^2 + \omega^2)} \tag{6.7}$$

expanding each part separately

$$\frac{1}{(\alpha + s)(s^2 + \omega^2)} = \frac{A}{(s + \alpha)} + \frac{Bs + C}{(s^2 + \omega^2)}. \tag{6.8}$$

Solving for A, B and C, we find

$$A = \frac{1}{(\omega^2 + \alpha^2)}, \quad B = \frac{-1}{(\omega^2 + \alpha^2)}, \quad \text{and} \quad C = \frac{\alpha}{(\omega^2 + \alpha^2)}$$

so that

$$\frac{1}{(\alpha + s)(s^2 + \omega^2)} = \frac{1}{(\omega^2 + \alpha^2)}$$
$$\left[\frac{1}{(s + \alpha)} - \frac{s}{(s^2 + \omega^2)} + \frac{\alpha}{(s^2 + \omega^2)} \right] \tag{6.9}$$

Taking the inverse Laplace Transform of the right side of Equation 6.9

$$\mathcal{L}^{-1} = \frac{1}{(\omega^2 + \alpha^2)} \left[e^{-\alpha t} - \cos(\omega t) + \frac{\alpha}{\omega} \sin(\omega t) \right]. \tag{6.10}$$

Now, for the remainder of Equation 6.8

$$\frac{s}{(\alpha + s)(s^2 + \omega^2)} = \frac{A}{(\alpha + s)} + \frac{Bs + C}{(s^2 + \omega^2)}. \tag{6.11}$$

If we simplify and solve Equation 6.11 for A, B and C, we find that

$$A = \frac{-\alpha}{(\omega^2 + \alpha^2)}, \quad B = \frac{\alpha}{(\omega^2 + \alpha^2)} \quad \text{and} \quad C = \frac{\omega^2}{(\omega^2 + \alpha^2)}$$

so that

$$\frac{s}{(\alpha + s)(s^2 - \omega^2)} = \frac{1}{(\omega^2 + \alpha^2)} \left[\frac{-\alpha}{(\alpha + s)} + \frac{\alpha s}{(s^2 + \omega^2)} + \frac{\omega^2}{(s^2 + \omega^2)} \right] \quad (6.12)$$

Taking the inverse Laplace Transform of the right side of Equation 6.12 yields

$$\mathcal{L}^{-1} = \frac{1}{(\omega^2 + \alpha^2)} \left[-\alpha e^{-\alpha t} + \alpha \cos(\omega t) + \omega \sin(\omega t) \right]. \quad (6.13)$$

Combining Equation 6.10 and Equation 6.13, we see that

$$i(t) = \frac{v_{max}}{L(\omega^2 + \alpha^2)} \quad (6.14)$$
$$[e^{-\alpha t}(\omega \cos(\theta) - \alpha \sin(\theta)) + \cos(\omega t)(\alpha \sin(\theta) - \omega \cos(\theta)) + \sin(\omega t)(\alpha \cos(\theta) + \omega \sin(\theta))].$$

Realizing that

$$\tan(\varphi) = \frac{X}{R} = \frac{\omega L}{R} \quad \text{and} \quad \alpha = \frac{R}{L}$$

so that

$$\tan(\varphi) = \frac{\omega}{\alpha}.$$

Then

$$\sin(\varphi) = \frac{\omega}{\sqrt{\omega^2 + \alpha^2}} \quad \text{and} \quad \omega = \sin(\varphi)\sqrt{\omega^2 + \alpha^2}$$

$$\cos(\varphi) = \frac{\alpha}{\sqrt{\omega^2 + \alpha^2}} \quad \text{and} \quad \alpha = \cos(\varphi)\sqrt{\omega^2 + \alpha^2}.$$

Substituting these expressions for ω and α in Equation 6.14

$$i(t) = \frac{V_{max}}{L(\omega^2 + \alpha^2)} [e^{-\alpha t}(\sin(\varphi)\sqrt{\omega^2 + \alpha^2}\cos(\theta)$$
$$- \cos(\varphi)\sqrt{\omega^2 + \alpha^2}\sin(\theta))$$
$$+ \cos(\omega t)(\cos(\varphi)\sqrt{\omega^2 + \alpha^2}\sin(\theta)$$
$$- \sin(\varphi)\sqrt{\omega^2 + \alpha^2}\cos(\theta))$$
$$+ \sin(\omega t)(\cos(\varphi)\sqrt{\omega^2 + \alpha^2}\cos(\theta)$$
$$+ \sin(\varphi)\sqrt{\omega^2 + \alpha^2}\sin(\theta))].$$

After some simplification

$$i(t) = \frac{v_{max}\sqrt{\omega^2 + \alpha^2}}{L(\omega^2 + \alpha^2)} [e^{-\alpha t}(\sin(\varphi)\cos(\theta) - \cos(\varphi)\sin(\theta))$$
$$+ \cos(\omega t)(\cos(\varphi)\sin(\theta) - \sin(\varphi)\cos(\theta))$$
$$+ \sin(\omega t)(\cos(\varphi)\cos(\theta) + \sin(\varphi)\sin(\theta))].$$

$$(6.15)$$

Using the trigonometric substitutions

$$\sin(A)\cos(B) = \frac{1}{2}\sin(A + B) + \frac{1}{2}\sin(A - B)$$

$$\sin(A)\sin(B) = \frac{1}{2}\cos(A - B) - \frac{1}{2}\cos(A + B)$$

$$\cos(A)\cos(B) = \frac{1}{2}\cos(A - B) + \frac{1}{2}\cos(A + B)$$

and

$$\cos(A)\sin(B) = \frac{1}{2}\sin(A + B) - \frac{1}{2}\sin(A - B)$$

Equation 6.15 becomes, after simplification

$$i(t) = \frac{V_{max}}{L\sqrt{\omega^2 + \alpha^2}} \left[\sin(\omega t + \theta - \varphi) - e^{-\alpha t}\sin(\theta - \varphi)\right]. (6.16)$$

Finally, we see that

$$L \sqrt{\omega^2 + \alpha^2} = \sqrt{\omega^2 L^2 + \alpha^2 L^2}$$

but since

$$\alpha = \frac{R}{L}$$

$$L \sqrt{\omega^2 + \alpha^2} = \sqrt{\omega^2 L^2 + R^2} = |Z|.$$

The final form of Equation 6.16 then becomes

$$i(t) = \frac{V_{max}}{|z|} [\sin(\omega t + \theta - \varphi) - e^{-\alpha t} \sin(\theta - \varphi)]. \tag{6.17}$$

Where:
   a) R, L and ω are resistance in ohms, inductance in henries and radians per second, respectively
   b) θ represents the closing angle of the fault on the voltage wave
   c) φ represents the familiar power factor angle
   d) $\alpha = R/L$

An examination of Equation 6.17 reveals several important points. First, the total current at any instant is a composite of two currents, one varying sinusoidally, the other decaying in an exponential manner. The sinusoidal term is actually the steady-state current flow following the transient period, while the second term is a decaying *dc* component.

Note that if the closing angle on the *ac* voltage wave equals the power factor angle, i.e., $\theta - \varphi = 0$, then the second term becomes zero for all t. This produces a completely symmetrical waveform. Conversely, if $\theta - \varphi = \pi/2$, the *dc* component reaches its maximum value for this specific value of R and L.

Also note that as L approaches zero, α will approach infinity, φ approaches zero, and the second term will then approach zero. This will reduce Equation 6.17 to the familiar form developed in Chapter 2 for the sinusoidal current flow in a purely resistive circuit.

Let us now utilize Equation 6.17 to explore the transient behavior of several *RL* circuits.

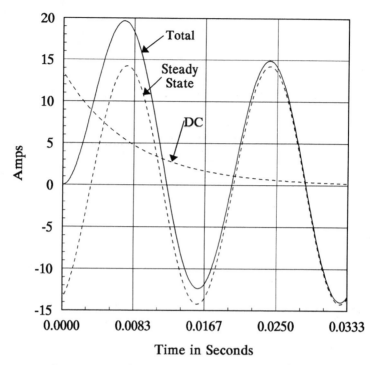

Figure 6.4 Asymmetrical fault current waveform in an *ac* circuit with both resistive and reactive components present.

Example 6.1: Given a 120 volt 60 Hz source connected to a circuit similar to that shown in Figure 6.3 with $R = 3.75\ \Omega$ and $L = 0.030h$. The closing angle on the voltage wave is zero degrees. Plot the resulting current waveforms.

Figure 6.4 shows a graph of the steady state and decaying *dc* components along with the composite waveform. Note that the composite current waveform begins at zero and reaches its highest peak during the first cycle, followed by a decay toward a symmetrical waveform as the *dc* component decays.

The key point is that the magnitude of this first peak is substantially greater than those in succeeding peaks. Remember that the force that tends to distort current-carrying components is proportional to the square of the peak current. Thus, the asymmetrical nature of the short-circuit current during the first cycle or so must be considered in the design of equipment in order for that equipment to safely withstand the resulting forces.

Let us now investigate a case in which the resistive component of the circuit is virtually zero. Actually, all circuits have some resistive component, even if it is only the resistance of the conductors themselves.

Example 6.2: Given a 120 Volt 60 Hz circuit connected as in Example 6.1 with $R = 0$ and $L = 0.04h$. Again the closing angle on the voltage wave is zero degrees. Plot the resulting current waveform.

Figure 6.5 shows the resulting plot of both the steady state, *dc* and composite waveform. Note that the composite current waveform is fully offset, and due to the lack of a resistive component, does not decay.

In summary, we should remember that the first cycle or so of fault current is likely to contain some degree of asymmetry. The degree of asymmetry depends on the value of R and L as well as the closing angle of the fault on the voltage wave. The electromagnetic forces involved are proportional to the square of the peak current. In practice, engineers calculate symmetrical short circuit currents, and equipment manufacturers

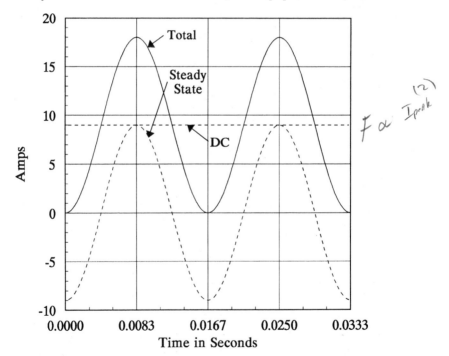

Figure 6.5 Fully offset fault current waveform in an *ac* system with only reactive component present.

design their equipment to withstand the higher forces associated with the first few cycles of asymmetrical current.

FAULT CURRENT ANALYSIS TECHNIQUES

There are two basic approaches to fault current analysis, the per-unit method and the direct method (IEEE Standard 242-1986). Both approaches can be applied and both yield similar results.

The per-unit approach is particularly applicable to systems which utilize several different voltage levels. In this approach, base values are established for current, voltage, ohms and kVA. The computed values are referred to the base values in terms of their per-unit quantity. For example, a voltage level of 0.9 per-unit on a 120 volt base would actually be 108 volts. The use of the per-unit system requires a fair amount of mathematical manipulation, using specialized equations. To those who do not work with fault analysis on a daily basis, per-unit quantities may be a bit abstract and perhaps more difficult to comprehend.

On the other hand, the direct method employs the data contained in the one-line diagram, along with the related conductor resistance and reactance values. The equations used are more familiar to most engineers, and the results of calculations are immediately available for application. In most building systems, there are only one or two voltage levels, so the required computations are not overly complex.

The latest edition of the IEEE Publication, *Protection and Coordination of Industrial and Commercial Power Systems*, commonly called the Buff Book, IEEE Standard 242-1986, presents both methods. We will utilize the direct method in this work. Those interested in pursuing the per-unit method can refer to the references at the end of this chapter.

In a professional design office environment that typically operates on a very tight time schedule, the more straightforward the method is, the better. For many years, short-circuit analysis was perceived as being somewhat difficult, and, all too often, received too little attention during the design process. In 1967, Russell Ohlson published an IEEE paper which he hoped would simplify fault analysis, and thereby make it more readily applicable in an engineering environment (Ohlson 1967). Ohlson's work was based on the direct method, and he later chaired the IEEE group that drafted the short-circuit analysis chapter in the 1986 edition of the Buff book.

FAULT CURRENT ANALYSIS
BY THE DIRECT METHOD

As discussed in previous sections, the basic goal of fault current analysis is to determine the prospective fault current at various points in the building's distribution system. This information is used to select appropriate circuit breakers, fuses and panel bracing. Actually, these items of equipment have two distinct current ratings. The first rating is the normal current-carrying capacity. The second rating specifies the highest abnormal current flow that the device can safely withstand during fault conditions.

In essence, the direct method involves the determination of the total impedance of the system at various locations of interest. The system phase-to-neutral voltage divided by the total system impedance at these points yields the available fault current. Since the typical ohmic values for most systems are quite small, it is easier to express them in terms of milliohms per phase.

The first step in the fault current analysis is to assemble the necessary data. The single-line diagram is the first item needed because it contains information on essential system elements such as service voltage, utility service type (overhead service or pad transformer), main distribution panel, lighting and power panels, feeders and step-down transformers. The single-line diagram will also contain information on conductor sizes and overcurrent device locations and types. In addition to this information we will need the length of all feeders, along with the conductor type (copper or aluminum), and the type of insulation material, as well as the type of raceway material (plastic, aluminum, or steel). In general, this data will be readily available to the engineer because such information is a necessary product of the normal design process.

Other information is of a more specialized nature, and is required primarily for the fault analysis study. For example, we will require information on the utility's available fault current level. This data must be supplied by the utility company. Other specialized information concerns the resistance and reactance values of various system components, such as conductors, as well as percent impedance and X/R ratios for step-down transformers.

Tables 6.1 and 6.2 contain resistance and reactance data for various conductor types. Note that this data is presented in terms of milliohms per 100 feet of conductor length at 25°C. It may be recalled that Table 9 of

| Cable Size | Copper Several Single-Conductor Cables | | | | | | Aluminum Several Single-Conductor Cables | | | | | |
| | Steel | | Aluminum | | Plastic | | Steel | | Aluminum | | Plastic | |
	R	X	R	X	R	X	R	X	R	X	R	X
14	257.0	5.60	257.0	4.48	257.0	4.48	422.0	5.60	422.0	4.48	422.0	4.48
12	162.0	5.23	162.0	4.18	162.0	4.18	266.0	5.23	266.0	4.18	266.0	4.18
10	101.8	4.90	101.8	3.92	101.8	3.92	167.0	4.90	167.0	3.92	167.0	3.92
8	64.04	5.14	64.04	4.12	64.04	4.12	105.0	5.14	105.0	4.12	105.0	4.12
6	41.00	5.04	41.00	4.03	41.00	4.03	67.40	5.04	67.40	4.03	67.40	4.03
4	25.90	4.77	25.90	3.82	25.90	3.82	42.40	4.77	42.40	3.82	42.40	3.82
3	20.50	4.58	20.50	3.66	20.50	3.66	33.60	4.58	33.60	3.66	33.60	3.66
2	16.40	4.49	16.40	3.59	16.20	3.59	26.60	4.49	26.60	3.59	26.60	3.59
1	13.03	4.58	13.03	3.66	12.90	3.66	21.10	4.58	21.10	3.66	21.10	3.66
1/0	10.40	4.46	10.40	3.56	10.20	3.56	16.80	4.46	16.80	3.56	16.80	3.56
2/1	8.35	4.35	8.35	3.48	8.12	3.48	13.30	4.35	13.30	3.48	13.30	3.48
3/0	6.68	4.22	6.68	3.37	6.43	3.37	10.60	4.22	10.60	3.37	10.50	3.37
4/0	5.34	4.14	5.34	3.31	5.11	3.31	8.44	4.14	8.44	3.31	8.38	3.31
250	4.57	4.23	4.57	3.38	4.33	3.38	7.22	4.23	7.22	3.38	7.09	3.38
300	3.85	4.14	3.85	3.31	3.62	3.31	6.02	4.14	6.02	3.31	5.92	3.31
350	3.33	4.07	3.33	3.25	3.11	3.25	5.20	4.07	5.20	3.25	5.07	3.25
400	2.97	4.04	2.97	3.23	2.73	3.23	4.60	4.04	4.60	3.23	4.44	3.23
500	2.44	3.96	2.44	3.17	2.20	3.17	3.75	3.96	3.75	3.17	3.56	3.17
600	2.09	4.01	2.09	3.21	1.85	3.21	3.19	4.01	3.19	3.21	2.98	3.21
750	1.74	3.94	1.74	3.15	1.50	3.15	2.64	3.94	2.64	3.15	2.40	3.15
1000	1.40	3.86	1.40	3.09	1.15	3.09	2.11	3.86	2.11	3.09	1.82	3.09

Reprinted from ANSI/IEEE Std. 242-1986, *IEEE Recommended Practice for Protection and Coordination of Industrial and Commercial Power Systems*, Copyright © 1986 by The Institute of Electrical and Electronic Engineers, Inc. with permission of IEEE.

Table 6.1 60 Hz Low-Voltage Cable in Conduit-Resistance (R) and Reactance (X) Data for Insulation Types THW, RHH, and Use in Milliohms per Conductor per 100 Feet at 25°C.

Handwritten margin notes:

- Specified at 25° C So That These values are more conservative
- T↗, E↓, ICA↑
- THWN → THHN
- Milliohms

Cable Size	Copper Several Single-Conductor Cables						Aluminum Several Single-Conductor Cables					
	Steel		Aluminum		Plastic		Steel		Aluminum		Plastic	
	R	X	R	X	R	X	R	X	R	X	R	X
14	257.0	4.93	257.0	3.94	257.0	3.94	422.0	4.93	422.0	3.94	422.0	3.94
12	162.0	4.68	162.0	3.74	162.0	3.74	266.0	4.68	266.0	3.74	266.0	3.74
10	101.8	4.63	101.8	3.71	101.8	3.71	167.0	4.63	167.0	3.71	167.0	3.71
8	64.04	4.75	64.04	3.80	64.04	3.80	105.0	4.75	105.0	3.80	105.0	3.80
6	41.00	4.37	41.00	3.49	41.00	3.49	67.40	4.37	67.40	3.49	67.40	3.49
4	25.90	4.41	25.90	3.53	25.90	3.53	42.40	4.41	42.40	3.53	42.40	3.53
3	20.50	4.30	20.50	3.44	20.50	3.44	33.60	4.30	33.60	3.44	33.60	3.44
2	16.40	4.20	16.40	3.36	16.20	3.36	26.60	4.20	26.60	3.36	26.60	3.36
1	13.03	4.27	13.03	3.42	12.90	3.42	21.10	4.27	21.10	3.42	21.10	3.42
1/0	10.40	4.17	10.40	3.34	10.20	3.34	16.80	4.17	16.80	3.34	16.80	3.34
2/0	8.35	4.09	8.35	3.37	8.12	3.27	13.30	4.09	13.30	3.27	13.30	3.27
3/0	6.68	4.00	6.68	3.20	6.43	3.20	10.60	4.00	10.60	3.20	10.50	3.20
4/0	5.34	3.93	5.34	3.14	5.11	3.14	8.44	3.93	8.44	3.14	8.38	3.14
250	4.57	3.99	4.57	3.19	4.33	3.19	7.22	3.99	7.22	3.19	7.09	3.19
300	3.85	3.93	3.85	3.14	3.62	3.14	6.02	3.93	6.02	3.14	5.92	3.14
350	3.33	3.83	3.33	3.11	3.11	3.11	5.20	3.88	5.20	3.11	5.07	3.11
400	2.97	3.85	2.97	3.08	2.73	3.08	4.60	3.85	4.60	3.08	4.44	3.08
500	2.44	3.79	2.44	3.03	2.20	3.03	3.75	3.79	3.75	3.03	3.56	3.03
600	2.09	3.82	2.09	3.05	1.85	3.05	3.19	3.82	3.19	3.05	2.98	3.05
750	1.74	3.76	1.74	3.01	1.50	3.01	2.64	3.76	2.64	3.01	2.40	3.01
1000	1.40	3.70	1.40	2.96	1.15	2.96	2.11	3.70	2.11	2.96	1.82	2.96

Reprinted from ANSI/IEEE Std. 242-1986, *IEEE Recommended Practice for Protection and Coordination of Industrial and Commercial Power Systems*, Copyright © 1986 by The Institute of Electrical and Electronic Engineers, Inc. with permission of IEEE.

Table 6.2 60 Hz Low-Voltage Cable in Conduit—Resistance (R) and Reactance (X) Data for Insulation Types THWN, and THHN in Milliohms per Conductor per 100 Feet at 25°C.

233

the *National Electrical Code* contains resistance and reactance data based on a conductor temperature of 75°C.

Short circuit analysis studies are based on this lower temperature because the results will be somewhat more conservative. This is because the lower temperature more closely represents unloaded, cooler conductors whose resistance values are commensurately lower, resulting in lower total impedance and higher prospective fault current levels. The inductive reactance component of the conductor's impedance is unaffected by temperature. For similar reasons, we use 75°C data in voltage drop analysis in order to simulate loaded conductors with higher resistance values. This results in higher predicted voltage drops and, again, in conservative results.

Table 6.3 contains data on copper and aluminum busway. This data is also presented in terms of milliohms per 100 feet at 25°C.

Table 6.4 contains data on single-phase and three-phase step-down transformers. This table gives representative values for both percent impedance and X/R values.

We will now discuss how we translate the assembled system and component data into the desired fault current study.

FAULT CURRENT ANALYSIS METHODOLOGY

We will use the simple system represented by the single-line diagram in Figure 6.6 to discuss the basic methodology involved in a fault current analysis. The first task is to review the single-line diagram in order to select appropriate points in the system where the available fault current must be determined. Most engineers begin with the available fault current level at the secondary of the utility transformer, indicated as Point A. The next point is normally at the location of the main distribution panel, Point B. The available fault current level at this point is necessary in order to specify adequate interrupting capacity for overcurrent devices located here. From this point we work through the system panelboard by panelboard, and switchboard by switchboard, determining the fault current levels for all other equipment, such as at Points C and D. The available fault current level is computed by dividing the phase-to-neutral voltage by the total impedance at each of these points in the system. Since we will work in milliohms, the result of this division must then be multiplied by 1000 to yield amps.

Table 6.3 60 Hz Low-Voltage Busway-Resistance (R) and Reactance (X) Data in Milliohms per 100 Feet at 25°C.

| | FEEDER | | | | PLUG-IN | | | |
| | Copper | | Aluminum | | Copper | | Aluminum | |
Ampere Rating	R	X	R	X	R	X	R	X
100	-	-	-	-	11.82	4.00	21.96	4.00
225	-	-	-	-	7.15	3.42	6.12	3.42
400	-	-	-	-	1.78	2.30	3.11	2.60
600	-	-	2.56	0.99	1.78	2.30	1.71	1.59
800	1.26	0.99	1.78	0.81	1.05	2.17	1.80	2.17
1000	1.05	0.82	1.59	0.50	1.05	2.17	1.20	1.43
1350	0.76	0.65	1.06	0.44	0.71	1.43	0.84	1.00
1600	0.70	0.53	0.89	0.40	0.49	1.00	0.75	0.90
2000	0.52	0.41	0.63	0.31	0.43	0.90	0.60	0.72
2500	0.38	0.32	0.47	0.25	0.35	0.72	0.42	0.50
3000	0.35	0.30	0.46	0.20	0.24	0.50	0.40	0.48
4000	0.25	0.20	0.31	0.16	-	-	-	-
5000	0.19	0.15	-	-	-	-	-	-

Reprinted from ANSI/IEEE Std. 242-1986, *IEEE Recommended Practice for Protection and Coordination of Industrial and Commercial Power Systems*, Copyright © 1986 by The Institute of Electrical and Electronic Engineers, Inc. with permission of IEEE.

The Utility Transformer Secondary

Our goal is to determine both the available fault current at the secondary of the utility transformer, as well as the corresponding system impedance. This is actually a two step process, and involves first, determining the impedance of the utility system referred to the utility transformer secondary, as well as the impedance of the utility transformer itself. The

Table 6.4 General Transformer Data.

GENERAL PURPOSE TRANSFORMERS STANDARD THREE-PHASE kVA TRANSFORMERS		
kVA	Avg. %Z	Avg. X/R
15.0	3.6	1.94
30.0	6.4	0.92
45.0	6.6	1.13
75.0	5.7	1.38
112.5	6.1	1.51
150.0	5.5	1.53
225.0	6.6	2.00
300.0	3.6	1.81
500.0	5.0	2.89
750.0	5.0	1.98
1000.0	5.8	2.38

GENERAL PURPOSE TRANSFORMERS STANDARD SINGLE-PHASE kVA RATINGS		
kVA	Avg. %Z	Avg. X/R
3.0	-	3.53
5.0	-	2.64
7.5	-	2.54
10.0	-	2.18
15.0	5.2	0.60
25.0	6.4	1.06
37.5	5.6	1.07
50.0	5.2	1.21
75.0	6.1	1.64
100.0	6.5	1.65
167.0	7.0	2.27

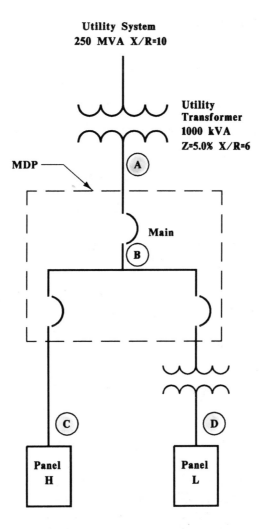

Utility System
250 MVA X/R=10

Utility
Transformer
1000 kVA
Z=5.0% X/R=6

MDP ———

Main

Panel H

Panel L

Figure 6.6 Basic distribution system for development of fault current methodology.

second step is to determine the available fault current level as discussed above.

Many utilities specify their system fault current capability in terms of MVA or kVA and *X/R* ratio. Most utilities can also supply data on the utility transformer percent impedance and *X/R* ratio. We first convert the available fault kVA into the corresponding impedance as seen from the utility transformer's secondary.

$$I_{sc} = \frac{kVA_{Sys}}{kV\sqrt{3}} \quad A \qquad (6.18)$$

The impedance on a per phase basis, using the secondary phase to neutral voltage $kV/\sqrt{3}$ is given by

$$Z_{Sys} = \frac{kV/\sqrt{3} \cdot 1000}{I_{sc}} \quad \Omega. \qquad (6.19)$$

because milliohms

Substituting I_{sc} and solving in terms of mΩ

$$Z_{Sys} = \frac{kV^2 \cdot 10^6}{kVA} \quad m\Omega. \qquad (6.20)$$

Using the utility system's X/R ratio, the values of resistance and inductive reactance will be given by

$$R_{Sys} = Z_{Sys} \cdot \cos\left[\arctan\left[\frac{X}{R}\right]\right] \quad m\Omega \qquad (6.21)$$

$$X_{Sys} = Z_{Sys} \cdot \sin\left[\arctan\left[\frac{X}{R}\right]\right] \quad m\Omega. \qquad (6.22)$$

Next, we find the impedance of the utility transformer itself.

Recall that the definition of the *percent impedance* of a transformer is the percentage of the rated primary voltage required to circulate full-rated current in the shorted secondary winding. If 5% of the rated primary voltage causes 100% of the rated secondary current to flow, then full-rated primary voltage will cause 100/5 or 20 times that current to flow under short-circuit conditions. The short circuit current at the secondary terminal of a transformer is then given by

$$\begin{aligned} I_{sc} &= I_{Rated} \cdot \frac{100}{\%Z} \quad A \\ &= \frac{kVA}{kV\sqrt{3}} \cdot \frac{100}{\%Z} \quad A. \end{aligned} \qquad (6.23)$$

The corresponding transformer impedance required to limit current to this level is given by

$$Z_{Trans} = \frac{V}{I_{sc}} \quad \Omega \tag{6.24}$$

since we wish to work in terms of kV and milliohms

$$Z_{Trans} = \frac{kV \cdot 10^6}{I_{sc}} \quad m\Omega. \tag{6.25}$$

Now substituting for I_{sc} and simplifying

$$Z_{Trans} = \frac{kV^2 \cdot \%Z \cdot 10,000}{kVA} \quad m\Omega \tag{6.26}$$

From the transformer's X/R ratio we can then determine values for the resistance and reactance.

$$R_{Trans} = \frac{kV^2 \cdot \%Z \cdot 10,000 \cdot \cos\left[\arctan\frac{X}{R}\right]}{kVA} \quad m\Omega \tag{6.27}$$

$$X_{Trans} = \frac{kV^2 \cdot \%Z \cdot 10,000 \cdot \sin\left[\arctan\frac{X}{R}\right]}{kVA} \quad m\Omega \tag{6.28}$$

The total utility system impedance at the secondary of the utility transformer is then

$$Z_{Total} = Z_{Sys} + Z_{Trans} \quad m\Omega \tag{6.29}$$

and the available fault current at the utility transformer secondary is

$$I_{sc} = \frac{V}{|Z_{Total}|} \cdot 1000 \quad A \tag{6.30}$$

where V is the secondary phase-to-neutral voltage.

Example 6.3: Let the utility system shown in Figure 6.6 have an available fault level of 250,000 kVA and an X/R ratio of 10. The utility transformer is rated 1000 kVA with an impedance of 5.0%, an X/R ratio of 6 and a secondary voltage of 208Y/120 V. Thus, the utility system impedance

$$Z_{Sys} = \frac{.208^2 \cdot 10^6}{250,000} = \boxed{0.173} \quad m\Omega$$

and

$$R_{Sys} = 0.173 \cdot \cos(\arctan(10)) = 0.017 \quad m\Omega$$

$$X_{Sys} = 0.173 \cdot \sin(\arctan(10)) = 0.172 \quad m\Omega$$

$$Z_{Sys} = 0.017 + j0.172 \quad m\Omega.$$

The transformer impedance itself is given by

$$R_{Transf} = \frac{0.208^2 \cdot 5.0 \cdot 10,000 \cdot \cos(\arctan(6))}{1000} = 0.356 \quad m\Omega$$

$$X_{Transf} = \frac{0.208^2 \cdot 5.0 \cdot 10,000 \cdot \sin(\arctan(6))}{1000} = 2.134 \quad m\Omega$$

$$Z_{Transf} = 0.356 + j2.134 \quad m\Omega.$$

The total impedance at the utility transformer secondary is then

$$Z_{Total} = (0.017 + j0.172) + (0.356 + j2.134)$$
$$= 0.373 + j2.306 \quad m\Omega.$$

Finally, the available fault current at this location is

$$I_{sc} = \frac{V}{|Z_{Total}|} \cdot 1000 = \frac{120}{2.336} \cdot 1000 = 51,370 \quad A.$$

The Main Distribution Panel - MDP

The total impedance at MDP is the sum of the total utility system impedance and the impedance of the main feeder from the transformer to the MDP. Tables 6.1 and 6.2 give resistance and reactance data for THW and THWN conductors, respectively. If the main service feeder is composed of N sets of parallel conductors each of length L

$$R_{Main} = \frac{L}{100} \cdot \frac{m\Omega}{100} \cdot \frac{1}{N} \quad m\Omega \tag{6.31}$$

$$X_{Main} = \frac{L}{100} \cdot \frac{m\Omega}{100} \cdot \frac{1}{N} \quad m\Omega \tag{6.32}$$

$$Z_{Main} = R_{Main} + jX_{main} \quad m\Omega. \tag{6.33}$$

The total impedance, then, at the MDP is the sum of the system impedance and the impedance of the service conductors. The available fault current at the MDP is then determined exactly as before.

Subfed Equipment

The impedance of the feeder for Panel H is determined in the same manner as for the main service. This impedance is added to the impedance value previously determined at the MDP to arrive at a new total impedance at Panel H. The available fault current level is then computed exactly as before.

If a step-down transformer is part of the system, as shown for Panel L, the task becomes a bit more complex. We must determine the total system impedance, including the utility system, main feeders and sub feeders up to the transformer primary. This impedance is then reflected through the transformer to an equivalent value on the secondary side by application of the square of the turns ratio, a^2, as follows:

$$Z_{Secondary} = \frac{1}{a^2} \cdot Z_{Primary} \quad m\Omega \tag{6.34}$$

where

$$a^2 = \left[\frac{V_{Primary}}{V_{Secondary}} \right]^2. \tag{6.35}$$

The impedance of the step-down transformer is computed exactly as before and added to the system impedance at the secondary to determine the total impedance at this point. The fault current is then computed as before, taking care to now use the secondary phase-to-neutral voltage.

We will now perform a short-circuit analysis on a small industrial power system.

Example System: Figure 6.7 shows the single-line diagram for a small industrial power system. Figure 6.8 shows the corresponding riser diagram. Our goal is to determine the available fault current level at points A through F.

Point A - Utility Transformer Secondary

a) Utility System - 500,000 kVA, X/R = 12

$$Z_{Sys} = \frac{.480^2 \cdot 10^6}{500,000} = 0.461 \quad m\Omega$$

and

$$R_{Sys} = 0.461 \cdot \cos(\arctan(12)) = 0.038 \quad m\Omega$$

$$X_{Sys} = 0.461 \cdot \sin(\arctan(12)) = 0.459 \quad m\Omega$$

$$R_{Transf} = \frac{.480^2 \cdot 5.5 \cdot 10,000 \cdot \cos(\arctan(12))}{2500} = 0.421 \quad m\Omega$$

$$X_{Transf} = \frac{.480^2 \cdot 5.5 \cdot 10,000 \cdot \sin(\arctan(12))}{2500} = 5.051 \quad m\Omega$$

and the total impedance

$$
\begin{aligned}
Z_{Total} &= Z_{Sys} + Z_{Transf} \\
&= (0.038 + j0.459) + (0.421 + j5.051) \\
&= 0.459 + j5.510 \quad m\Omega.
\end{aligned}
$$

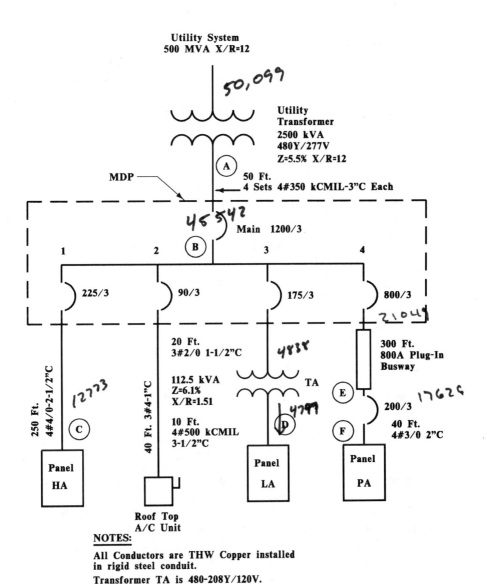

Figure 6.7 Single-line diagram of a small industrial power system for fault current analysis.

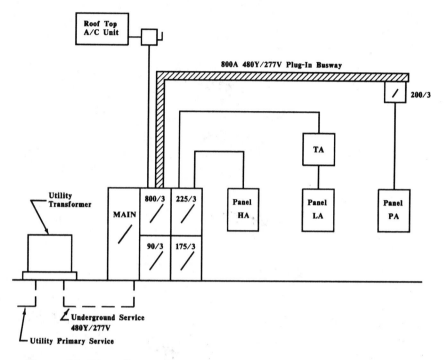

Figure 6.8 Riser diagram for industrial power system shown in Figure 6.7.

The available fault current level at this point is then

$$I_{sc} = \frac{277}{5.529} \cdot 1000 = 50,099 \; A.$$

Let us now consider the main building service feeder from the utility transformer to the main switchboard. This feeder will consist of either conduit and wire or busway. The specified conductor type will have X and R values, as shown in Table 6.1 and 6.2 for conductors or 6.3 for busway.

The main feeder consists of 50 feet of 4 sets of 4 #350 kCMIL THW Copper conductors in steel raceway. From Table 6.1, the appropriate values are $R = 3.33$ and $X = 4.07$ mΩ per 100 feet. For the total feeder then

$$R = \frac{Length}{100} \cdot \frac{m\Omega}{100} \cdot \frac{1}{Number \; of \; Runs}$$

$$= \frac{50}{100} \cdot 3.33 \cdot \frac{1}{4} = 0.416 \quad m\Omega$$

$$X = \frac{50}{100} \cdot 4.07 \cdot \frac{1}{4} = 0.509 \quad m\Omega.$$

This feeder then adds an additional $Z = 0.416 + j0.509$ mΩ to the total already computed for the utility system. The total impedance at the main distribution panel would be

$$\begin{aligned}
Z_{MDP} &= Z_{Sys} + Z_{Feeder} \\
&= (0.459 + j5.510) + (0.416 + j0.509) \\
&= 0.875 + j6.019 \quad m\Omega.
\end{aligned}$$

The available fault current level at the main distribution panel will then be given by

$$I_{sc} = \frac{V}{|Z|} \cdot 1000 = \frac{277}{6.082} \cdot 1000 = \boxed{45,542} \quad A.$$

Thus, the available fault current level at the utility transformer, 50,099 A, has been reduced to 45,542 A at the main distribution panel by the impedance of the feeder conductors. We know that the fault bracing of the panel and the interrupting rating of the circuit breakers must be at least this value in order for the equipment to safely operate.

We will next turn our attention to Panel HA. This panel is served by a single run of 4#4/0 THW, copper in steel raceway. The length of the run is 250 feet. Referring again to Table 6.1, we find that $R = 5.34$, $X = 4.14$ mΩ per 100 feet.

$$R = \frac{250}{100} \cdot 5.34 = 13.350 \quad m\Omega$$

$$X = \frac{250}{100} \cdot 4.14 = 10.350 \quad m\Omega$$

The total impedance at Panel HA will then be

$$\begin{aligned}
Z_{Total} &= Z_{MDP} + Z_{Feeder} \\
&= (0.875 + j6.019) + (13.350 + j10.350) \\
&= 14.225 + j16.369 \quad m\Omega.
\end{aligned}$$

and the available fault current level at Panel HA will be

$$I_{sc} = \frac{V}{|Z|} \cdot 1000 = \frac{277}{21.686} \cdot 1000 = 12,773 \quad A.$$

We now turn our attention to Panel LA. It will be necessary to determine the total impedance of the utility system, plus all conductors up to transformer TA. This impedance must then be reflected through to the secondary side and added to the impedance of the transformer itself. The secondary phase-to-neutral voltage, divided by this impedance, will yield the available fault current at the secondary connections. Finally, the impedance of the conductors from the transformer to Panel LA must be added to determine the total impedance at this point.

The feeder from MDP to transformer TA is 3#2/0 THW, copper in steel raceway and the length of the feeder is 20 feet.

$$R = \frac{20}{100} \cdot 8.35 = 1.670 \quad m\Omega$$

$$X = \frac{20}{100} \cdot 4.35 = 0.870 \quad m\Omega$$

The total impedance at the primary of TA will then be

$$
\begin{aligned}
Z_{Total} &= Z_{MDP} + Z_{Feeder} \\
&= (0.875 + j6.019) + (1.670 + j0.870) \\
&= 2.545 + j6.889 \quad m\Omega.
\end{aligned}
$$

To reflect this impedance through the transformer to the secondary side

$$Z_{Sec} = \frac{1}{a^2} Z_{Primary}$$

$$a = \frac{v_P}{v_S} = \frac{480}{208} = 2.308 \quad \text{and} \quad a^2 = 5.325 \quad \frac{1}{a^2} = 0.188$$

$$Z_{Sec} = 0.188 \cdot (2.545 + j6.889) = 0.478 + j1.295 \quad m\Omega.$$

Transformer TA is rated 112.5 kVA, 480 to 208Y/120 V and from Table 6.4, we see that a typical percent impedance is 6.1% and X/R ratio is 1.51. Using the technique developed previously

$$R = \frac{kV^2 \cdot \%Z \cdot 10{,}000 \cdot \cos\left[\arctan\left[\frac{X}{R}\right]\right]}{kVA}$$

$$= \frac{0.208^2 \cdot 6.1 \cdot 10{,}000 \cdot \cos\left(\arctan\left(1.51\right)\right)}{112.5} = 12.953 \quad m\Omega.$$

Similarly

$$X = \frac{0.208^2 \cdot 6.1 \cdot 10{,}000 \cdot \sin\left(\arctan\left(1.51\right)\right)}{112.5} = 19.559 \quad m\Omega.$$

The total impedance at the secondary of the transformer is then

$$
\begin{aligned}
Z_{Total} &= Z_{Sys} + Z_{Transf} \\
&= (0.478 + j1.295) + (12.953 + j19.559) \\
&= 13.431 + j20.854 \quad m\Omega.
\end{aligned}
$$

At the secondary transformer, TA, the available fault current will be

$$I_{sc} = \frac{V}{|Z|} \cdot 1000 = \frac{120}{24.805} \cdot 1000 = 4838 \quad A.$$

Note that the correct voltage for the last step is now the phase-to-neutral voltage at the secondary of the transformer. Also note that the step-down transformer has had a dramatic effect on lowering the available fault current level.

Finally, we must consider the impedance of the feeder from the transformer to Panelboard LA. This feeder is 10 feet in length and consists of 4#500 kCMIL, THW in 3-1/2" steel raceway. For this feeder

$$R = \frac{10}{100} \cdot 2.44 = 0.244 \quad m\Omega$$

$$X = \frac{10}{100} \cdot 3.96 = 0.396 \quad m\Omega.$$

The total impedance at Panel LA is then

$$
\begin{aligned}
Z_{Total} &= Z_{Sys} + Z_{Feeder} \\
&= (13.431 + j20.854) + (0.244 + j0.396) \\
&= 13.675 + j21.250 \quad m\Omega
\end{aligned}
$$

and the available fault current level at Panelboard LA

$$
I_{sc} = \frac{V}{|Z|} \cdot 1000 = \frac{120}{25.270} \cdot 1000 = 4749 \quad A.
$$

The last steps in this fault current analysis involve the busway serving Panel PA. This copper plug-in busway is 300 feet in length and is rated 800 A. For this busway

$$
R = \frac{300}{100} \cdot 1.05 = 3.150 \quad m\Omega
$$

$$
X = \frac{300}{100} \cdot 2.17 = 6.510 \quad m\Omega.
$$

The total impedance at the end of the busway

$$
\begin{aligned}
Z_{Total} &= Z_{MDP} + Z_{Busway} \\
&= (0.875 + j6.019) + (3.150 + j6.510) \\
&= 4.025 + j12.529 \quad m\Omega.
\end{aligned}
$$

The available fault current at the 200 A plug-in circuit breaker

$$
I_{sc} = \frac{V}{|Z|} \cdot 1000 = \frac{277}{13.160} \cdot 1000 = 21{,}049 \quad A.
$$

The short feeder from the 200 A breaker to Panel PA is 40 feet in length and consists of 4#3/0, THW copper in 2" steel raceway. The impedance of this segment is

$$
R = \frac{40}{100} \cdot 6.68 = 2.672 \quad m\Omega
$$

$$
X = \frac{40}{100} \cdot 4.22 = 1.688 \quad m\Omega.
$$

The total impedance at Panel PA is then

$$
\begin{aligned}
Z_{PA} &= Z_{Sys} + Z_{Feeder} \\
&= (4.025 + j12.529) + (2.672 + +j1.688) \\
&= 6.697 + j14.217 \quad m\Omega
\end{aligned}
$$

and the available fault current level at Panel PA is

$$
I_{sc} = \frac{V}{|Z|} \cdot 1000 = \frac{277}{15.715} \cdot 1000 = \boxed{17,626} \ A.
$$

Table 6.5 shows a summary of the fault current calculations, along with the appropriate circuit breaker interrupting ratings based on the standard breaker interrupting ratings discussed in Chapter 3. It is crucial that the selected breaker interrupting rating be at least equal to the available fault current level at the point in the system at which the breaker is installed. Some engineers allow a 10% safety margin between the breaker rating and the available fault current level to offset any slight error in feeder length or conductor characteristics.

Table 6.5 Summary of Results of Fault Analysis.

POINT	LOCATION	AVAILABLE FAULT	STANDARD BREAKER RATING
A	Utility Trans. Secondary	50,099 A	65,000 A
B	MDP	45,542 A	50,000 A
C	Panel HA	12,773 A	14,000 A
D	Panel LA	4,749 A	10,000 A
E	200 A Busway Plug	21,049 A	25,000 A
F	Panel PA	17,626 A	25,000 A

PHASE-TO-PHASE AND
PHASE-TO-GROUND FAULTS

As discussed previously, traditional fault current analysis focuses on the three-phase bolted fault because this type of fault normally produces the highest fault current flow and, hence, subjects the system to maximum stress. The actual occurrence of bolted three-phase faults is somewhat rare. Phase-to-phase faults and faults involving ground are far more common.

Phase-to-phase faults can reach fault current levels of 87% of the bolted three-phase value. Faults of this magnitude will normally cause overcurrent devices to operate very rapidly and, therefore, interrupt the current flow in a manner similar to the bolted three-phase fault.

Phase-to-ground faults may be either bolted faults or higher impedance faults, such as arcing faults. In either case, the added impedance of the ground path results in a current flow that may be substantially lower than the bolted three-phase value.

Figure 6.9(a) shows a bolted ground fault in which the current is limited only by the utility system impedance, the impedance of the conductors within the building and the impedance of the ground path. A reduction in the level of fault current flow will tend to introduce time-delay in the operation of overcurrent devices which have inverse time-current trip characteristics. As a result, ground faults may persist for longer periods of time, with an increased potential for damage.

Arcing Ground Faults

Another type of ground fault involves arcing. In this type of fault the total impedance is the sum of the utility system impedance, the building conductor impedance, the impedance of the fault itself and the impedance of the ground path. The arcing in such cases can become self-sustaining, and enormous system damage is possible. The arc itself can be initiated a number of ways, such as the dropping of a tool, or physical damage to a panelboard or switchboard. The arc is initiated in air, but rapidly becomes an arc through the vapors of the conductor material itself. The arc operates at temperatures which are far beyond the operative capacity of any known material. As a result, the conductor material involved, as well as equipment housings, can be rapidly vaporized. Because of the potential for property damage and personal injury associated with this type of fault, we will discuss it in some detail.

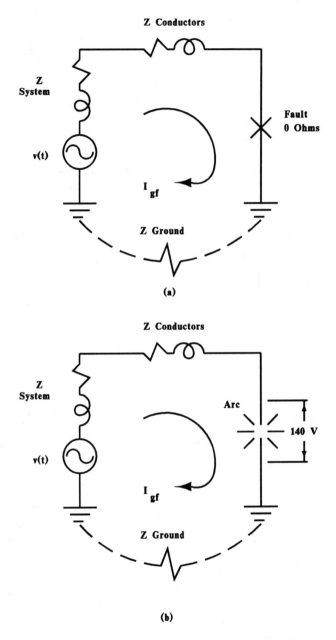

Figure 6.9 *AC* system with ground fault (a) bolted (b) arcing.

The problem of arcing ground faults in building electrical systems first came to widespread attention in the early 1960s with the increased use of 480Y/277 volt systems. This system voltage was developed in response to the need for a higher voltage system in order to more economically serve larger loads than the 208Y/120 volt system. Unfortunately, the 480Y/277 volt system also proved to be more susceptible to arcing faults. Research has shown that an instantaneous voltage level of 375 volts is required to establish a self-sustaining arc. The system *rms* level of 277 V (392 V peak) was just high enough to initiate such an arc. Most faults of this kind were found to either begin as phase-to-ground faults, or to begin as phase-to-phase or phase-to-neutral faults, that rapidly escalate to involve ground as well.

In Figure 6.9(b) we see that in an arcing fault involving ground, the total impedance now includes the impedance at the point of fault. It has been observed that bolted faults occur rarely, and that most faults involve some level of impedance and, often, arcing. The arc itself appears to be purely resistive in nature, and the arc voltage drop is relatively flat, topped at approximately 140 volts. This suggests that the arc resistance is inversely proportional to the current.

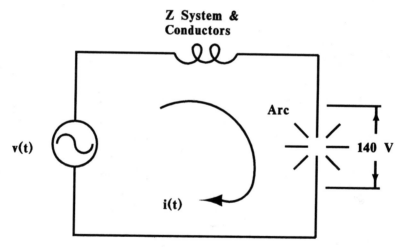

Figure 6.10 *AC* system with arcing ground fault - idealized with reactive component only.

The Discontinuous Current Waveform of Arcing Faults

Let us now discuss the nature of the arcing ground fault current waveforms. In order to do this, let us use Figure 6.10 that shows a simplified circuit, containing only inductive reactance.

Writing an equation for the voltage drop around the circuit

$$v(t) = L\frac{di}{dt} + V_{arc} \qquad (6.36)$$

where

$$v(t) = v_{max} \sin(t) \qquad V \qquad (6.37)$$

and V_{arc} represents the voltage drop across the arc.

Equation 6.36 can then be simplified to

$$\frac{di}{dt} = \frac{1}{L}\left(v_{max} \sin(t) - V_{arc}\right) \qquad (6.38)$$

now, integrating

$$i = \frac{1}{L}\int_{t_a}^{t_d} \left[v_{max} \sin(t) - V_{arc}\right] dt \qquad A \qquad (6.39)$$

where t_a represents the time at which the arc initiates and t_d the time at which the arc extinguishes. Note that both t_a and t_d are expressed in radians.

We know that the arc will strike when the system voltage reaches the restrike voltage V_r, 375 V, so t_a will be given by

$$t_a = \arcsin\left(\frac{V_r}{v_{max}}\right) \qquad Radians. \qquad (6.40)$$

We also know that the arc current will reach its maximum value when the system voltage $v(t)$ equals V_{arc}. We will call this point in time t_c.

$$t_c = \pi - \arcsin\left(\frac{V_{arc}}{v_{max}}\right) \qquad Radians \qquad (6.41)$$

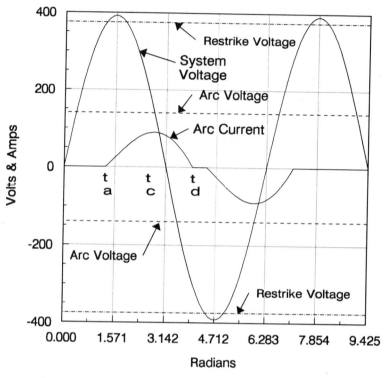

Figure 6.11 Arcing ground fault current waveform for circuit shown in Figure 6.10.

The current will then decrease, reaching zero when the integral of the volt-time area, shown in the right side of Equation 6.39, equals zero. We will call this time t_d.

At time t_d, the arc extinguishes and does not restrike until the system voltage reaches the negative restrike voltage $-V_r$. At this point, the arc restrikes, resulting in the negative loop of current shown in Figure 6.11. This process repeats itself cycle after cycle, resulting in a current flow that is discontinuous, and that contains significant harmonics. The *rms* value of the current can also be substantially lower than the three-phase fault current values. It is this lower *rms* value of current that causes many overcurrent devices to have significant time delays in opening or, in some cases, causing the devices to not open at all.

Figure 6.12 shows the damage to an industrial busway system resulting from a low level arcing ground fault. The fault was initiated by workmen

Figure 6.12 480Y/277 V busway system damaged by arcing ground fault (Photo: Mary Matthews).

255

who attempted to install a plug-in device on the energized busway. Unfortunately, the plug-in device was defective, and the resulting fault rapidly escalated to an arcing ground fault. Note the damage to both the busway system conductors and the metal housing. The arcing persisted for several seconds. The building system contained protective features to detect such faults, however, the equipment had not yet been connected at the time of the accident.

Ground Fault Detection Systems

The solution for the problem of ground faults and arcing ground faults is a ground fault detection system that rapidly detects the fault, and removes it from the system. There are several approaches to detecting ground faults, but the basic idea is quite simple. The ground fault detection system monitors the current in the phase conductors as well as the current in the neutral. If the vector sum of these currents is not zero, some portion of the current is returning through a path involving ground. Upon detecting this condition, the ground fault system signals the overcurrent device to open. The necessary current-sensing devices can be built into a circuit breaker, or supplied as separate components installed in the panel or switchboard. The trip point is frequently adjustable, so as to avoid nuisance tripping due to normal system leakage current. In recognition of the danger of ground faults, the *National Electrical Code* now requires such ground fault detection systems to be installed on all 480 volt system overcurrent devices rated at 1000 A or greater.

Safety Hazards Due to Arcing Ground Faults

In addition to the damage to system components discussed above, there is also severe risk of injury to nearby personnel. This hazard results from the blast pressure produced by the arc, as well as the thermal energy liberated by the arc.

The blast pressure occurs within a few milliseconds of the initiation of the arc. This blast pressure can range from a few pounds/ft^2 to several hundred pounds/ft^2, and can be something of a mixed blessing to nearby personnel. Although they may be injured by being propelled into nearby objects, the movement away from the arc source can reduce the degree of burn injury. The shock of blast pressures is often sufficient to result in short-term memory loss. As a result, personnel close to arcing ground faults often report finding themselves knocked to the floor, with no recollection of how they came to be there.

Risk of burn injury results from both the ejection of molten material as the arc erodes conductor material, and thermal burns caused by the very high temperatures associated with the arc. The arc itself is known to involve temperatures of up to 35,000°F. The ejected droplets of molten metal may reach temperatures of 1000°C. The combination of molten material and very high arc temperatures often ignites clothing, resulting in even more severe injury. As you can see, the combination of blast and burn injury makes arcing ground faults one of the most hazardous and particularly lethal types of system faults.

LIMITING THE AVAILABLE
FAULT CURRENT LEVEL

Perhaps one of the most effective methods of limiting the available fault current in a system is the installation of current-limiting fuses. In Chapter 3 we discussed the characteristics of this device, and it might be appropriate to review this material at this time. Current-limiting fuses are capable of detecting high-level faults, and opening well before the current reaches its first cycle peak. In essence, this severely limits the actual level of available fault current on the load side of the fuse. This in turn permits us to utilize components with lower interrupting ratings. For example, in Figure 6.7, the circuit breakers shown in Panel MDP might be changed to switches with current-limiting fuses. Depending on the fuses selected, it might be possible to lower the interrupting rating of devices located in Panels HA and PA.

A second method of limiting the available fault current in a system is through the use of current-limiting circuit breakers. These devices are designed to operate at speeds sufficient to produce the desired current-limiting effect.

Following the completion of the fault current study, the engineer is in a position to determine the best system-protective device combination. For example, the main device might be a fused switch or circuit breaker, while the feeder overcurrent devices might also be either fused switches or circuit breakers. The final decision is often based upon a careful economic analysis between alternative system configurations.

COMPUTER FAULT CURRENT ANALYSIS

As you can imagine, a manual fault current analysis using only a calculator can become very time consuming and tedious, even for a moderate-sized building system. Typically, design budgets are very tight, making time even more scarce. Unfortunately, tedious tasks are often overlooked or approximate solutions sought. In the case of fault current analysis, such neglect can have negative consequences of epic proportions.

For many years the cost and availability of computers and software for fault current analysis made manual solutions a necessity. The software that was available often required such massive preparation and complex data entry that most engineers found it difficult to justify the time. As late as 1978, the author found it necessary to personally develop software for his company's use, due to the lack of effective professional software. Even today, there are few options available, simply due to the limited market, and the large expense of software development. Even with excellent software, it is critical that the engineer fully comprehend every step in the fault current analysis in order to detect errors which may surface during data entry.

Evaluating Software for Fault Current Analysis

When evaluating software for fault current analysis, several points should be considered:

1. We live in a very demanding profession in which design time is at a premium, and the margin for error quite small. The software should be very interactive and input data should be requested in the commonly used terms of normal fault current analysis. It should be possible to quickly, and painlessly, modify the system after the data is entered. Few building systems escape several design changes, often before the project is even initially released. The ability to rapidly re-configure the system permits the engineer to realize one of the most valuable advantages of the computer, the ability to play "what if?" This ability to analyze several possible alternatives is simply not possible in the case of manual calculations.
2. The software should be selected with the nature of your firm's projects in mind. A professional acquaintance once told the author of a

wondrous new fault analysis program his firm had just purchased. Several months later, when asked how the use of the software was progressing, he responded that the software was only used on very large projects with commensurately larger design fees, due to the enormous complexity of data entry. If your firm does only seventy-story office buildings then, by all means, select powerful and versatile software. If your firm normally has smaller commercial and industrial projects, then a somewhat simpler and faster package might be in order. In short, pick the right tool for the job.

3. The final output of the software should be rendered in clearly comprehensible terms. It is helpful if the computer run bears a time and date since several different versions of the system may have been run during the design process. It is also important for all input data to be clearly repeated on the output, along with the source of other data, such as conductor resistance and reactance values. Remember that you may have to submit the results of the fault current study to the local building plans review department or other inspection authorities. These people do not take kindly to the results of fault current analysis which seem to have appeared to the engineer in a vision.

4. If at all possible, it is desirable for the fault current study to utilize data already entered into the software for other purposes, such as load studies, panelboard and switchboard schedules, or voltage drop studies. In this way, the complete design, including all computations, can be readily reviewed and revised. Such integrated software also has the advantage of reducing the ever present danger of the failure to make system modifications to all affected data files.

Electrical System Analysis Programs

Figure 6.13 shows a portion of the output data for one electrical system analysis program. Notice that the input data is clearly referenced, and that results are presented in an itemized and readily understandable manner. This particular package allows easy entry of system loads, circuit by circuit. The system determines the appropriate conductor sizes, overcurrent device ratings and panelboard ratings, based on the input load data. The final output includes panelboard and switchboard schedules ready to import directly into CAD drawings, or for inclusion on manually prepared drawings. Finally, the software performs a complete fault current and voltage drop study, a portion of which is shown in the referenced figure.

```
E/CALC DISTRIBUTION SYSTEM REPORT - Project: Sample System

for

John Matthews & Associates, Inc. - Atlanta, Georgia

Mon Jan 27 11:21:22 1992

Data drive:\path  C:\ECALC\EX1\

Wire resistance and reactance values are from 1990 NEC(r) Chapter 9 Table 9.

Voltage drop computations use ohmic values.
NEC(r) resistance is adjusted to Centigrade value shown on "Phase wires" line.

Fault current computations use per unit values - base = 10,000.
NEC(r) resistance is adjusted to Centigrade value shown on "Ambient temp" line.

Utility available fault current = 50000 RMS symmetrical amps     X/R = 9.0
Utility line-to-neutral voltage = 277

Service entrance equipment: MDP        480Y/277V 3P 4W

Downstream equipment = 2
HA    LA
```

```
-------------------------------------------------------------------
MDP    480Y/277V 3P 4W Switchboard         - ckt br - Feeder UTILITY

Load Load  Load      Voltage at      Voltage drop %   Fault contrib - RMS Sym
     FLA   PF%       equipment      Cond Trans Total  Utility Motors    Total
NL    -     -     479.8Y/277.0       -     -    -      37102    -        37102
Ht   662  98.8    478.3Y/276.1      0.33   -   0.33    37102   489       37591
A/C  606  96.1    478.2Y/276.1      0.34   -   0.34    37102   916       38018

                  Feeder
OCP Amps           800        (minimum =   800)
Type-Loc-Len      THW Copper  - OH    75ft
Conduit(s)        (2) 3-1/2"  EMT
Phase wires       3#500KCM    (75C - reduced 1)
Neutral wire      #500KCM
Ground wire        -
Isolated Gnd       -

Ambient temp       30C
```

```
-------------------------------------------------------------------
HA    480Y/277V 3P 4W Lighting panel          - Feeder MDP

Load Load  Load      Voltage at      Voltage drop %   Fault contrib - RMS Sym
     FLA   PF%       equipment      Cond Trans Total  Utility Motors    Total
NL    -     -     479.8Y/277.0       -     -    -      30600    -        30600
Ht   101  95.0    477.9Y/275.9      0.39   -   0.39    30600   332       30932
A/C  122  93.7    477.8Y/275.8      0.42   -   0.42    30600   646       31246

                  Feeder
OCP Amps           225        (minimum =   175)
Type-Loc-Len      THW Copper  - OH    25ft
Conduit(s)        (1) 2-1/2"  EMT
Phase wires       3#4/0       (75C - 100% load)
Neutral wire      #4/0
Ground wire       #4          (NEC(r) 250-95)
Isolated Gnd       -

Ambient temp       30C
-------------------------------------------------------------------
LA    208Y/120V 3P 4W Lighting panel            - Feeder MDP
Transformer KVA=    75    Z%= 5.2    X/R=  1.4   Taps= Normal

Load Load  Load      Voltage at      Voltage drop %   Fault contrib - RMS Sym
     FLA   PF%       equipment      Cond Trans Total  Utility Motors    Total
NL    -     -     208.0Y/120.0       -    0.00   -      3627    -         3627
Ht   107 100.0    203.9Y/117.7      0.38  1.59  1.97    3627    2         3629
A/C  131  85.0    200.4Y/115.7      0.59  3.05  3.64    3627   89         3716

                  Secondary                    Primary
OCP Amps           225        (minimum =   150)   125        (minimum =    70)
Type-Loc-Len      THW Copper  - OH    10ft      THW Copper  - OH    40ft
Conduit(s)        (1) 2-1/2"  EMT              (1) 1-1/4"  EMT
Phase wires       3#4/0       (75C - 100% load)  3#2        (75C - reduced 1)
Neutral wire      #4/0
Ground wire       #2          (NEC(r) 250-94)    #6         (NEC(r) 250-95)
Isolated Gnd       -                             -

Ambient temp       30C                           30C
-------------------------------------------------------------------
End of report
```

Figure 6.13 Typical fault current analysis software output. (Data courtesy of E/CALC, Robert S. Atkinson, 3716 Woodmont Lane, Nashville, TN 37215.)

As a final point, the presence of a complete and professional fault current study is not only important data but often distinguishes the product of a truly dedicated design professional.

References

Atkinson, Robert S., P.E. 1987. *ECALC Software Manual*. Nashville: Atkinson & Associates, P.C.

Beeman, Donald. 1955. *Industrial Power Systems Handbook*. New York: McGraw Hill Book Company, Inc.

Dunki-Jacobs, Jr. 1986. Escalating Arcing Ground Fault Phenomenon. *Institute of Electrical and Electronic Engineers Transactions: Industry Applications*. pp. 1156-1161.

Dunki-Jacobs, Jr. 1972. Effects of Arcing Ground Faults on Low Voltage System Design. *Institute of Electrical and Electronic Engineers Transactions: Industry Applications*.

Greenwood, Alan. 1971. *Electrical Transients in Power Systems*. New York: Wiley Interscience.

Institute of Electrical and Electronic Engineers, Inc. 1990. *IEEE Recommended Practice for Power Systems Analysis. IEEE Std. 399-1990*. New York: Institute of Electrical and Electronic Engineers, Inc.

Institute of Electrical and Electronic Engineers, Inc. 1990. *IEEE Recommended Practice for Electric Power Systems in Commercial Buildings. IEEE Std. 241-1990*. New York: Institute of Electrical and Electronic Engineers, Inc.

Institute of Electrical and Electronic Engineers, Inc. 1986. *IEEE Recommended Practice for Electric Power Distribution for Industrial Plants. ANSI/IEEE Std. 141-1986*. New York: Institute of Electrical and Electronic Engineers, Inc.

Institute of Electrical and Electronic Engineers, Inc. 1986. *IEEE Recommended Practice for Protection and Coordination of Industrial and Commercial Power Systems. ANSI/IEEE Std. 242-1986*. New York: Institute of Electrical and Electronic Engineers, Inc.

Lee, Ralph H. 1987. Pressure Developed by Arcs. *Institute of Electrical and Electronic Engineers Transactions: Industry Applications*. pp. 760-764.

Lee, Ralph H. 1982. Other Electrical Hazard: Electric Arc Blast Burns. *Institute of Electrical and Electronic Engineers Transactions: Industry Applications*. pp. 246-251.

National Fire Protection Association. 1990. *National Electrical Code*. Massachusetts: National Fire Protection Association.

Ohlson, Russell O. 1967. Procedure for Determining Maximum Short Circuit Values in Electrical Distribution Systems. *Institute of Electrical and Electronics Engineers' Transactions on Industry and General Applications*. IGA-3(2).

7

Utility Systems

There is an old saying about looking into the future—"you cannot understand where you are going unless you understand where you have been." This saying may be somewhat simplistic, but there is more than a little truth in that statement, especially as it relates to the utility industry. The utility industry today is a product of all the dynamic forces to which it has been exposed since its inception in about 1879. The growth of our nation's electrical requirements was precipitated by the increasing energy use and sophistication of building electrical systems. The utility industry has gone through periods of steady growth and stable regulatory constraints, as well as volatile periods of public environmental concerns, fuel shortages and rapidly changing regulatory policies. Utilities have sometimes seen their public image shift from that of the "good guys" to one that may be somewhat less flattering. Through this time they have successfully met their goal of reliably providing the nation's power requirements. These requirements have gone through significant change. It is important for the engineer to understand how the various forces which shape and mold the utility industry have combined to produce today's utility operating environment. This environment has direct bearing on the design and planning of building electrical systems.

So, let's take a few moments to look backward at "where we've been" so that perhaps we will be in a better position to understand tomorrow's electrical energy challenges.

HOW IT ALL BEGAN

We often think of the utility industry as having originated with Edison's work around 1879, and to a major extent this is true. Before Edison, however, a lot of basic work had been done by others, such as Britain's William Gilbert, who in 1600 published works concerning the fundamentals of magnetism. In 1808, a British chemist, Sir Humphry Davy, used a battery and two pieces of charcoal to create an arc light. Later, in 1831, Michael Faraday, another British subject, and Joseph Henry, an American, independently discovered the principle of electromagnetic induction, that was to play such an important role in the electrical industry. In 1859, another American inventor, Moses G. Farmer, experimented with lamps which used strips of platinum connected to batteries. In 1860, Joseph Swan, in England, employed a partial vacuum in a bell jar and a carbonized paper filament to demonstrate an incandescent lamp (Hyman 1985). All of these experiments were significant because they demonstrated that important applications of electricity, such as electrical lighting, were feasible.

By the time Edison turned his attention to the electric lamp, he had already made significant contributions in the area of telegraphy, and had invented the phonograph and the telephone. After considerable research and experimentation, Edison was ready to demonstrate the first practical incandescent lamp at his laboratory in Menlo Park, New Jersey in 1879. After this demonstration, gas stocks plummeted as rumors of the new light source spread. Edison's demonstrations caused an immediate sensation, and the public flocked to see the new invention. Shortly afterward, Henry Villard, a wealthy railroad owner of the day, ordered the first shipboard lighting system for the S. S. Columbia. This 115-lamp installation created much public attention when the brightly lighted Columbia sailed from New York in May, 1880 (Cox 1979). It might be said that the sailing of the Columbia also marked the "sailing" of the fledgling electric utility industry.

During this period, Edison clearly realized the technical and production challenges facing the widespread use of electricity. In effect, a new industry would have to be created. Someone would have to build the dynamos which provided the electricity. Still others would have to produce the many different electrical devices necessary to convey the electricity to the point of utilization which was, at this time, primarily the incandescent lamp. Finally, the incandescent lamp itself would have to be refined to the point that mass production was possible. All of this would have to occur at costs which would make the entire venture economically feasible (Cox 1979). It is a tribute to Edison and those who followed that these challenges were overcome and the new industry began to flourish.

These early utility systems were essentially stand-alone (isolated) affairs, and were built to serve either a specific industrial site or perhaps a small area of a city (Vennard 1979). There was no interconnection of systems, and one system might be designed to supply power to an electric trolley system, primarily a daytime operation, while another system was installed for electric lighting, primarily a nighttime operation. Demand for power for most industrial process and motor loads occurred between the time of peak demand for the trolley and the lighting system. It was soon realized that one generation system could serve the electric trolley, electric lighting and the industrial systems (Hyman 1985). As *ac* power became more widely used, higher voltages became necessary in order to transmit power over greater distances. The concept of central generating plants then evolved, based on the realization that large generating facilities could produce electricity far more efficiently, and at less cost, than many smaller facilities.

The early days of the utility industry were chaotic. Electric light companies were formed and often franchised the use of their equipment, sometimes only in a given area or part of a city. A city might have a number of electric companies operating for different purposes, such as street lighting, home lighting, industrial systems and electric trolley systems.

One of the utility leaders of the day was Samuel Insull of Chicago. By 1892, he had realized several crucial facts regarding the fledgling utility industry. He realized that the operation of a utility system was extremely expensive from the standpoint of fixed costs. The required equipment, although long-lived, was quite expensive to purchase and install. He also recognized that, once installed, the generating facilities were not terribly expensive to operate. Finally, he realized that systems had to be designed and constructed so that they were capable of providing the customer's peak demand requirements, which often occurred for only a few hours a day, thus leaving the system lightly loaded the remainder of the time. These were formidable challenges, and remain so, even today.

Between 1902 and 1932, the generating capacity of isolated systems serving individual industrial plants decreased. From 1902 to 1932 the generating capacity of utilities grew from 1.2 million kW to 34.4 million kW. At the same time industrial generation grew from 1.8 million kW to only 8.5 million kW (Hyman 1985). The movement toward central station generation of power was underway.

THE GROWTH OF THE ELECTRIC UTILITY INDUSTRY

The electric utility industry grew along three basic lines of ownership and operation, which were the investor-owned, the government-owned and the cooperative-owned systems. Investor-owned systems were essentially corporations established to produce and deliver power at a profit. The capital requirements for construction and operation of these systems were derived from the sale of stocks and bonds to the public. The continuing revenues from the sale of electric power to the utilities' customers repaid the cost of construction and operation, and also provided the stockholders with a return on their investment. Government-owned systems consisted of systems owned and operated by federal, state and local agencies. There was considerable political pressure early in the century for the establishment of such government-owned systems. Cooperative ownership differs from investor-owned systems in that the system's customers are also its owners. Cooperative systems raise capital in the same manner as investor-owned systems. The federal government, through the Rural Electrification Administration, or REA, has played a major role in the development of the cooperatives (Schap 1986).

During this time, the first regulation of utilities occurred as a result of public pressure as well as pressure from within the utility industry itself. The basic concept of regulation had its roots in the fact that the utility industry was something of a monopoly (Hyman 1985). It was hoped that the regulation of the industry by an independent state agency would prevent exploitation of the customer. The regulatory agency would be empowered to monitor and approve the rates charged by the utility. In many states these agencies are called Public Service Commissions. Today, essentially all states have such organizations.

During this early period many utilities were bought by holding companies. These holding companies often owned several utilities, and provided engineering as well as financial services to their operating units. The state regulatory agencies controlled the actual rates charged by utilities, but the costs associated with the services provided by the utilities' holding companies were not as well regulated. This led to excesses on the part of some holding companies, who charged inflated amounts for technical and financial services, which were then passed on to the utility's customers as part of the utility's operating costs. In 1928, the Federal Trade Commission began what was to be a long investigation of the utility industry and its holding companies. The investigation resulted in the *Public Utility Holding Act of 1935* which, in effect, severely curtailed the activities of the holding companies.

During this same period, presidential candidate, and then president, Franklin Roosevelt, voiced support for government power projects. He was especially interested in hydroelectric projects which offered both power generation and flood control. His interest eventually led to the establishment of government power projects such as the Tennessee Valley Authority and the Bonneville Power Administration.

Shortly after the end of World War II, the utility industry had evolved into much the same kind of configuration we see today. It was composed of investor-owned, government-owned and cooperative-owned utilities. It was heavily regulated by state agencies, and employed generation, transmission, and distribution equipment which was both expensive to build and long-lived. The cost of electric power decreased as more efficient generation equipment was developed and the resulting demand for low cost electrical power steadily increased.

The period of time between World War II and the mid-1960s can perhaps be best characterized as a period of stability and steady load growth. It had long been understood that utility generation and transmission equipment was very costly and required a significant length of time to construct. This made effective projections of future electrical demand important in order to have adequate generating capacity in place when needed. Investor-owned utilities routinely projected load growth, planned new facilities and financed construction from the sale of stocks and bonds. Utilities were viewed as "good guys." The utility industry and regulatory agencies were not involved in significant levels of conflict, and the nation's economy developed at a rate which required ever increasing amounts of electrical power. If a utility happened to slightly overbuild generating capacity, it was certain that the excess capacity would be required within a few years, so no harm was done.

The primary fuels for power generation were coal and oil. Hydroelectric generation eventually provided a smaller part of the expanding needs, due, in part, to the limited number of suitable dam locations. By the mid-1950s the potential of nuclear power was clearly visible on the horizon. Finally, the available reserves of fossil fuel seemed limitless.

These were probably the most worry-free days the electric utility company has ever experienced, but storm clouds were just over the horizon. In fact, the foundation for future problems was already in place. As the efficiency of generation systems increased, the utilities were able to maintain the low cost of power year by year, despite increases in the cost of most other goods and services. One measure of generation efficiency is *heat rate*. *Heat rate* is the energy in British Thermal Units, BTU's, required to generate one kilowatt-hour of energy. This heat rate had shown essentially constant improvement up through the mid-1960s, but has remained fairly constant thereafter. The low cost of electrical power prompted many industries to shift ever increasing amounts of process load, as well as other energy requirements, to the electrical utilities. Electric power was a "good buy." Increased use of electrical power was also encouraged by the fact that power was priced on a declining block basis, with the cost of the first several hundred kWh being more expensive than succeeding energy blocks. This declining scale had great economic advantage for many industries. As a final point, the utility industry was very conservative, and tended to make decisions carefully, and only after careful deliberation. They may have also been somewhat slow in recognizing that fundamental changes in their operating environment were about to occur.

A PERIOD OF UNCERTAINTY FOR THE ELECTRIC UTILITY INDUSTRY

The period from about 1965 to the present has been one of literally constant change and challenge for the utility industry. These changes can perhaps be best viewed in the context of some of the watershed events which precipitated them.

On November 2, 1965, a failure in the Ontario Hydro system plunged a large portion of the northeast United States into darkness. The "Northeast Blackout of 1965," as it has been called, was symptomatic of problems within the utility industry. Perhaps the power transfer capabilities between regions were not adequate. Perhaps control and early detection of system overloads were not adequate. In any event, the ensuing public outcry placed the utility industry under close scrutiny. The industry focused considerable attention on determining the cause of the problem, and in taking steps to lessen the likelihood of its reoccurrence.

It is interesting to review the utility's *reserve margin* situation in conjunction with this and later periods of time. The *reserve margin* is based on the utility's peak demand during the summer and winter months, respectively, and also on the overall system capability during these periods. The expression for percent reserve margin then becomes:

$$\% \ Reserve \ Margin \ = \ \frac{(Capability \ - \ Peak \ Load) \cdot 100}{Peak \ Load}. \quad (7.1)$$

Figure 7.1 shows the summer and winter percent reserve margin for the nation's utility system during the period from 1960 through 1989. Notice that both summer and winter reserve margins were lower in the mid-1960s than in subsequent years. Also note that the summer reserve margin is

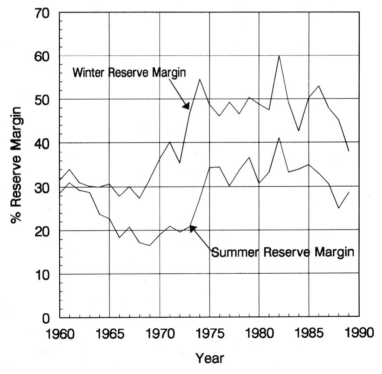

Figure 7.1 Utility industry reserve margins for the period 1960 to 1989. (Data from Table 7 *Statistical Yearbook of the Utility Industry*, Edison Electric Institute, Various Years.)

somewhat less than the winter reserve margin. The figure further reveals that both summer and winter reserve margins showed improvement from about 1970 through the mid-1980s. It will be interesting to watch these trends during the next few years.

During the late 1960s the utilities continued to offer power at very low cost compared to the overall cost increase in other areas of the economy. They also may have underestimated the degree of escalation of their own facility construction costs as well as fuel costs.

By the mid-1960s there was increased public awareness of the potential negative environmental impact of power generation. This environmental concern prompted many utilities to shift to the use of oil to replace some coal-fired plants with their higher release of carbon dioxide as well as other pollutants. Additional emphasis was also placed on nuclear power generation as a source of more pollution-free power.

By the late 1960s escalating construction costs and fuel costs finally forced relatively rapid increases in the cost of power. Regulatory agencies were somewhat slow in allowing rate increases, so utilities were caught between rapidly increasing costs and more slowly changing rates. The result was a squeeze on utility profits, and a subsequent decline in investor interest in utility stocks. As their financial condition worsened, it became more difficult for them to obtain necessary financing. Figure 7.2 shows the pattern of utility cost changes for residential, industrial, and commercial customers, as well as composite data for the entire electric utility industry. The data covers the period from 1960 through 1989.

It was an already weakened utility industry that confronted the Arab Oil Embargo of 1973 and 1974. The Oil Producing and Exporting Countries, OPEC, had cut off or severely curtailed oil exports to the United States in the wake of the Yom Kippur War. This subjected the utility industry to several pressures concurrently. The cost of electric power had historically been based on a system that included fuel costs as an integral part of the cost per kWh of energy. The utilities suddenly found that the necessary fuel might not be available, and the fuel that could be obtained was subject to sudden and unpredictable cost variations. The regulatory agencies were slow to permit rate increases to cover these changes in fuel costs and the resulting squeeze further crippled many utilities. One solution to this problem eventually emerged as the *fuel cost recovery* system, whereby a utility could add a separate amount to the customer's bill to reflect unforeseen changes in fuel costs. This amount could be an addition or a credit, depending on the trend in fuel costs. The fuel cost recovery, or *FCR*, is a component of almost all modern day utility bills.

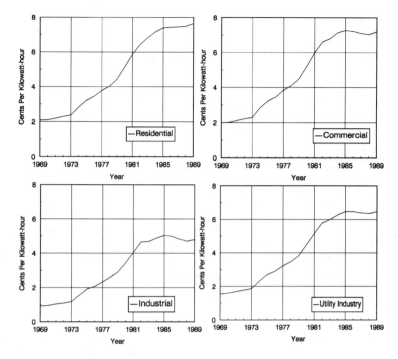

Figure 7.2 Cost per kilowatt-hour for several utility customer classes for the period 1969 to 1989. (Data from Table 63, *Statistical Yearbook of the Utility Industry*, Edison Electric Institute, Various Years.)

During the same period of time there was pressure to reduce the use of oil and return to more coal-based generation. This required expensive changes in generation plants, and the new pollution control systems favored by environmentalists added further costs. Increased use of gas-fired generation facilities also occurred. Nuclear plants were already being built, but were subject to both growing anti-nuclear opposition and unexpected increases in construction costs. To some there appeared to be increased public opposition to almost all practical fuel sources (Cott and Spinrad 1985). Figure 7.3 presents data on the percentage of the total utility generation represented by coal, oil, gas, nuclear and hydro for the period 1960 through 1989. Notice the decrease in coal use and increase in oil use until about 1973, the time of the Arab Oil Embargo. Also note the slow but steady rise in nuclear power's share of the total generation and the slowly decreasing role of hydro generation.

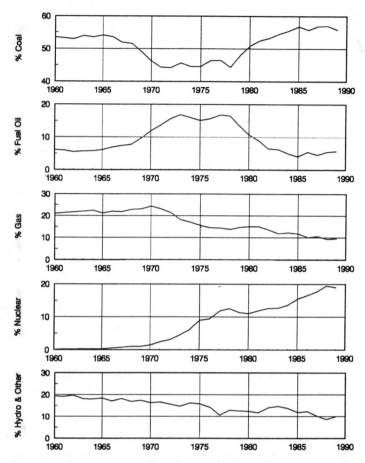

Figure 7.3 Percentage of utility generation by various fuel types for the period 1960 to 1990. (Data from Table 22, *Statistical Yearbook of the Utility Industry*, Edison Electric Institute, Various Years.)

As a result of the Arab Oil Embargo and attendant electric power rate increases, the public became more sensitive to the need for energy conservation and, hence, the historically steady rate of increase in the use of electric power lessened. This exacerbated the utilities' situation, since the decrease in energy usage decreased the revenue that was necessary to finance projects already being constructed in anticipation of future needs.

To many utility executives the situation probably appeared as if it could not be worse. Unfortunately, it could, as events at Three Mile Island were to prove. On March 28, 1979, at a nuclear generating plant at Three Mile Island near Harrisburg, Pennsylvania, owned by General Public Utilities, a cooling system problem rapidly escalated to an emergency shutdown of the reactor. Over the following days public attention was riveted to the news media as details of the accident emerged (Stoler 1985). Sadly, it seemed very little attention was paid to the simple fact that despite the nature of the accident and associated operator errors, the overall construction safety features worked. There was no serious release of radioactivity to the environment. A power generation reactor had been subjected to a serious emergency condition for the first time, and the countless thousands of hours of careful engineering paid off — a catastrophe was averted. It was clear, however, that both operational and design flaws existed.

In the wake of Three Mile Island, the public anti-nuclear contingent, which had never been particularly reticent about expressing its views, became even more vocal. Under tremendous pressure, the Nuclear Regulatory Commission temporarily suspended permits for new nuclear plants, and over a period of time, instituted sweeping design and construction changes to all nuclear plants. This included existing facilities, those under design, and those under construction. The effects on construction schedules and costs for nuclear plants were devastating. Many plants were delayed for years, while others had construction budgets increase many times over original estimates. A good example is Georgia Power's Plant Vogtle. The construction of this facility was started in the early 1970s in southeast Georgia. Design and construction changes, as well as other cost overruns, saw this facility, originally estimated at $660 million, escalate to a final cost of over $8.7 billion by the time it began operation in 1988 (Georgia Power Company 1988). Many plants were simply abandoned at their current state of construction resulting in an enormous financial loss.

The utility industry carefully watched General Public Utilities, Three Mile Island's owner, effectively pilloried by the public. General Public Utilities had to purchase power elsewhere to replace the lost generation at Three Mile Island, and this power was more costly than it would have been if generated as originally planned. The regulatory agency having control did not allow the utility to pass the increased cost along to the customer. As a result, the financial position of the utility eroded and it became difficult to sell securities. Finally, it had to omit stock dividends, which only worsened its position with investors (Hyman 1985).

In order to understand the effects of these events on the utility industry, it is necessary to understand traditional utility planning. The historically stable utility operating environment had prompted utilities to make decisions based on the best interests of the rate-paying public over a long period of time. Power generation alternatives such as nuclear power, which promised long term rewards in the form of lower power costs, were embraced despite high initial costs. The utility industry suddenly witnessed this decision-making process go awry. The industry had made decisions it felt were in the public interest, and suddenly found that these decisions exposed utility stockholders to much increased risk. Utility management was to understandably become much more reluctant to make future decisions from which only the customer might benefit, while at the same time exposing the utility stockholders to increased risk (Long-Range System Planning Working Group, Power Engineering Society 1980).

RECENT YEARS

The period since the early 1980s has seen what many hope is the beginning of a more stable environment for the utilities, despite the fact that concern over the availability of natural gas prompted government pressure on utilities to reduce dependence on this fuel. The very costly and often delayed nuclear plants are coming on line, and will continue to do so through the 1990s, but as of 1991 there appears to be little interest in constructing additional nuclear plants. As you can imagine, the utility industry would be very careful in initiating any new nuclear construction. Recently, the trend seems to be toward increased use of coal as a fuel source, with nuclear power shouldering about 20% of the generation burden.

Looking toward the future, we see the potential of nuclear fusion, but this is, by most accounts, several decades away. Although there is much interest in "soft technologies" such as wind power, photovoltaic systems and other techniques, there appears to be little hope that these will provide substantial amounts of power at their current level of development.

The 1991 Persian Gulf War has again brought home the clear danger of our over-dependence on foreign energy sources. Following the war, there has been renewed public discussion of increased use of nuclear fission power. Discussion has centered around small nuclear units in the 500 MW range. These could be constructed to standardized specifications, and licensed under more clearly defined terms to lessen potential construction cost overruns and changes by regulatory agencies. While few expect nuclear power to reach the early predictions of "power too cheap to meter," it may yet assume a major role in our energy requirements until the day of fusion power arrives.

In summary, we see that the utility industry, born in 1879, maturing during the early and mid-1900s and subjected to successive environmental, fuel supply, economic and regulatory challenges since the 1960s, is a product of its sometimes volatile evolution. Some utilities are investor-owned, some are owned by federal, state, or local governments and some by cooperatives. Utilities are regulated by state agencies who play an important role in regulating the cost of power to the consumer. Utilities have faced formidable challenges since the 1960s with economic, environmental and fuel source availability problems affecting the construction and operating costs. Utilities have become more conservative in their decision making, partly in order to protect their investors (Matthews 1989). In a broader context, the utility situation must be viewed against the overall national energy and environmental situation. The earth's resources are quite finite, and we must carefully balance the indisputable need for technological advancement with equal concern for maintaining our "home." The utilities are clearly a part of this picture and will remain so. Finally, they also represent a significant part of the cost of running our homes and businesses.

Figure 7.4 shows the relative percentages of energy billing to industrial, commercial, residential and other customers for 1989. During this year the total utility industry billing was $169.1 billion dollars.

CUSTOMER LOAD PATTERNS

From the earliest days of utility systems, several points were quite clear. First, the utility generation, transmission and distribution systems must be constructed so as to supply the maximum demand of each customer. It was also clear that a given customer would use this maximum demand for only a few hours each day, and that the load pattern would depend on the nature of the business involved. Once the system was installed, the customer should be charged for each increment of energy, kWh, he consumed. Customers who used energy relatively evenly throughout their operating cycle were desirable, because this provided more revenue for the same

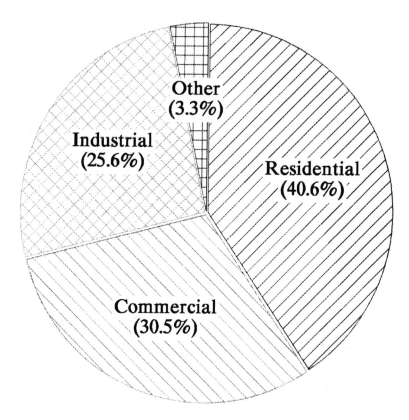

Figure 7.4 Percentage of utility company revenue for 1989 by customer class. (Data from Table 58, *Statistical Yearbook of the Utility Industry*, Edison Electric Institute, Various Years.)

capital expenditure for equipment, such as distribution lines and transformers. In fact, the actual costs of providing each kWh above a certain value, which depended on both the load pattern and the total power consumed, decreased. As a result, power could logically be priced on a declining-scale basis.

Today's most common billing systems take these same factors, plus several others, into account. In addition to peak demand and total kWh consumed, today's bills also include a category for fuel cost recovery, or *FCR*, as discussed in previous sections. The *FCR* rate provision permits utilities to respond in a timely manner to unforeseen changes in fuel costs by either increasing the *FCR* to offset increased fuel costs, or by decreasing it should fuel costs decline. The impetus for the *FCR* sprang from the days of the Arab Oil Embargo of 1973-1974, during which the delays involved in regulatory approval of rate increases severely hurt many utilities.

Another factor to be considered is seasonal variation in utility load. For example, in some areas of the United States, the utility experiences its peak annual load during the winter, while in other areas, the peak occurs during the summer. In the southeast, many utilities experience their peak during the summer, due to air conditioning systems, while their winter peaks are lower, due to the widespread use of gas for heating. In contrast, many parts of the northeast experience winter peaks, due to the more widespread use of electric heat. Most utilities discourage energy use during their peak period by charging more for each kWh than during seasons when system demand is less. Figure 7.5 shows the profile for a typical day in both February and August for a southeastern utility, Georgia

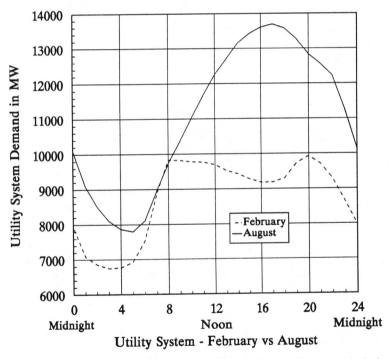

Figure 7.5 Demand profile for Georgia Power System for a typical day in February and in August. (Data courtesy of Georgia Power Company.)

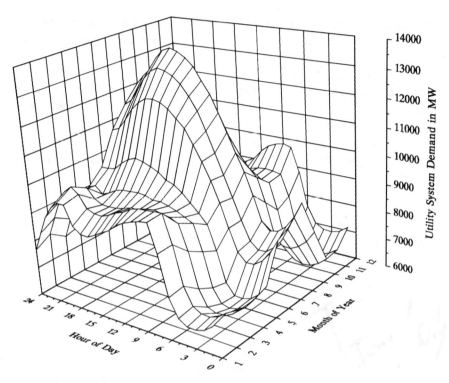

Figure 7.6 Three dimensional annual load profile for Georgia Power System. (Data courtesy of Georgia Power Company.)

Power Company. Note that the summer peak is considerably higher than the winter peak, and occurs during the afternoon, when air conditioning requirements are at a maximum, making this utility a summer peaking system.

If we represent the annual load pattern for this utility in three dimensions with the X and Y axis in a horizontal plane representing the months of the year and the hours of the day respectively, and the vertical Z axis representing total system demand in MW, we see a three dimensional shape as shown in Figure 7.6. Note that the total demand is at its lowest during the early hours of the day, and again late at night, during the early months of the year—January, February and March. The same pattern occurs in the fall during October, November and December. The high ridge shown during the mid-day time period during August reveals that the total system load peaks during this period at about 14,000

MW. The system must therefore be capable of supplying this peak level of power despite the fact that the average demand is considerably less.

Finally, we must understand that the actual utility system load at any instant is actually a composite of all the utility customers' requirements at that point in time. Not all industrial or commercial customers have the same load profile. For example, schools may reach a peak load around noon, while grocery stores might reach their peak during the middle of the afternoon. Hospitals might reach their peak during the morning hours, while churches might reach their peak during the early evening. To complicate matters even further, each of these customer classes might have a somewhat different load profile in summer months than in winter months. Figures 7.7 through 7.10 show typical normalized load profiles for several classes of utility customers during a typical day in both February and in August. Take a few moments to compare these and see if you can explain why each customer class might have the load profile shown.

In Figure 7.7 we see that, during February, church loads typically begin to increase during the morning and decrease during the afternoon. Around 6:00 P.M. the load again increases, reaching a peak around 7:00 P.M. during evening services. During the summer, the pattern is much the same, except the early afternoon load does not drop off as sharply, perhaps due to air conditioning. Figure 7.7 also shows load profile data for colleges. Note the February peak around 10:00 A.M., with declining load during the remainder of the day. During August, the peak occurs from late morning until mid-afternoon, a broader, longer peak.

Figure 7.8 shows data for grocery store type customers. Note the relatively flat demand during the night followed by a rapid buildup during morning shopping hours, with a fairly broad time of near maximum demand during the afternoon and early evening. During August, the period of near maximum demand continues further into the evening, again perhaps due to air conditioning requirements. Hospitals, as shown in Figure 7.8, have a demand profile characterized by rapid buildup of load early in the morning, and a decrease beginning before noon in February, but delayed until late afternoon during August.

Figure 7.9 shows load profile data for hotels. During February, note the rapid load buildup during early morning checkout hours, and the steady decline until late in the afternoon when customer registration and checkin begins. The August profile reveals that a midday load decrease does not occur, probably again due to air conditioning. Finally, Figure 7.9 also contains load profile data for office buildings. During February, note the

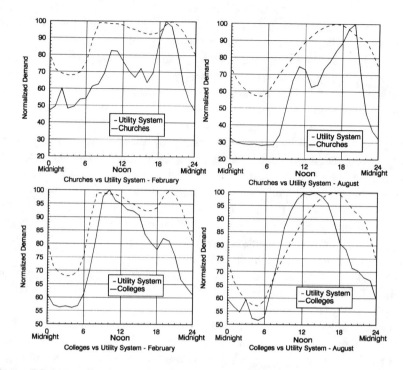

Figure 7.7 Normalized demand profile for churches and colleges for a typical day in February and in August. (Data courtesy of Georgia Power Company.)

relatively modest, flat demand during the nighttime hours, followed by a rapid increase beginning about 7:00 A.M. as workers arrive. By 6:00 P.M. the loads are decreasing toward the nighttime levels, as workers leave for home. The profile for summer months is remarkably similar except for the longer time of near peak demand, occurring during the afternoon hours of heavy air conditioning use.

Figure 7.10 shows load profile data for restaurants and schools. Note that restaurant load begins to increase during the morning, during preparation for noon and evening customer activity. A relatively significant portion of the load, 70%-80%, continues well into the evening hours. Schools are unique in their very rapid load buildup during early morning hours followed by an equally rapid decrease during the mid afternoon as they close for the day.

Again, the utility's overall load as shown in Figure 7.5 is a composite of its thousands of individual customers. These customers may be grouped

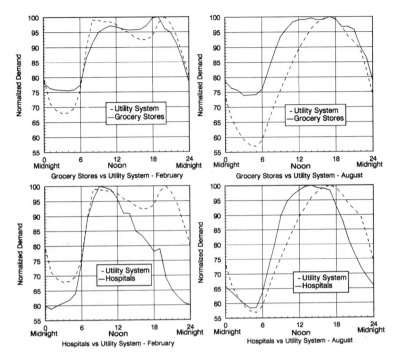

Figure 7.8 Normalized demand profile for grocery stores and hospitals for a typical day in February and in August. (Data courtesy of Georgia Power Company.)

into several classes, a few of which we have reviewed. The utility system must be capable of responding rapidly to these changing load conditions, while at the same time maintaining adequate reserve capacity to deal with unforeseen generating plant outages, or for normal maintenance shutdowns. While providing the required level of generation, the utility also continuously monitors the use of its various plants, so as to optimize the mix of generation between coal, oil, nuclear and hydro in order to achieve the lowest cost of total generation. As you may imagine this entire process is quite complicated. It is carried out 24 hours a day, 365 days a year, and historically, with great reliability.

Let us now turn our attention to the all important financial aspects of utility operation—the customer billing system.

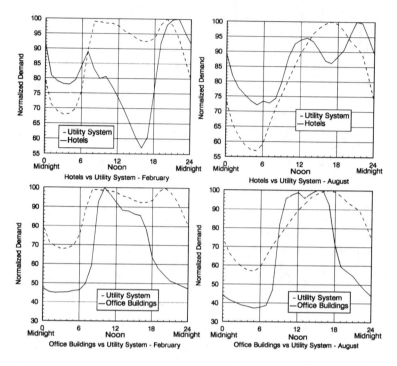

Figure 7.9 Normalized demand profile for hotels and office buildings for a typical day in February and in August. (Data courtesy of Georgia Power Company.)

UTILITY BILLING

As discussed in earlier sections, utilities have a number of factors which must be addressed as part of the billing process. Electrical utility bills have historically been based on the number of kilowatt-hours of energy consumed. This is natural because this is a measure both of the useful work performed by electric power in the customer's facilities, and a measure of the generation commitment made by the utility in providing the power. kWh is also a measure of the utility fuel requirements and is, therefore, the basis for any fuel cost recovery factor, *FCR*, applied by the utility.

In addition to the kWh consumption, the utility is also concerned with the customer's peak kW demand. You will recall from the discussion earlier in this chapter that utilities must provide generation, transmission and distribution systems capable of supplying this peak load. The full

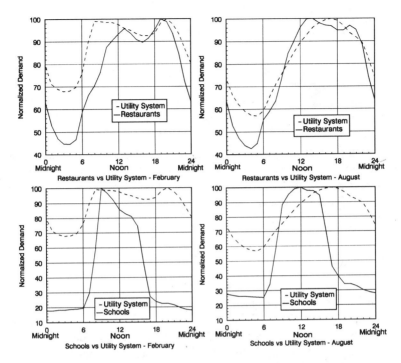

Figure 7.10 Normalized demand for restaurants and schools for a typical day in February and in August. (Data courtesy of Georgia Power Company.)

capacity of this equipment is frequently utilized only a few hours per day, as we have seen in the profile of various load classes.

Another utility concern relates to the fact that the system peak load occurs either in summer or winter, depending primarily on the extent of the area's electric versus gas heating. Many southern utilities are summer-peaking, while many northern utilities are winter-peaking. Utilities would naturally like to encourage power usage during periods when their system is more lightly loaded, and discourage use during periods of heavy system loads. As a result, a customer's peak demand during the system's heaviest load season is a significant factor in the calculation of the customer's bill.

Most utilities are also concerned with the overall power factor of the customer's load. The higher the power factor, in other words, the closer to unity, the higher the utilities' efficiency. This is due to factors such as lower transmission and distribution losses.

Let us summarize the factors which must be considered by utilities in the determination of a customer's bill. Some of the most important are:

1. The customer's kWh consumption
2. The customer's peak kW demand
3. The time of year of occurrence of the customer's peak demand
4. The customer's power factor

The method of incorporating these factors into a customer's bill varies from utility to utility. We will review the approach taken by a representative utility, Georgia Power Company. Remember that all utility rates are subject to review and approval by the governing public service commission.

Rate Schedules

Georgia Power Company currently has available at least 19 different rate schedules, each tailored to the specific needs of different customer classes. In addition, it has several additional rates, often called riders, for charges such as fuel cost recovery. We will concern ourselves here with only the most common rate schedules, for residential and commercial/industrial customers.

Residential Rates

This particular utility employs only one basic residential rate, which it refers to as its R rate schedule. The R is followed by a number indicating its current revision. For example, the current residential rate schedule is called R-12. Figure 7.11 shows details of this rate schedule. Note that the full text of the schedule contains other contract options and requirements which are not included here. This rate schedule contains a minimum charge that covers the basic cost of providing and maintaining the equipment necessary to serve the customer. The next portion of the schedule covers the actual cost of service based on the number of kWh consumed. There are actually two categories for this part of the schedule, depending on whether the bill represents a summer or winter month. Note the differences between the two. Note that the cost of each increment, or block of energy usage, changes. The total bill is subject to an additional charge for fuel cost adjustment. The current fuel cost adjustment rider is entitled FCR-9, and this schedule is shown in Figure 7.12. Finally, the total bill is subject to any applicable taxes.

Let us look at a representative residential bill under the R-12 rate schedule.

Example 7.1: A Georgia Power residential customer consumes a total of 2400 kWh during the month of July. If the prevailing tax rate is 5%, and the bill is subject to the fuel cost recovery factor rider R-9, what is the customer's total bill?

Based on the rate schedule R-12 and fuel cost recovery rate R-9:

Base Charge	$ 7.50
First 650 kWh @ $0.04551 per kWh	$ 29.58
Next 350 kWh @ $0.07566 per kWh	$ 26.48
Next 1400 kWh @ $0.07789 per kWh	$109.05
Total	$172.61
FCR - 2400 kWh @ $0.016896 per kWh	$ 40.55
Total	$213.16
Taxes-5% of Total Bill	$ 10.66
Grand Total	$223.82

Compare the cost of the same total energy usage during a winter month to compare differences between summer bills and winter bills for equal kWh usage.

Commercial and Industrial Rates

Some utilities have billing classifications for commercial and industrial customers, based on their peak demands. Georgia Power Company is such a utility, and currently provides several such categories. Its PLS rate schedule applies to customers whose peak demand is 30 kW or less. The PLM and PLL schedules apply to customers with demands of between 30

GEORGIA POWER COMPANY

Residential Service

Schedule "R-12"

AVAILABILITY:

   Throughout the Company's service area from existing lines of
   adequate capacity.

APPLICABILITY:

   For all domestic uses of a Residential Customer in a
   separately or commonly metered dwelling unit. A Residential
   Customer hereunder is defined in Rules and Regulations for
   Electric Service.

TYPE OF SERVICE:

   Single or three-phase, 60 hertz, at a standard voltage.

MONTHLY RATE:

   Winter - October through May
   Base Charge...........................$7.50
   First 650 kWh......................@.....4.551¢ per kWh
   Next 350 kWh.......................@.....3.907¢ per kWh
   Over 1000 kWh......................@.....3.845¢ per kWh

   Summer - June through September
   Base Charge...........................$7.50
   First 650 kWh......................@.....4.551¢ per kWh
   Next 350 kWh.......................@.....7.566¢ per kWh
   Over 1000 kWh......................@.....7.789¢ per kWh

FUEL COST RECOVERY:

   The amount calculated at the above rate will be increased
   under the provisions of the Company's effective Fuel Cost
   Recovery Schedule, including any applicable adjustments.

Minimum Monthly Bill: $7.50

Figure 7.11 Residential rate schedule R-12 for Georgia Power Company. (Data courtesy of Georgia Power Company.)

kW and 500 kW, and 500 kW or greater, respectively. The same fuel cost recovery schedule applies to all utility customers.

Figures 7.13, 7.14 and 7.15 show the requirements of these three rate structures. There are several important distinctions in these schedules. First, there is no difference in the rate structure between summer and

GEORGIA POWER COMPANY
Fuel Cost Recovery
Schedule "FCR-9"

APPLICABILITY:

This schedule is applicable to and becomes a part of each
retail rate schedule in which reference is made to the Fuel
Cost Recovery Schedule.

RATE:

All bills rendered subject to the Fuel Cost Recovery
Schedule shall be respectively increased in an amount equal
to 1.6896¢ per kWh.

Fuel Costs shall be the cost of:

(1) fossil and nuclear fuel consumed in the Company's own
     plants, and the Company's share of fossil and nuclear
     fuel consumed in jointly owned or leased plants.

(2) the identifiable fossil and nuclear fuel costs
     associated with energy purchased for reasons other than
     identified in (3) below.

(3) the net energy cost of energy purchases, exclusive of
     capacity or demand charges (irrespective of the
     designation assigned to such transaction) when such
     energy is purchased on an economic dispatch basis.
     Included therein may be such costs as the charges for
     energy purchases and the charges as a result of
     scheduled outages, all such kinds of energy being
     purchased by the buyer to substitute for its own higher
     cost of energy; and less

(4) the cost of fossil and nuclear fuel recovered through
     intersystem sales when sold on an economic dispatch
     basis.

Service hereunder subject to Rules and Regulations for
Electric Service on file with the Georgia Public Service
Commission.

Figure 7.12 Fuel cost recovery rate R-9 for Georgia Power Company. (Data courtesy of Georgia Power Company.)

winter. Next, note that the billing kWh increments are determined by the number of hours times the billing demand. The billing demand itself should not be confused with the actual peak demand during the billing month, as the two may or may not be the same. The billing demand is generally determined by selecting the higher of either the current month,

GEORGIA POWER COMPANY
Determination of Billing Demand for
PLS, PLM & PLL Rate Schedules

The Billing Demand shall be based on the highest 30-minute kW measurements during the current month and the preceding eleven (11) months. For the billing months of June through September, the Billing Demand shall be the greatest of (1) the current actual demand or (2) ninety-five percent (95%) of the highest actual demand occurring in any previous applicable summer month or (3) sixty percent (60%) of the highest actual demand occurring in any previous applicable winter month (October through May). For the billing months of October through May, the Billing Demand shall be the greatest of (1) ninety-five percent (95%) of the highest summer month (June through September) or (2) sixty percent (60%) of the highest winter month (including the current month). In no case shall the Billing Demand be less than the greatest of (1) the contract minimum, (2) fifty percent (50%) of the total contract capacity, or (3) 5 kW (PLS), 30 kW (PLM) and 500 kW (PLL).

Figure 7.16 Method of determining billing demand for small, medium, and large commercial/industrial customers - Georgia Power Company. (Data courtesy of Georgia Power Company.)

an agreed contract minimum, or a stipulated fraction of the higher of winter or summer months during a specified period. In the case of Georgia Power Company rates PLS, PLM and PLL, this fraction is 95% for summer months and 60% for the winter months. This method considers the current month as well as the previous 11 months. Figure 7.16 shows the actual statement of the method of determining billing demand. This requirement is applicable to all three rate schedules, along with minimum contract requirements for each classification.

Example Rate Problem

Perhaps the easiest way to understand these somewhat involved schedules is by means of an example.

Example 7.2: The Aardvark Manufacturing Company is a Georgia Power Company industrial customer. The company has consumed 260,000 kWh during the current month of May. During this time, its peak utility demand was 670 kW. The utility bill is subject to both FCR rates and a local tax of 4%. The maximum utility demand over the preceding 11 months were as follows:

Month	Peak Utility Demand	
June	710 kW	Summer Month
July	750 kW	Summer Month
August	760 kW	Summer Month
September	700 kW	Summer Month
October	675 kW	Winter Month
November	655 kW	Winter Month
December	590 kW	Winter Month
January	580 kW	Winter Month
February	595 kW	Winter Month
March	590 kW	Winter Month
April	610 kW	Winter Month
May	670 kW	Current Winter Month

(a) After reviewing the criteria for determining billing demand, what is the appropriate billing demand for the current month?
(b) What is the appropriate utility rate schedule for this customer?
(c) What is the total utility bill for the current month?

(a) Billing Demand: Since the current billing period is May, we refer to the part of the determination method shown in Figure 7.16. The billing demand will be the highest of:

    1) 95% of the highest summer month's demand

$$.95 \cdot 760 = 722 \text{ kW}$$

    2) 60% of highest winter month's demand
       (Including current month) $.60 \cdot 675 = 405 \text{ kW}$

(b) Rate Schedule: The appropriate rate schedule is PLL. (Actually the applicable rate schedule is established at the time the customer first requests service). The appropriate billing demand is 722 kW.

(c) Monthly Bill:

1) **Base Charge**	$ 16.00
2) **1st 200 Hr. times Billing Demand**	
$200 \cdot 722 = 144{,}400$ kWh	
First 3,000 kWh @ $0.10570 per kWh	$ 317.10
Next 7,000 kWh @ $0.09630 per kWh	$ 674.10
134,400 kWh @ $0.08200 per kWh	$1,020.80
3) **All consumption in Excess of 200 hours and not greater than 400 hours times billing demand**	
$200 \cdot 722 = 144{,}400$ kWh - but we have only	
$260{,}000 - 144{,}400 = 115{,}600$ kWh left, so	
115,600 kWh $\cdot$ $0.01093 per kWh	$1,263.51
4) **Fuel Recovery Charge**	
260,000 kWh @ $0.016896 per kWh	$4,392.96
Total	$7,684.47
Taxes @ 4%	$ 707.38
Grand Total	$8,391.85

THE UTILITY AND THE BUILDING DESIGN ENGINEER

During the design phase of a building, the design engineer and the utility must interact on several important matters.

GEORGIA POWER COMPANY
Power and Light
Small
Schedule "PLS-1"

AVAILABILITY:

Throughout the Company's service area from existing lines of
adequate capacity.

APPLICABILITY:

To all electric service of one standard voltage required on
the customer's premises, delivered at one point and metered
at or compensated to that voltage for any customer with a
demand, as determined under the Special Applicability
Provisions, of less than 30 kW.

TYPE OF SERVICE:

Single or three-phase, 60 hertz, at a standard voltage.

MONTHLY RATE - Energy Charge Including Demand Charge:

Base Charge (includes the first 25 kWh or less) $16.00

All consumption (kWh) not greater than 200
200 hours times the billing demand:

First 25 kWh.................included in the Base Charge
Next 2,975 kWh.....................@......10.430¢ per kWh
Next 7,000 kWh.....................@...... 9.530¢ per kWh
Over 10,000 kWh....................@...... 8.150¢ per kWh

All consumption (kWh) in excess of
200 hours and not greater than 400 hours times
the billing demand.................@...... 1.050¢ per kWh

All consumption (kWh) in excess of
400 hours and not greater than 600 hours times
the billing demand.................@...... 0.750¢ per kWh

All consumption (kWh) in excess of
600 hours times the billing demand @...... 0.650¢ per kWh

Figure 7.13 Small load commercial/industrial rate schedule PLS-1 for Georgia
Power Company. (Data courtesy of Georgia Power Company.)

1. The engineer must establish contact with the utility very early in the
   design process, in order to reach agreement on:
   a) The available service voltages
   b) The available service method (for example, overhead service or
      service from a pad-mounted transformer)

```
                    GEORGIA POWER COMPANY
                      Power and Light
                          Medium
                    Schedule "PLM-1"
```

AVAILABILITY:

   Throughout the Company's service area from existing lines of
   adequate capacity.

APPLICABILITY:

   To all electric service of one standard voltage required on
   the customer's premises, delivered at one point and metered
   at or compensated to that voltage for any customer with a
   demand, as determined under the Special Applicability
   Provisions, of not less than 30 kW but less than 500 kW.

TYPE OF SERVICE:

   Single or three-phase, 60 hertz, at a standard voltage.

MONTHLY RATE - Energy Charge Including Demand Charge

   Base Charge...16.00

   All consumption (kWh) not greater than
   200 hours times the billing demand:

   First   3,000 kWh....................@........10.450¢ per kWh
   Next   7,000 kWh....................@........ 9.550¢ per kWh
   Next   190,000 kWh....................@........ 8.200¢ per kWh
   Over   200,000 kWh....................@........ 6.300¢ per kWh

   All consumption (kWh) in excess of
   200 hours and not greater than
   400 hours times the billing demand..@....... 1.0650¢ per kWh

   All consumption (kWh) in excess of
   400 hours and not greater than
   600 hours times the billing demand..@........ 0.800¢ per kWh

   All consumption (kWh) in excess of
   600 hours times the billing demand..@........ 0.700¢ per kWh
```

**Figure 7.14** Medium load commercial/industrial rate schedule PLM-1 for Georgia Power Company. (Data courtesy of Georgia Power Company.)

   c) A mutually acceptable location for the service entrance to the building, including utility pole or pad-mounted transformer location

   d) Method of utility metering (meter on building, on transformer or in building switchgear)
```

GEORGIA POWER COMPANY
Power and Light
Large
Schedule "PLL-1"

AVAILABILITY:

Throughout the Company's service area from existing lines of adequate capacity.

APPLICABILITY:

To all electric service of one standard voltage required on the customer's premises, delivered at one point and metered at or compensated to that voltage for any customer with a demand, as determined under the Special Applicability Provisions, of 500 kW or greater.

TYPE OF SERVICE:

Single or three-phase, 60 hertz, at a standard voltage

MONTHLY RATE - Energy Charge Including Demand Charge:

Base Charge...16.00

All consumption (kWh) not greater than
200 hours times the billing demand

First   3,000 kWh.......................@......10.570¢ per kWh
Next   7,000 kWh.......................@...... 9.630¢ per kWh
Next 190,000 kWh.......................@...... 8.200¢ per kWh
Over 200,000 kWh.......................@...... 6.380¢ per kWh

All consumption (kWh) in excess of
200 hours and not greater than 400
hours times the billing demand........@...... 1.093¢ per kWh

All consumption (kWh) in excess of
400 hours and not greater than 600
hours times the billing demand........@...... 0.865¢ per kWh

All consumption (kWh) in excess of
600 hours times the billing demand....@...... 0.755¢ per kWh

Figure 7.15 Large load commercial/industrial rate schedule PLL-1 for Georgia Power Company. (Data courtesy of Georgia Power Company.)

2. As the design process continues, the nature of the building's various loads will become clearer, and the engineer must relay this data to the utility. The utility will in turn use this information to make decisions on the appropriate equipment size and type to serve the building. One of the most crucial items is the selection by the utility of the correct transformer kVA rating. The utility may have the transformer in stock, or may have to place a special order. Ordering a transformer can take

from several weeks to several months, depending on the manufacturer's delivery schedule. The utility may also find that new distribution lines, or the rerouting of existing lines, is necessary. Such changes require considerable time as well as careful coordination with other customers on the same part of the system.

3. During the latter phases of the design process, the engineer will have more completely defined the building's distribution system. This will provide information on panelboard and switchgear locations, step-down transformer locations and sizes, and feeder lengths and conductor sizes. Also, the utility will have determined the kVA rating and other characteristics of its service transformer. At this point, the utility should be formally requested to provide information on the utility's available fault current level at the service location. The response may be in the form of the utility transformer's percent impedance, a computed available fault current level, or parameters of a Thevenin equivalent of the utility system at the point of service. In any event, the engineer will need this data in order to perform the very important fault current study. Without this analysis, it is not possible to assure the proper building equipment interrupting ratings.

From the above discussion, it should be clear that prompt and effective communication between the utility and the design engineer is vital to the success of the design effort. Unfortunately, the failure of this interface is far too often the source of conflict. For example, imagine the position of an engineer who must explain to a building owner that the delay in the opening of a project is due to ineffective communication that left the utility insufficient time to provide the necessary service equipment. Another all too common assumption on the part of some engineers is that any desired service voltage or service method is readily available. This can lead to very expensive redesign, or even delay in completion of the project should these assumptions turn out to be false.

In summary, the engineer must contact the serving utility early in the design. Most utilities have very efficient engineering groups who will work with the engineer to assure that all necessary information required by both parties is available. Plainly put, early contact benefits everyone, except the occasional attorney who may have to straighten things out if adequate coordination does not occur.

References

Cassidy, Frank and Gerald W. Schirra, 1977. Treatment of Inflation in Engineering Economic Analysis. *IEEE Transactions on Power Apparatus and Systems*. May/June: pp. 1027-1035.

Cox, James A. 1979. *A Century of Light*. New York: Larousse & Company, Inc.

Georgia Power Company. 1988. *Vogtle Fact Sheet*. Internal Publication of Georgia Power Company.

Hyman, Leonard S. 1985. *America's Electric Utilities: Past, Present and Future*. Arlington, Virginia: Public Utilities Reports, Inc.

Long-Range System Planning Working Group, Power Engineering Society. 1980. The Significance of Assumptions Implied in Long-Range Electric Utility Planning Studies. *IEEE Transactions on Power Apparatus and Systems*. May/June: pp. 1047-1056.

Matthews, John H., P.E. 1989. *The Application of Interval Mathematics to Utility Revenue Requirement Theory*. Dissertation. Tennessee Technological University, Cookeville, Tennessee.

Ott, Karl O. and Bernard I. Spinrad. 1985. *Nuclear Energy - A Sensible Alternative*. New York: Plenum Press.

Schap, David. 1986. *Municipal Ownership in the Electric Utility Industry*. New York: Praeger Special Studies.

Stoler, Peter. 1985. *Decline and Fail: The Ailing Nuclear Power Industry*. New York: Dodd, Mead & Company.

Vennard, Edwin. 1979. *Management of the Electric Energy Business*. New York: McGraw-Hill Book Company.

● 50 KVA Xformer 4160/208 volts connected Y-Y with
6% Z And $X/R = 5$ calc. Resistance For both primary And sec
And 3ϕ copper losses when full load current flows

Pri. Y

$I_p = \dfrac{50KVA /3}{4160/\sqrt{3}} = 6.94A$

$100\% \ Z = \dfrac{4160/\sqrt{3} = 2400v}{6.94A} \dfrac{}{6.94A} = 345.6\,\Omega$

$Z_{spi} = 3\% \cdot (345.6\,\Omega) = 10.36\,\Omega$

3% because of 2 transformers $6\% = 3\%$

$Z = \sqrt{R^2 + x^2}$: $\dfrac{x}{R} = 5$

$10.36 \ \sqrt{R^2 + (5R)^2}$: $10.36^2 = 26R^2$

$R = \sqrt{\dfrac{(10.36)^2}{26}} = \boxed{2.03}$

Sec Y

$I_s = \dfrac{500kv/3}{120v} = 138.89$

$100\% \ Z = \dfrac{120v}{138.89} = 0.864$

$Z_{s \ sec} = 3\% \cdot (0.864) = 0.02512$

$R = \sqrt{\dfrac{0.02512^2}{26}} = \boxed{0.00508}$

Load Loss

Pri x 3 because Y

$I^2R = 6.9A^2 (2.03) \times 3 = 289.9\,w$

Sec

$I^2R = (138.89)^2 (0.00508) \times 3 = 293.9\,w$

$\boxed{P_{tot} = 583.85\,w}$

3ϕ 500 KVA xformer rated @ 12,500 - 480 connected Δ-Y And
has 5.75% Z per phase calc per-phase Z

Pri. Δ

$I_p = \dfrac{500KVA/3}{12500/\sqrt{3}} = 23.09A$

$100\% \ Z = \dfrac{12500/\sqrt{3}}{23.09} = 312.5\,\Omega$

$Z_{sp} = 0.0575 (312.5) = 17.97$

$Z_{s \Delta} = 17.97\,\Omega$ ←

$Z_s \ Y = 17.97/3 = 5.99\,\Omega$

Secondary Y

use secondary
ΔR Y conversion
and use a^2 to
get Δ side.

8

Power Quality

During recent years the quality of power itself has become an issue of great concern. In order to effectively design building electrical systems the engineer must understand the origins of this concern, and also the nature of the various types of power quality related problems. When power quality is first mentioned, many people are somewhat puzzled and may remark "What's the problem? The utility generates power and delivers it to us. Where's the power quality problem?" In actuality, the problems are very real and are increasing in severity.

In broad terms, we can break power quality down into two subcategories. The first is the *availability of power*. Our utilities have done such an excellent job over the years of providing high quality, reliable power that many of us simply take power reliability for granted. As one utility company executive said, "People are not surprised when their car doesn't start, but they assume that the lights at home will come on whenever they throw the switch." There's a lot of truth in this comment. In Chapter 7 we discussed the effects of the last twenty years on the utility industry. It is a sad but true fact that utilities are now operating in an environment that is far less conducive to maintaining adequate power reserves than in the past. Construction of new generation facilities has been hampered by environmental concerns, cost overruns and regulatory constraints. This simply means that our existing utility systems are operating closer and closer to their maximum capacity, leaving less spare capacity for unexpected outages. Our existing power generation facilities are also aging, and they will require careful maintenance to avoid reliability problems. Many parts of the United States already feel the effects of limited generation capacity, and the situation can become much worse during the present decade.

The second area of concern deals with the *characteristics* of the power we receive. Our increased sensitivity to events such as normal utility line switching, lightning and harmonics stems from basic changes in building equipment and how we operate. During the past two decades, we have seen an enormous increase in the use of computers and computer controlled manufacturing, as well as in power electronic applications such as the variable speed control of motors. As systems like these have increased, we have seen a reduction in our overall tolerance for power outages or transients, such as those discussed above. This is because of the sensitivity of solid state devices to power interruptions, transients or distortions of the voltage or current waveform. In years past, equipment was not nearly as sensitive to such problems, and brief service interruptions were not a severe problem. Today, many companies are almost totally dependent on power that is very reliable. For example, companies that process credit card charge slips, arrange airline reservations or operate processes such as plastic molding, are extremely sensitive to any power disruption, no matter how brief. As a final point, building loads themselves are now a major source of potential disruption, due to the nonsinusoidal current drawn by much of the newer solid state equipment. This equipment often generates significant levels of harmonic distortion which can affect nearby equipment, and even be transmitted back into the utility system where other customers can be affected.

In summary, we have long enjoyed reliable power and the minor disruptions which did occur were not a severe problem. Today, the reliability of power is becoming an increasing concern. Finally, the source of many power quality related problems is within the building itself.

In order to put these concerns into perspective, we will begin by reviewing some of the more important operating characteristics of the utility system.

THE UTILITY AND POWER QUALITY

Consider for a moment the "product" of the utility company—clean, undistorted and reliable power. Over the years, they have delivered this "product" with tremendous reliability. In fact, most utilities have well documented reliability records of 99.8% or better. Even with this excellent record, events can occur which affect utility system operations. Examples of such events are trees striking overhead lines, construction workers accidentally digging into underground utility cables, or cars colliding with utility poles. In any of these events, protective devices within the utility system automatically remove power from the affected circuit in order to minimize disruption, and to protect both equipment and human lives. As a result of the action of these protective devices, however, facilities on the same circuit are often temporarily without power.

Let's take a few moments to review how a typical segment of a utility system operates. Figure 8.1 shows a utility substation served directly from a transmission system. At the substation, the transmission voltage, often 46 to 500 kV, is transformed to a lower voltage for overhead or underground distribution, often at 4 to 35 kV. Typically, the substation transformer might serve several distribution circuits, each with its own protective circuit breaker. The circuit breaker, unlike the normal molded case or power circuit breaker we have discussed in previous chapters, is capable of automatically reclosing after being tripped. This is very important because in many cases, automatic reclosing helps quickly restore service and minimizes the length of the outage. If the short circuit, or overload, was caused by a tree limb momentarily contacting a overhead line, it may well have been blown or burned clear, thus permitting automatic restoration of utility service. This reclosing cycle is often programmed to repeat itself several times (three is a common number) before locking the breaker in the open position. At this point, the action of operating personnel is required to restore service.

Moving further from the substation, we see various pole-mounted reclosers which function in a similar manner to the substation breaker recloser just discussed. These pole-mounted reclosers are intended to isolate smaller segments of the distribution circuit and, as such, are of lower current capacity than the substation breaker. This recloser also goes through up to three cycles before locking in the open position.

The third protective device shown in Figure 8.1 is the fuse, and might be installed to protect an even smaller segment of the utility circuit, perhaps an individual transformer serving a specific customer. The pole-mounted recloser is intentionally set so that its first tripping cycle will occur prior to the downstream fuse melting point. Thus, the recloser affords an opportunity for the fault or overload to clear. After two such cycles, the recloser is set to delay long enough for the fuse to melt, thus isolating the faulted segment of the circuit, but also allowing the recloser

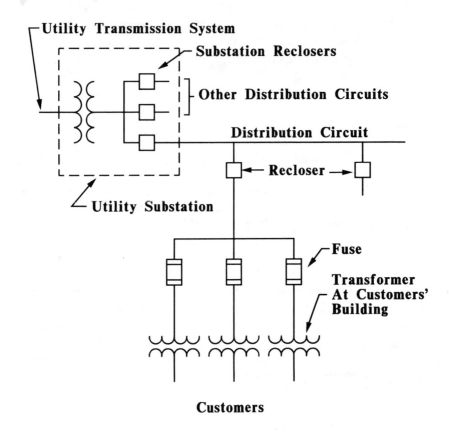

Figure 8.1 Segment of a typical utility system.

to successfully close the third time, and restore service to the unaffected portion of the circuit.

It should be clear that this recloser/fuse combination effectively reduces the length of time of many outages, but subjects other customers on the circuit to a higher number of shorter outages. At the present time, there are few devices under development which promise to offer substantial improvements in the operational sequence just discussed.

It should be noted that utilities engage in very aggressive programs to lessen the likelihood of service interruptions. For example, periodic tree trimming programs lessen the likelihood of outages due to tree contact. The proper use of lightning arresters and effective grounding practices lessens the likelihood of damage due to lightning strikes, but not the danger of momentary outages. Early generations of direct burial cable have significantly shorter service life than originally anticipated. Replacement programs are underway, using newer cable that is expected to offer much improved service life, but such programs will require some length of time for implementation.

Animal contact with overhead lines or substation components is a more serious problem than one might expect. For example, squirrels often come into contact with energized components of pole-mounted transformers. This is generally very stressful for the utility system components, as well as the squirrel. The danger of automobiles colliding with utility poles is well recognized, and programs are underway by many utilities to relocate poles in high-risk locations to better protected areas, or at least further from the roadway. In areas of heavy construction activity, underground cables are frequently damaged by earth-moving equipment. Some utilities have initiated programs to encase in concrete the cable in such high threat areas. Additionally, many utilities make underground cable location and marking services readily available to contractors and others planning excavation activities.

In summary, the utilities are making substantial efforts to reduce the frequency and severity of outages, but it is equally clear that such interruptions are a fact of life, and must be anticipated and appropriate precautions planned. Unfortunately, adequate power reserves may pose a longer-range problem due to the very long lead time required for the construction of generation facilities.

TYPES OF DISTURBANCES

It should be clear from the above discussion that the building electrical system may be subjected to a variety of power disturbances, ranging from complete power outages, low or high voltage conditions and transients, to excessive harmonic content. It should also be remembered that many of these conditions can occur due to disturbances generated within the building. As mentioned in the introductory section of this chapter, many building loads are becoming increasingly sensitive to even minute power system disturbances. The utilities are very aware of this increased sensitivity, and many are attacking the problem from two directions. First, they are actively seeking methods of reducing utility-generated disturbances, as well as outright outages, by the improved construction and maintenance practices discussed previously. Secondly, they are aware that they are often subject to criticism for some types of problems which are totally beyond their control, such as building generated transients and harmonics. To help counter this potential criticism, many utilities have established very aggressive customer educational programs sponsored by newly formed power quality departments. These groups provide technical information on power quality related matters for the utility's customers, help determine the source of specific customer complaints and, in some cases, even arrange for the lease or purchase of power conditioning equipment. Forward looking utilities realize that power quality related problems will be a very real problem for many years to come.

Figure 8.2 shows the sensitivity of typical building computer loads to disturbances of various durations. The proper operating band is the area between the upper and lower curves. Note that the equipment can withstand high or low voltage conditions for very brief periods of time.

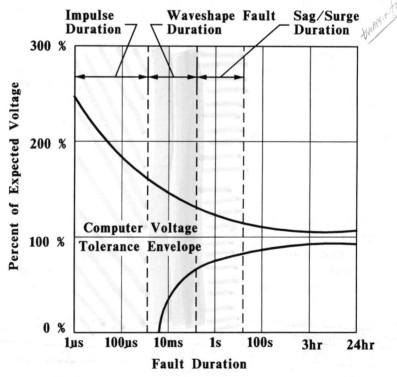

Figure 8.2 Computer equipment voltage tolerance.

However, once the duration of the disturbance exceeds the operating bounds shown, erratic operation, or outright failure, can occur. In Figure 8.2 note that disturbances of very short duration are called *impulses*, while disturbances of slightly longer duration are classified as *wave shape faults*. Low or high voltage conditions which persist for longer periods are called *sags* or *surges*.

↳ #1 problem

Disturbance Measurement

Before reviewing several typical building power disturbances, we will briefly discuss how we measure and record power quality information. Figure 8.3 shows a typical power analyzer that is designed to automatically acquire data on power characteristics, such as voltage, current, power flow, power factor, harmonic spectrum, and Total Harmonic Distortion. Modern equipment can operate unattended, and data can be recorded on printed tape, stored on a magnetic disk for later analysis, or downloaded by an internal modem to a remote location. Such power analyzers are capable of simultaneously monitoring voltage and current in all three phases of three-phase systems. Voltages are recorded using clip-on voltage probes, as shown, and currents are recorded using clamp-on probes which fasten around the conductors.

Many power analyzers are capable of performing Fourier Waveform Analysis on the digitally sampled voltage and current waveforms. This data is then presented as a harmonic spectrum. In the next section, we will discuss harmonics and will review several graphs captured by a power analyzer. Similar equipment is available to monitor disturbances, such as impulses, wave shape faults, surges, sags and outages. These disturbance analyzers are similar in appearance to the power analyzer shown in Figure 8.3. The disturbances we are about to review were recorded using such a device.

Typical Disturbances

As discussed in earlier sections, there are a number of possible sources of building power disturbances which originate within the utility system. Figure 8.4 shows a graph of a typical utility system power outage. Note that the voltage and frequency decay over a period of several cycles. The decay is often exponential in nature. A failure of this type can occur due to a number of causes, and there is very little the building owner or

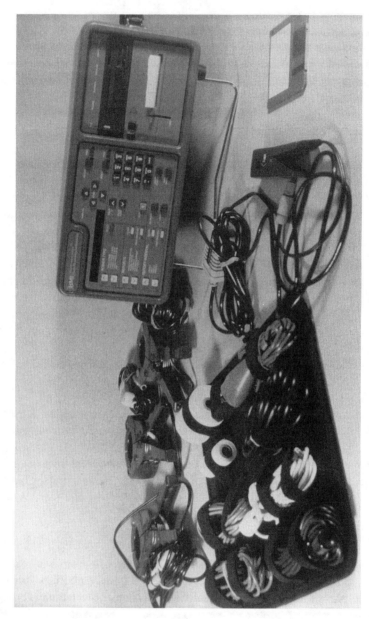

Figure 8.3 Microprocessor-based power system analyzer. (Photo: William C. Kennedy.)

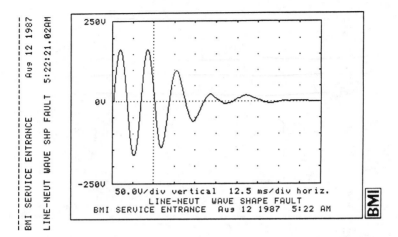

Figure 8.4 Disturbance graph for a typical utility power outage. (Graphic courtesy of Basic Measuring Instruments, Foster City, California.)

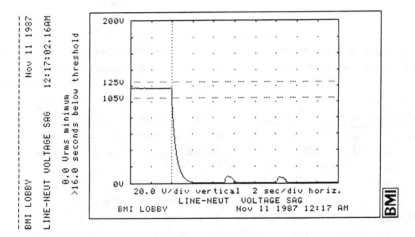

Figure 8.5 Disturbance graph for a typical utility power outage and recloser operation. (Graphic courtesy of Basic Measuring Instruments.)

operator can do to mitigate the effect, except install a battery backup system for critical loads. Note that this graph shows only a few cycles of the voltage wave.

A similar utility outage is shown in Figure 8.5, except that the events extend over several seconds instead of a few cycles. This is due to the fact

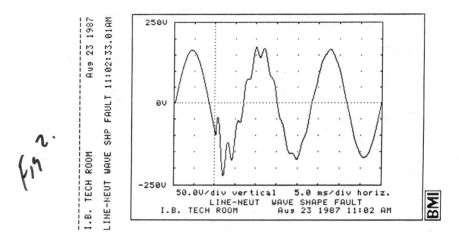

Figure 8.6 Disturbance graph for switching transient produced by a utility system power factor correction capacitor bank. (Graphic courtesy of Basic Measuring Instruments.)

that this utility's reclosers make two attempts to re-energize the system, as evidenced by the slight voltage increases. As discussed earlier, such reclosers may automatically attempt to restore service as many as three times.

Figure 8.6 shows the effects of energizing a bank of utility system power factor correction capacitors. Note that the disturbance decays over a period of a cycle or so. This damped ringing typically occurs at frequencies of 200 Hz to 1 kHz, and always moves, initially, toward the zero axis. In general, the lower the ringing frequency, the further back in the distribution system the capacitor bank is located. This type of disturbance can be very disruptive to sensitive electronic loads. It is interesting to note that capacitors are normally switched on in response to increases in the utilities' inductive load, such as industrial motors coming on line. The capacitor switching can be either manual or automatic, so it should be remembered that this type of transient can occur any time.

Figure 8.7 shows the effects of a recloser on a nearby utility branch. The overload or short circuit, that prompted the recloser to open, results in a decrease in system voltage, visible as the notch in the voltage waveform. When the recloser opens, the voltage returns to its former level. This disturbance will be repeated each time the recloser operates. This type of disturbance may cause erratic operation of electronic loads or unexpected triggering of power failure circuits.

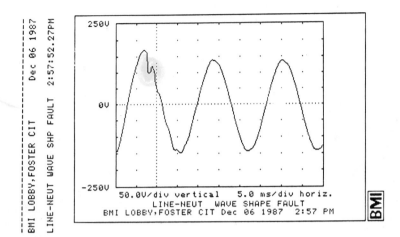

Figure 8.7 Disturbance graph for recloser operation on a nearby utility circuit. (Graphic courtesy of Basic Measuring Instruments.)

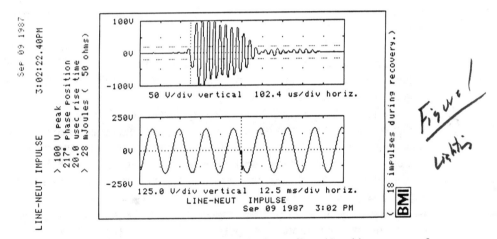

Figure 8.8 Disturbance graph for a typical lightning strike. (Graphic courtesy of Basic Measuring Instruments.)

Figure 8.8 shows the effects of a lightning strike. The surge can be of the same, or opposite, polarity from the voltage sine wave shown. Note the trigger point on the lower graph of the voltage wave. Lightning impulses typically ring at a frequency of 10 to 100 kHz. Several impulses of this kind might be observed during a lightning storm. This type of disturbance

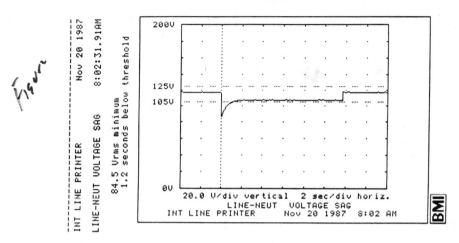

Figure 8.9 Disturbance graph for a typical motor starting transient. (Graphic courtesy of Basic Measuring Instruments.)

can cause damage to power supplies. Data can be disrupted, fuses may open, and ground fault circuits can be triggered. Lightning transients are a simple fact of life, especially in areas of high thunderstorm activity. It is often possible to install surge suppression systems within the building to reduce the magnitude of lightning transients, and thereby provide some degree of protection for sensitive equipment.

Let us now turn our attention to a very common disturbance generated within a building—the starting of a motor. Motors may require four to six times their normal running current during startup. This large current flow can cause a temporary reduction in voltage, as shown in Figure 8.9. Note that the voltage dips below the trigger threshold of 105 volts, and then exponentially rises to a value just below its initial value. Should this motor be on the same circuit as a sensitive load, it is very likely that it will disrupt normal operation. The only solution to the problem is to either move the motor to another circuit, or to increase the size of the conductors serving the circuit. Preferably, a motor should not share the same circuit as a sensitive load. This type of problem occurs frequently in office buildings, where receptacle circuits might serve a number of tenants in different areas of the building, making it difficult to locate the source of the disturbance.

From this brief discussion, it should be clear that power disturbances stem from a number of causes, and can be quite challenging to locate and correct. Understanding the characteristic shape or "signature" of these transient events can be a very important diagnostic tool. Recent studies, such as those by Alexander McEachern, have added to our knowledge of these kinds of events by carefully and systematically recording and documenting a variety of common disturbances. The previous graphs have been gratefully reprinted from the *Handbook of Power Signatures* with permission from Alexander McEachern.

Not all power disturbances are transient in nature. The increased use of power electronic devices has resulted in increased voltage and current waveform distortion. This harmonic distortion can lead to a variety of building power system problems. We will now discuss the basic concepts of harmonic analysis, sources of harmonics and their effect on building systems.

HARMONIC CONSIDERATIONS

Power system harmonics are of great concern today due to the rapid increase in the number of distortion producing loads. These loads are primarily electronic in nature, and their increase is due to continuing advances in the design and application of solid state power conversion devices. In the case of linear loads, the relationship between the sinusoidal voltage and current is a straight line, resulting in load currents which are also sinusoidal in nature. Nonlinear loads, on the other hand, often draw system current, and hence power, during only a portion of the voltage wave. In this case, the current might be nonsinusoidal and even discontinuous in nature.

In years past, most system loads were linear, and harmonic related problems were minimal. In fact, most building power systems and components are designed for operation on sinusoidal voltages and currents. Distorted voltages and currents can cause very serious system problems ranging from overheating in transformers to damage to conductors. The rapid expansion of power electronic applications, ranging from uninterruptible power supply (*UPS*) systems and variable frequency speed control systems for motors to computer power supplies, has increased the percentage of the building's overall load represented by such nonlinear loads.

In this section we will discuss the nature of harmonics, how they are studied and evaluated, their effects on building systems, and recent trends in attempts to develop relevant harmonic standards.

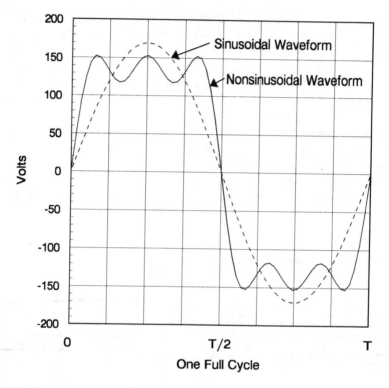

Figure 8.10 Sinusoidal and distorted periodic waveforms.

Harmonic Analysis

Figure 8.10 shows a sinusoidal waveform, as well as a distorted waveform. These waveforms could represent either voltage or current. Conventional circuit analysis techniques do not work in the case of nonsinusoidal waveforms. An effective way to study system response to distorted waveforms is through the use of Fourier waveform analysis. Fourier found that a periodic waveform can be seen as being composed of a number of sinusoidal waveforms of various magnitudes and phase angles. These are the harmonic components of the waveform. The system's response to the waveform can then be evaluated by studying the response of the system to each of the individual harmonic components, using the superposition theorem. The individual harmonic components occur at integer multiples of the basic system frequency. Since the United States' power system operates at 60 Hz, relevant harmonics occur at 120, 180, 240, 300 Hz and higher frequencies. These are called the 2nd, 3rd, 4th, and 5th harmonics, respectively. Reactance Increase with f, => V becomes more non-sinusoidal

Fourier waveform analysis techniques allow us to separate nonsinusoidal voltages and currents into their harmonic components. Fourier showed that a periodic waveform, $i(t)$, can be described by

$$i(t) = \frac{1}{2}a_0 + \sum_{n=1}^{\infty}\left[a_n \cos(n \cdot \omega \cdot t) + b_n \sin(n \cdot \omega \cdot t)\right] \quad (8.1)$$

where a_n and b_n represent the Fourier coefficients given by

$$a_n = \frac{2}{T}\int_0^T i(t)\cos(n \cdot \omega \cdot t)\,dt \quad (8.2)$$

$\cos\theta \;+\; \frac{1}{3}\cos 3\theta + \frac{1}{5}\cos 5\theta + \frac{1}{7}\cos 7\theta$
$-$ approaches a square wave

and

$$b_n = \frac{2}{T}\int_0^T i(t)\sin(n \cdot \omega \cdot t)\,dt. \quad (8.3)$$

Here T represents the period of the waveform, and ω the angular frequency. The magnitude of the individual harmonic components, c_n, is given by

$$c_n = \sqrt{a_n^2 + b_n^2}. \quad (8.4)$$

Equation 8.1 can now be written in terms of the c_n components as

$$i(t) = \frac{1}{2}a_0 + \sum_{n=1}^{\infty} c_n \cos(n \cdot \omega \cdot t - \theta_n) \quad (8.5)$$

where the harmonic phase angle, θ_n, is given by

$$\theta_n = \arctan\left[\frac{b_n}{a_n}\right]. \quad (8.6)$$

The magnitudes of the individual harmonic components, c_n, are generally normalized by dividing each by the magnitude of the fundamental, or first harmonic. The harmonic percentages, h_n, can then be expressed by

$$h_n = \frac{c_n}{c_1} \cdot 100\%. \qquad (8.7)$$

Finally, to evaluate the relative degree of waveform distortion, we must introduce the concept of Total Harmonic Distortion, *THD*. *THD* is normally expressed as a percentage, and is given by

Relative Degree
of waveform
Distortion

$$THD = \frac{\sqrt{\sum_{n=2}^{\infty} h_n^2}}{h_1} \cdot 100\%. \qquad (8.8)$$

Now let's look at an example.

Example 8.1: A periodic, sawtooth waveform is shown in Figure 8.11, and is given by

$$i(t) = 100 \cdot \frac{t}{T} \quad Amps \qquad (8.9)$$

where $f = 60$ Hz and $T = 1/f$.

Using Equations 8.2, 8.3, 8.4, and 8.7, we find that the average value of the waveform

$$\frac{1}{2} a_0 = 50$$

and for harmonics 1 through 9:

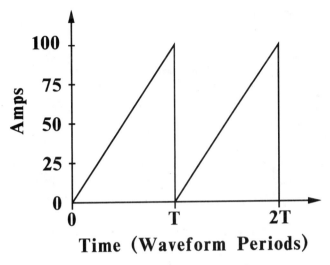

Figure 8.11 Sawtooth current waveforms.

n	a_n	b_n	c_n	$h_n\%$
1	0	-31.83	31.83	100.0%
2	0	-15.92	15.92	50.0%
3	0	-10.61	10.61	33.3%
4	0	- 7.96	7.96	25.0%
5	0	- 6.37	6.37	20.0%
6	0	- 5.31	5.31	16.7%
7	0	- 4.55	4.55	14.3%
8	0	- 3.98	3.98	12.5%
9	0	- 3.54	3.54	11.1%

Using Equation 8.8

$$ THD = \frac{\sqrt{\sum_{n=2}^{9} h_n^2}}{h_1} \cdot 100 = 73.5\%. $$

If we wish to reconstruct the original waveform from its harmonic components, we can use either Equation 8.1 or 8.5. Figure 8.12 shows the

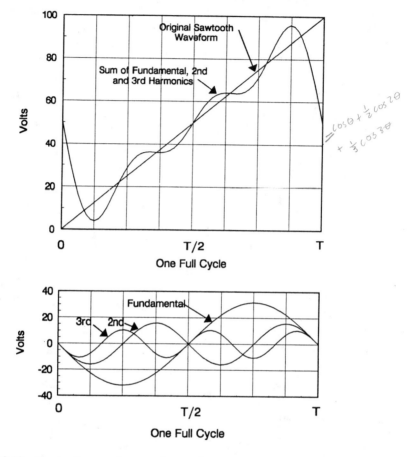

Figure 8.12 Sawtooth waveform and waveform synthesis using fundamental, second and third harmonics.

original sawtooth waveform, the reconstruction using the first three harmonics, as well as the individual harmonics 1, 2, and 3. If we had expanded the waveform synthesis to include the first 9 harmonics, we would have more closely modeled the original waveform.

Unfortunately, we rarely have the luxury of having sufficient information to express distorted waveforms in equation form. This makes practical application of the Fourier waveform analysis techniques difficult. We generally acquire actual waveform data by digital means, resulting in waveform data at discrete intervals. Figure 8.13 shows the sawtooth

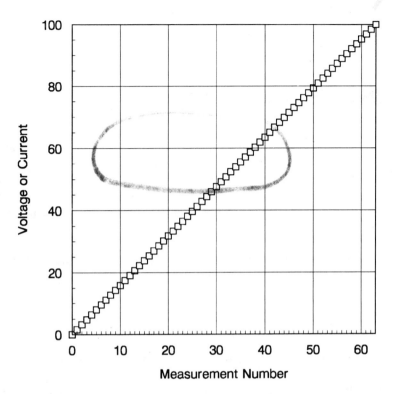

Figure 8.13 Digitally sampled sawtooth waveform for discrete Fourier Transform.

waveform of the previous example in terms of a number of values at incremental intervals.

According to the Fourier Transform for discrete data, the c_n components are given by

$$c_n = \left[\frac{1}{\sqrt{N}}\right] \cdot \sum_{k=0}^{N-1} i_k e^{2\pi j \left(\frac{n}{N}\right) k} \tag{8.10}$$

where N is the total number of data samples, i_k is the value of the waveform at the kth measurement, and j is the imaginary unit.

In the case of the Discrete Fourier Transform, we also normalize the magnitude of the c_n components by dividing each by the magnitude of the fundamental. These harmonic percentages are given by

$$h_n = \frac{|c_n|}{|c_1|} \cdot 100\% \qquad n = 1 \ldots N - 1. \qquad (8.11)$$

The total harmonic distortion of the waveform is given by

$$THD = \frac{\sqrt{\sum_{n=2}^{N-1} h_n^2}}{h_1} \cdot 100\%. \qquad (8.12)$$

Let's apply the Discrete Fourier Transform technique to the previous example and compare the results.

Example 8.2: For the sawtooth wave of Example 8.1, the value of i at N data points beginning at 0 will be given by

$$i_k = 100 \cdot \frac{k}{(N - 1)} \qquad for\ k = 0 \ldots (N - 1). \qquad (8.13)$$

Applying Equation 8.13, with $N = 64$ and $k = 0$ to 63, yields a series of discrete values of i as shown in Figure 8.13.

Applying Equation 8.10 for $n = 0\ldots9$, $k = 0\ldots63$, and $N = 64$ yields the following values of c_n, $|c_n|$, and h_n. Corresponding values of c_n and h_n from Example 8.1 are also shown for comparison.

n	c_n	$|c_n|$	h_n	c_n	h_n
	Discrete Fourier Transform			*Fourier Transform* (Example 8.1)	
1	-6.35 - j129.24	129.40	100.0	31.83	100.0
2	-6.35 - j 64.46	64.78	50.1	15.92	50.0
3	-6.35 - j 42.80	43.27	33.4	10.61	33.3
4	-6.35 - j 31.92	32.54	25.2	7.96	25.0
5	-6.35 - j 25.35	26.13	20.2	6.37	20.0
6	-6.35 - j 20.93	21.87	16.9	5.31	16.7
7	-6.35 - j 17.74	18.85	14.6	4.55	14.3
8	-6.35 - j 15.33	16.59	12.8	3.98	12.5
9	-6.35 - j 13.42	14.85	11.5	3.54	11.1

Using Equation 8.12, *THD* = 73.9%. From Example 8.1, we found the *THD* to be 73.5%.

Note that the individual values of h_n for the Discrete Fourier Transform and the Fourier Transform agree fairly well. In fact, had we sampled the waveform more frequently, i.e., increased the value of N to perhaps 128, 256, 512, or 1024, we would have found even closer agreement.

Finally, it is worth noting that the magnitudes of the c_n values for the Discrete Fourier Transform can be related to the c_n values of the Fourier Transform by the coefficient α, given by

$$\alpha = \frac{1}{2}\sqrt{N}.$$

In the present case, $N = 64$, so $\alpha = 4$. If we divide the magnitude of the c_n components for the Discrete Fourier Transform by α, we can compare them with the values of c_n found by using the Fourier Transform.

n	Discrete Fourier Transform $|c_n|$	Fourier Transform $|c_n|$
1	32.35	31.83
2	16.19	15.92
3	10.82	10.61
4	8.14	7.96
5	6.53	6.37
6	5.47	5.31
7	4.71	4.55
8	4.15	3.98
9	3.71	3.54

In summary, we see that the Fourier Transform and Discrete Fourier Transform yield similar results. Almost all modern power analyzers use a variation of the Discrete Fourier Transform called the Fast Fourier Transform or *FFT*.

Hopefully, the reader is now convinced that:

a. It is possible to break a periodic waveform down into its constituent harmonic components by either the Fourier Transform or Discrete Fourier Transform.
b. The individual harmonic components, expressed as a percentage of the fundamental, agree quite well using either technique.
c. Using the individual harmonic components, we can compute the Total Harmonic Distortion (*THD*) of the waveform. *THD* is useful in evaluating the degree of waveform distortion.
d. The Discrete Fourier Transform lends itself well to waveforms which cannot be readily expressed mathematically, but which can be measured using digital sampling techniques.

As we will see, information on the individual harmonic components of a distorted waveform, h_n, as well as its Total Harmonic Distortion (*THD*), will be very useful in understanding the effects of distorted currents and voltages on building power systems.

✳ Measurement Errors

The increase in the number of harmonic producing loads has also created errors in current and voltage measurements. The occurrence of these errors stems from the fact that traditional measurement techniques are based on sinusoidal voltages and currents. In the case of nonlinear voltages and currents, the waveforms are periodic, but not sinusoidal.

Most measurement instruments are based on one of two basic design principles. The first type of instrument senses the peak value of the voltage or current waveform, and converts this to an *rms* value by multiplying by 0.707 (or dividing by $\sqrt{2}$). If we recall the expression for computing the *rms* value of a waveform discussed in Chapter 2, we see that the assumption that the peak and *rms* values are related by $\sqrt{2}$ is always true for only sinusoidal waveforms. In the case of periodic non-sinusoidal waveforms, this relationship may or may not be valid. As a result, peak sensing, *rms* calibrated instruments often give erroneous readings on circuits serving nonlinear loads. For example, an instrument might indicate a lower value of *rms* current than is actually present. Those making such measurements are often baffled by the inexplicable fact that conductors are clearly overheating, despite the fact that readings indicate they are operating well within their current-carrying capabilities. The second type of instrument senses the average value of the voltage or current waveform and multiplies it by 1.11 to arrive at the *rms* value. Here again, the assumption is made that the waveforms are sinusoidal in nature. These meters are called average sensing, *rms* calibrated instruments, and are widely used today. In fact, both peak sensing and average sensing instruments are subject to considerable error. Fortunately, instruments are now available which are capable of making accurate *rms* measurements on sinusoidal and non-sinusoidal waveforms alike.

There are three basic approaches to obtaining true *rms* measurements. These are the *thermal, analog* and *digital* techniques. Thermal instruments have fairly slow response time. Analog meters offer better response time, but may possibly introduce errors due to their limited ability to accurately measure the crest factor of the waveform. Digital instruments are capable of sampling the signal at very high frequencies, often 2 MHz or faster, so accuracy is vastly improved. Such instruments are generally used to analyze brief samples of data.

Figure 8.14 shows a comparison of the results obtained from peak sensing, averaging and true *rms* type instruments. The current values have been normalized to 100 amperes, as measured on an averaging instrument. Note that the three instruments yield similar results only in the case of a sinusoidal waveform.

True *rms* instruments are generally much more expensive than peak sensing or averaging instruments. However, the increased accuracy can be well worth the additional money.

Harmonic Sources

Harmonic currents result from an increase in the number of nonlinear building loads. In this section, we will characterize several such harmonic sources.

Before reviewing these harmonic sources, we should first examine how harmonic data is presented. Figure 8.15 shows the power input, current and voltage waveforms, as well as a current spectrum for an incandescent lamp. This data, along with that of the following examples, was acquired by the author using the power analyzer shown in Figure 8.3. The incandescent lamp is a very linear load, so we would expect very little

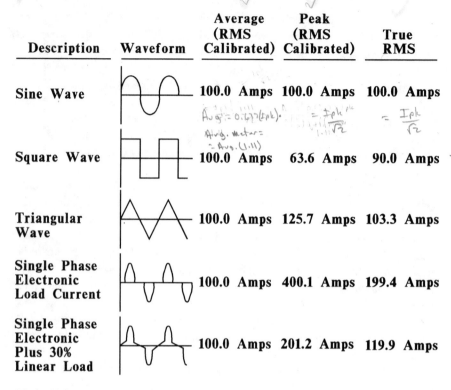

Description	Waveform	Average (RMS Calibrated)	Peak (RMS Calibrated)	True RMS
Sine Wave		100.0 Amps	100.0 Amps	100.0 Amps
Square Wave		100.0 Amps	63.6 Amps	90.0 Amps
Triangular Wave		100.0 Amps	125.7 Amps	103.3 Amps
Single Phase Electronic Load Current		100.0 Amps	400.1 Amps	199.4 Amps
Single Phase Electronic Plus 30% Linear Load		100.0 Amps	201.2 Amps	119.9 Amps

Figure 8.14 Measurement errors associated with peak sensing and average sensing instruments for several waveforms.

harmonic distortion. The graph shows two full cycles of current and voltage. Note that the power frequency is double the voltage and current frequency, so four full power cycles occur during two cycles of current and voltage. Note that the current waveform is very sinusoidal and is, as we would expect, in phase with the voltage. The harmonic spectrum includes the first 50 harmonics. The h_n components are expressed as a percentage of the fundamental, as discussed previously. The Total Harmonic Distortion (*THD*) is only 1.1%. Take a few moments to familiarize yourself with the presentation of this information before proceeding.

Figure 8.15 will serve as a useful comparison for the following examples.

Let us now examine several common sources of harmonics.

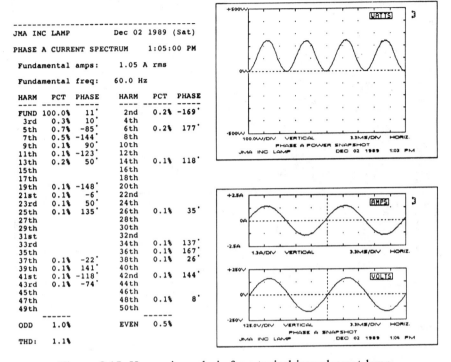

Figure 8.15 Harmonic analysis for a typical incandescent lamp.

- **Speed Control of AC Induction Motors.** During the past few years, we have seen significant advances in power electronic applications for speed control of *ac* induction motors. In years past, such speed control systems were quite expensive, and required very specialized equipment. As a result, speed control was reserved for critical applications only. Today, speed control is feasible for motor loads ranging from air conditioning compressors to supermarket refrigeration systems.

  Systems of this kind can represent a significant portion of the total facility load. Figure 8.16 shows the power input, current and voltage waveforms for a retail food store refrigeration system. Note that the power input and current waveforms are quite distorted. The current waveform Total Harmonic Distortion (*THD*) was found to be 117.2%. The speed control system used equipment similar to that discussed in the induction motor speed control section of Chapter 3.

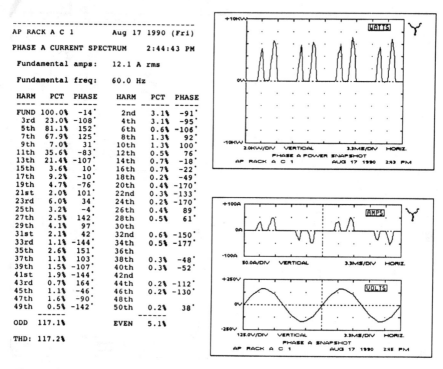

```
----------------------------------------
AP RACK A C 1          Aug 17 1990 (Fri)

PHASE A CURRENT SPECTRUM    2:44:43 PM

 Fundamental amps:   12.1 A rms

 Fundamental freq:   60.0 Hz

HARM    PCT   PHASE    HARM    PCT   PHASE
----  ------  -----    ----  ------  -----
FUND  100.0%   -14°     2nd   3.1%   -91°
 3rd   23.0%  -108°     4th   3.1%   -95°
 5th   81.1%   152°     6th   0.6%  -106°
 7th   67.9%   125°     8th   1.3%    92°
 9th    7.0%    31°    10th   1.3%   100°
11th   35.6%   -83°    12th   0.5%    76°
13th   21.4%  -107°    14th   0.7%   -18°
15th    3.6%    10°    16th   0.7%   -22°
17th    9.2%   -10°    18th   0.2%   -49°
19th    4.7%   -76°    20th   0.4%  -170°
21st    2.0%   101°    22nd   0.3%  -133°
23rd    6.0%    34°    24th   0.2%  -170°
25th    3.2%    -4°    26th   0.4%    89°
27th    2.5%   142°    28th   0.5%    61°
29th    4.1%    97°    30th
31st    2.1%    42°    32nd   0.6%  -150°
33rd    1.1%  -144°    34th   0.5%  -177°
35th    2.6%   151°    36th
37th    1.1%   103°    38th   0.3%   -48°
39th    1.5%  -107°    40th   0.3%   -52°
41st    1.9%  -144°    42nd
43rd    0.7%   164°    44th   0.2%  -112°
45th    1.1%   -46°    46th   0.2%  -130°
47th    1.6%   -90°    48th
49th    0.5%  -142°    50th   0.2%    38°
      ------                 ------
ODD   117.1%           EVEN   5.1%

THD: 117.2%
```

Figure 8.16 Harmonic analysis of a variable speed drive for a refrigeration compressor.

- **HID and Fluorescent Ballasts**. Both fluorescent and high intensity discharge lamps require ballasts for proper operation. The ballast, a transformer-like device, provides the proper starting voltage, and also limits the current flow through the lamp. Such ballasts can produce a significant degree of current waveform distortion. Figure 8.17 shows the power, current and voltage waveforms for a high pressure sodium lamp ballast. The Total Harmonic Distortion (*THD*) is shown to be 31.8%.
- **Electronic Power Supplies**. During recent years a new type of power supply has been introduced that has rapidly replaced traditional power supplies in many applications. This new device is called a switchmode power supply. Figures 8.18(a) and (b) show details of a linear regulator power supply and a switchmode power supply. In the case of the linear regulator power supply, the input voltage is transformed, and then rectified and regulated to provide the necessary *dc* output. In

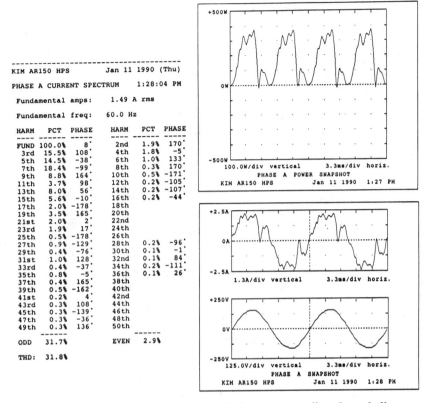

```
----------------------------------------
KIM AR150 HPS          Jan 11 1990 (Thu)

PHASE A CURRENT SPECTRUM        1:28:04 PM

Fundamental amps:      1.49 A rms

Fundamental freq:      60.0 Hz

HARM   PCT   PHASE      HARM   PCT   PHASE
----  -----  -----      ----  -----  -----
FUND  100.0%    8°       2nd   1.9%   170°
 3rd   15.5%  108°       4th   1.8%    -5°
 5th   14.5%  -38°       6th   1.0%   133°
 7th   18.4%  -99°       8th   0.3%   170°
 9th    8.8%  164°      10th   0.5%  -171°
11th    3.7%   98°      12th   0.2%  -105°
13th    8.0%   56°      14th   0.2%  -107°
15th    5.6%  -10°      16th   0.2%   -44°
17th    2.0%  -178°     18th
19th    3.5%  165°      20th
21st    2.0%    2°      22nd
23rd    1.9%   17°      24th
25th    0.5%  -178°     26th
27th    0.9%  -129°     28th   0.2%   -96°
29th    0.4%  -76°      30th   0.1%    -1°
31st    1.0%  128°      32nd   0.1%    84°
33rd    0.4%  -37°      34th   0.2%  -111°
35th    0.8%   -5°      36th   0.1%    26°
37th    0.4%  165°      38th
39th    0.5%  -162°     40th
41st    0.2%    4°      42nd
43rd    0.3%  108°      44th
45th    0.3%  -139°     46th
47th    0.3%  -36°      48th
49th    0.3%  136°      50th
       ------                 ------
ODD    31.7%             EVEN   2.9%

THD:   31.8%
```

Figure 8.17 Harmonic analysis of a high pressure sodium lamp ballast.

the case of the switchmode power supply, the input voltage is first rectified and the power stored in the capacitor, c_1. The switchmode controller switches this *dc* voltage on and off at a high frequency, 10 to 100 kHz, that is stepped down by the transformer, and rectified by diodes D_1 and D_2. The rectified output is then smoothed by the output capacitor. The input current waveform is both discontinuous and non-sinusoidal. Figure 8.18(c) also shows the input voltage and current for a switchmode power supply.

This type of power supply has gained wide acceptance due to its excellent ride-through capability, low parts count, light weight, good voltage regulation and last, but far from least, its low cost. Switchmode power supplies are used in many electronic devices, ranging from PC computers to fax machines. Recently, interest has

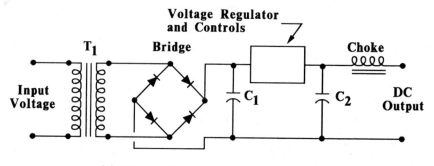

(a) Linear Regulator Power Supply

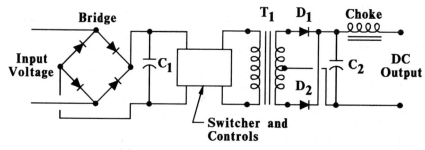

(b) Switchmode Power Supply

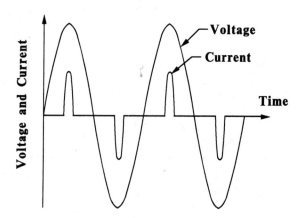

(c) Voltage and Current Input For Switchmode Power Supply

Figure 8.18 Comparison of linear regulator and switchmode power supplies.

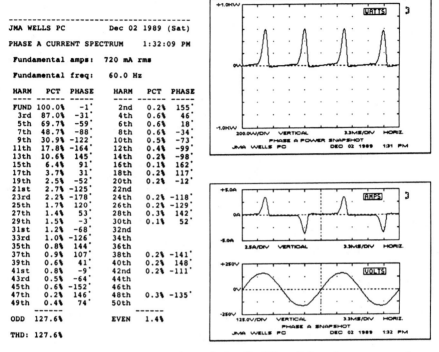

HARM	PCT	PHASE	HARM	PCT	PHASE
FUND	100.0%	-1˙	2nd	0.2%	155˙
3rd	87.0%	-31˙	4th	0.6%	46˙
5th	69.7%	-59˙	6th	0.6%	18˙
7th	48.7%	-88˙	8th	0.6%	-34˙
9th	30.9%	-122˙	10th	0.5%	-73˙
11th	17.8%	-164˙	12th	0.4%	-99˙
13th	10.6%	145˙	14th	0.2%	-98˙
15th	6.4%	91˙	16th	0.1%	162˙
17th	3.7%	31˙	18th	0.2%	117˙
19th	2.5%	-52˙	20th	0.2%	-12˙
21st	2.7%	-125˙	22nd		
23rd	2.2%	-178˙	24th	0.2%	-118˙
25th	1.7%	120˙	26th	0.2%	-129˙
27th	1.4%	53˙	28th	0.3%	142˙
29th	1.5%	-3˙	30th	0.1%	52˙
31st	1.2%	-68˙	32nd		
33rd	1.0%	-126˙	34th		
35th	0.8%	144˙	36th		
37th	0.9%	107˙	38th	0.2%	-141˙
39th	0.6%	41˙	40th	0.2%	148˙
41st	0.8%	-9˙	42nd	0.2%	-111˙
43rd	0.5%	-64˙	44th		
45th	0.6%	-152˙	46th		
47th	0.2%	146˙	48th	0.3%	-135˙
49th	0.4%	74˙	50th		
ODD	127.6%		EVEN	1.4%	

JMA WELLS PC   Dec 02 1989 (Sat)

PHASE A CURRENT SPECTRUM   1:32:09 PM

Fundamental amps: 720 mA rms

Fundamental freq: 60.0 Hz

THD: 127.6%

Figure 8.19 Harmonic analysis of a PC switchmode power supply.

been expressed in the use of larger switchmode power supplies in mainframe computer applications. Figure 8.19 shows the power, current and voltage input for a PC equipped with a switchmode power supply. Note the power and current pulse near the peak of the *ac* voltage wave. The current is very rich in third, fifth, and seventh harmonics. The Total Harmonic Distortion is 127.6%.

As mentioned previously, the switchmode power supply is also used extensively in a wide range of other office equipment applications, including fax machines. Figure 8.20 shows the power, current and voltage input for a typical fax machine. Again, note the switchmode power input profile, and resulting current waveform distortion. The Total Harmonic Distortion is 134.5%. Other items of equipment which exhibit nonlinear characteristics are copy machines. Figure 8.21 shows the power, current and voltage characteristics of a standard office copier.

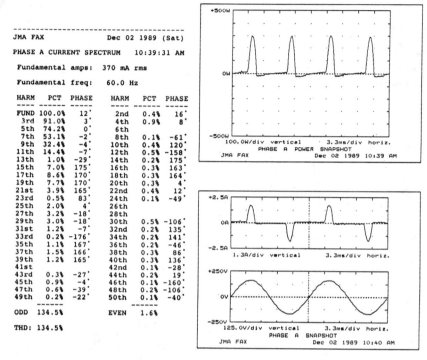

Figure 8.20 Harmonic analysis of a fax machine switchmode power supply.

- **Other Sources of Harmonics**. Another type of nonlinear system load is the Uninterruptible Power Supply (*UPS*) system. A schematic for a typical *UPS* system is shown in Figure 8.22. Note that the input voltage is rectified and used to charge a battery bank, that in turn, supplies power to a *dc/ac* inverter. Should normal power fail, the battery simply continues to supply uninterrupted power to the inverter, and its load, for a period of time limited only by the battery capacity. Typically, such systems are designed to allow operation for 10-15 minutes to facilitate the orderly shut-down of the sensitive load. If unlimited operational time is required, a generator can be added to serve the *UPS* system through a transfer switch, as shown in Figure 3.1. A bypass switch is often provided to permit continued operation of the critical load on normal power during maintenance on the *UPS* system. The input current to the *UPS* system battery charger can be

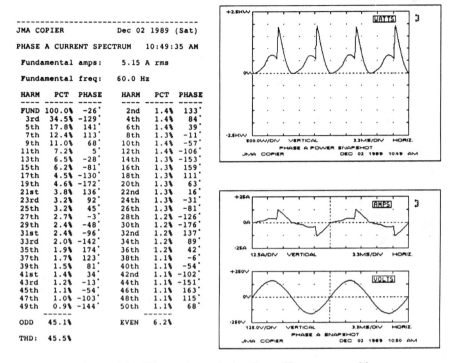

```
-------------------------------------
JMA COPIER              Dec 02 1989 (Sat)

PHASE A CURRENT SPECTRUM    10:49:35 AM

  Fundamental amps:     5.15 A rms

  Fundamental freq:     60.0 Hz

 HARM   PCT  PHASE    HARM   PCT  PHASE
 ----  ----- -----    ----  ----- -----
 FUND 100.0%  -26'     2nd  1.4%   133'
  3rd  34.5% -129'     4th  1.4%    84'
  5th  17.8%  141'     6th  1.4%    39'
  7th  12.4%  113'     8th  1.3%   -11'
  9th  11.0%   68'    10th  1.4%   -57'
 11th   7.2%    5'    12th  1.4%  -106'
 13th   6.5%  -28'    14th  1.3%  -153'
 15th   6.2%  -81'    16th  1.3%   159'
 17th   4.5% -130'    18th  1.3%   111'
 19th   4.6% -172'    20th  1.3%    63'
 21st   3.8%  136'    22nd  1.3%    16'
 23rd   3.2%   92'    24th  1.3%   -31'
 25th   3.2%   45'    26th  1.3%   -81'
 27th   2.7%   -3'    28th  1.2%  -126'
 29th   2.4%  -48'    30th  1.2%  -176'
 31st   2.4%  -96'    32nd  1.2%   137'
 33rd   2.0% -142'    34th  1.2%    89'
 35th   1.9%  174'    36th  1.2%    42'
 37th   1.7%  123'    38th  1.1%    -6'
 39th   1.5%   81'    40th  1.1%   -54'
 41st   1.4%   34'    42nd  1.1%  -102'
 43rd   1.2%  -13'    44th  1.1%  -151'
 45th   1.1%  -54'    46th  1.1%   163'
 47th   1.0% -103'    48th  1.1%   115'
 49th   0.9% -144'    50th  1.1%    68'
       ------                ------
 ODD    45.1%         EVEN    6.2%

 THD:   45.5%
```

Figure 8.21 Harmonic analysis of an office copy machine.

quite distorted, and is often a source of considerable system harmonic distortion. Most manufacturers offer optional filters for the *UPS* input, but these represent additional costs and are often omitted.

The aggregate of the various linear and nonlinear loads in a building determine the extent of the harmonic distortion seen by the utility system. Figure 8.23 shows the input power, current and voltage for a portion of the author's office building. This load represents lighting and receptacles, as well as miscellaneous equipment other than air conditioning or water heating. Note that the *THD* is 30%, and that the third harmonic is 24.5%, indicating that a significant portion of the load is probably electronic in nature. The percentage of electronic load in this kind of office building is expected to increase significantly in the future.

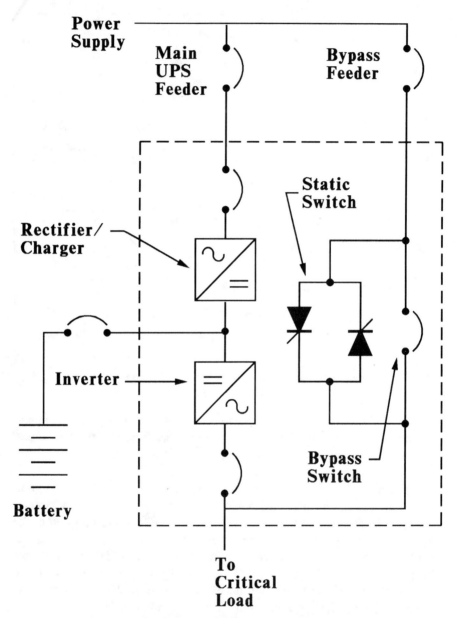

Figure 8.22 Schematic of an uninterruptible power supply (*UPS*) system.

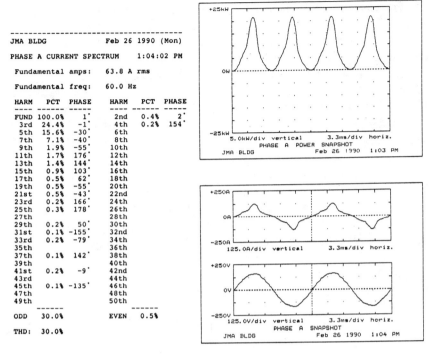

```
----------------------------------------
JMA BLDG                Feb 26 1990 (Mon)

PHASE A CURRENT SPECTRUM        1:04:02 PM

  Fundamental amps:    63.8 A rms

  Fundamental freq:    60.0 Hz

HARM   PCT  PHASE     HARM   PCT  PHASE
----  -----  -----    ----  -----  -----
FUND 100.0%    1˚     2nd   0.4%    2˚
 3rd  24.4%   -1˚     4th   0.2%  154˚
 5th  15.6%  -30˚     6th
 7th   7.1%  -40˚     8th
 9th   1.9%  -55˚    10th
11th   1.7%  176˚    12th
13th   1.4%  144˚    14th
15th   0.9%  103˚    16th
17th   0.5%   62˚    18th
19th   0.5%  -55˚    20th
21st   0.5%  -43˚    22nd
23rd   0.2%  166˚    24th
25th   0.3%  178˚    26th
27th                 28th
29th   0.2%   50˚    30th
31st   0.1% -155˚    32nd
33rd   0.2%  -79˚    34th
35th                 36th
37th   0.1%  142˚    38th
39th                 40th
41st   0.2%   -9˚    42nd
43rd                 44th
45th   0.1% -135˚    46th
47th                 48th
49th                 50th
      ------                ------
ODD    30.0%         EVEN   0.5%

THD:   30.0%
```

Figure 8.23 Harmonic analysis of the secondary load current for a 75 kVA transformer serving a representative office building.

Effects of Harmonics on Building Systems

Excessive levels of harmonic current can have a very serious effect on several types of building power system components. Step-down transformers, such as those commonly used in building systems, are specifically designed for operation on 60 Hz. Load currents which contain higher frequency harmonics increase the eddy current and hysteresis losses associated with transformer operation. The final result is often an overheated transformer and shortened insulation life. This overheating can occur despite the fact that the transformer may be operating well within its nameplate operating current range. When overheating is noted, current readings made with an averaging or peak sensing ammeter can exacerbate the problem by yielding erroneous values for current flow.

Fortunately, guidance for proper transformer selection for nonlinear loads is available in the form of ANSI/IEEE C57.110-1986. The IEEE recommended practice guidelines establish procedures for the appropriate derating factor for transformers serving nonsinusoidal loads.

Another building component frequently at risk due to nonsinusoidal currents is the neutral conductor. In the case of balanced linear loads, the neutral of a wye connected system carries no current due to the inherent current cancellations resulting from the three-phase system. In the past, reduced size neutral conductors were often installed in recognition of this fact. In the case of nonsinusoidal load currents, the third harmonics add, instead of cancel, in the neutral. As a result, we have seen a number of neutral overheating problems during the past several years. Also, many branch circuits are designed with this same neutral cancellation in mind, and three circuits are combined on one neutral conductor that is typically the same size as the phase conductor. Here again, we have seen increased problems with neutral conductor overheating. The *NEC* no longer permits reduced neutral conductors where nonsinusoidal load currents are expected. Unfortunately, we still have thousands of existing buildings which may be at risk due to the older design approach. Higher frequency harmonics can also interfere with telecommunications circuits in the form of high frequency noise.

The load current drawn by nonlinear loads can cause distorted voltages in other parts of the system. These distorted voltages can cause overheating of motors. This is simply due to the fact that the motor will draw harmonic current when exposed to supply voltages containing harmonic distortion. The resulting distorted current will cause the motor to experience increased losses in the motor laminations, and also higher losses due to skin effect in the windings.

Harmonic distortion can have negative effects on overcurrent protective devices, such as fuses, thermal magnetic circuit breakers and electronic trip circuit breakers. In the case of fuses, the higher frequencies cause the fuse to operate at a higher temperature than normal, thus increasing the likelihood of nuisance melting. In the case of thermal magnetic circuit breakers, the bimetal strip that senses overloads tends to operate at a higher temperature when exposed to harmonic currents, which may well result in unexpected tripping. The sensing coil in the instantaneous trip portion of the breaker can also overheat due to the higher frequencies present in harmonic currents. This increased heat can be transmitted to the bimetal strip, further adding to its temperature. There have also been indications that higher frequency harmonics may reduce the accuracy of the breaker's trip response. Even the new electronic trip circuit breakers are not immune to harmonic related problems. Most modern electronic trip circuit breakers employ microprocessors which compute the *rms* current based on a fairly high sampling rate. Unfortunately, the accuracy of this device depends on the proper frequency response of the current transformers. If the response of the current transformers is linear throughout the range of present harmonic frequencies, accurate tripping will occur. If this is not the case, then tripping errors can occur.

Harmonic Standards

Almost all industrialized nations now have some form of regulatory requirements which limit power system harmonics. In the United States, regulation guidelines have evolved through the IEEE Harmonic Standard 519, *Recommended Practices and Requirements for Harmonic Control in Electric Power Systems*. This document has been in almost constant review and revision for a number of years. IEEE-519 recognizes that harmonics themselves originate in the utility customer's premises in the form of nonlinear loads. The resulting nonsinusoidal current flow can create harmonic voltage problems within the utility system, which can, in turn, create harmonic problems for other utility customers. As a result, IEEE-519 seeks to place limits on the harmonic currents drawn by customers at the point of common coupling between the utility and the customer's system. The utility, on the other hand, is required to maintain its own system voltages within specified harmonics limits.

For utility customers, the maximum current *THD* is a function of the ratio of the available short circuit current, I_{sc}, at the point of common coupling (*PCC*), to the customer's maximum fundamental load current, I_L. The basic concept is to limit the percentage of harmonic current injected into the utility system. This helps protect other customers on the same part of the utility system. The current *THD* permitted ranges from 5% for an I_{sc}/I_L of less than 20, to 20% for an I_{sc}/I_L of over 1000.

In the case of utility system voltages, harmonic distortion limits are established for individual harmonics, as well as for the *THD*. For example, on systems of 2.3 to 69 kV there is a limit of 3% on individual harmonic voltages and a *THD* limit of 5%. For systems of over 138 kV, the individual harmonic voltage limit is 1%, and the voltage *THD* is limited to 1.5%.

Each year, utilities are becoming more and more aware of the necessity of monitoring their levels of voltage and current distortion. Many have established their own customer harmonic limit requirements. As non linear loads assume a larger and larger percentage of the total utility load, the problems associated with power system harmonics will also require increased attention by the building design engineer.

References

Arrillaga, J., D.A., Bradley and P.S. Bodger. 1985. *Power System Harmonics.* New York: John Wiley & Sons.

Basic Measuring Instruments, Inc. 1989. *End-Use Power Line Harmonics.* Foster City, California: Basic Measuring Instruments, Inc.

Basic Measuring Instruments, Inc. 1989. *The PowerProfiler: Use and Applications.* Foster City, California: Basic Measuring Instruments, Inc.

Brozek, James P. 1990. The Effects of Harmonics on Overcurrent Protection Devices. *Conference Record of the 1990 IEEE Industry Applications Society Annual Meeting - Part II.* Seattle: pp. 1965-1967.

Carter, Wendell W. 1989. *Power Quality.* Paper presented at Professional Program Session 21. 30 March 1989 at Georgia World Congress Center, Atlanta, Georgia.

Freund, Arthur. 1988. Double the Neutral and Derate the Transformer - Or Else. *Electrical Construction and Maintenance.* December, 1988: pp. 81-85.

Freund, Arthur. 1988. Nonlinear Loads Mean Trouble. *Electrical Construction and Maintenance.* March, 1988:83-90.

Institute of Electrical and Electronic Engineers. 1990. *IEEE Recommended 2Practice for Electric Power Systems in Commercial Buildings. IEEE Std. 241-1990.* New York: Institute of Electrical and Electronic Engineers, Inc.

Institute of Electrical and Electronic Engineers. 1990. *IEEE Recommended Practice for Power Systems Analysis. IEEE Std. 399-1990.* New York: Institute of Electrical and Electronic Engineers.

Institute of Electrical and Electronic Engineers. 1988. *IEEE Recommended Practice for Establishing Transformer Capability When Supplying Nonsinusoidal Load Currents. ANSI/IEEE C57.110-1986.* New York: Institute of Electrical and Electronic Engineers, Inc.

Long, Leland, and John H. Matthews. 1991. Electrical Power Distortion Produced by Microprocessor-Based Office Equipment. *Association for Computing Machinery - Proceedings - 29th Annual Southeast Regional Conference.* April, 1991.

McEachern, Alexander. 1989. *Handbook of Power Signatures.* Foster City, California: Basic Measuring Instruments, Inc.

McEachern, Alexander. 1987. *Voltage, Current, Power Factors, and Spectrum Measurements on Non-sinusoidal AC Power Circuits*. Foster City, California: Basic Measuring Instruments, Inc.

Mathsoft, Inc. 1987. *Mathcad 2.0 Users' Manual*. Cambridge: Mathsoft, Inc.

Calculate Voltage Drop (if Reactance is neg.)   At unity pf — 2 wire

$$Vd = 2 \times R_{cond} \times \ell \times I$$

Ex. 230v, 2 wire system,
24 Alead, 200ft, #10wire $Vd = 2 \times \dfrac{1.21\Omega}{1000ft} \times 200ft \times 24A = \boxed{11.62V}$

$R_{\#10} = 1.21\Omega/1000ft$

(use #8 so Reduce Voltage Drop)

NEC suggest 5% MAX V.D

OR use "F" Values;

Ex. I = 24A,   ∴ F = 0.210 ·(1.155) = 0.24255
    ℓ = 200ft
    V = 230   % Vd = $\dfrac{F \times \ell \times I}{V}$ = $\dfrac{0.24255 \times 24A \times 200}{230}$ = 5.06%
    #10
                Vd = 5.06% (230v) = 11.63

Example of Conduit and Decl.   3φ induction motor rated @ 200 hp, 460v, 240A, is to
                                 be installed in a plant. A 200ft, 3φ, 3 wire circuit is needed
                                 Assume the motor operates at a pf of 0.87 And Amb. temp = 40°C

a). Select conductor size and type. Assume the C.B for motor has lugs rated at 75°C.

• Use RH with a copper conductor — Rubber conductor that resist up to 75°C

•   Size => pg. 70.   240A, RH.

    correction Factor for 40° = 0.88

    ∴ $\dfrac{2.400}{0.88}$ = 273v

    ∴ use A conductor At 285v ⟹ $\boxed{300\text{kcmil}}$

o select proper size of conduit.

    ✗ RH @ 300kcmil, 3 wire for 3φ
    + use 2½ in (Fro-table)

o Assume source is 480, And There is A 3% Voltage loss in Motor
  Circuit. will the motor receive At least rated voltage?

    3% Vd = 14.4v, ∴ Voltage At MCC = 465.6v

    Vd = $\sqrt{3} \times 200ft \times \dfrac{0.0515}{1000ft} \times 240A$ = 4.28

    Voltage to Motor = 480 - 14.4 - .428 = 461.3v ⟹ √3

9

Introduction to Illumination Engineering

Perhaps one of the most crucial aspects of the design of a building is its illumination system. It is this system that determines how well the building's occupants will be able to perform their visual tasks. The lighting system also helps establish the mood or feeling of a space, and as such becomes an important architectural tool. In fact, the building lighting system is often the only part of the electrical system noticed by many people. In this, and succeeding chapters, we will examine some of the important aspects of illumination engineering.

In this chapter we will introduce the fundamentals of lighting, beginning with Edison's work in the late 1800s. We will also discuss the nature of light as visible radiant energy, as well as vision and human response to color and variations in illumination intensity. Illumination engineering has its own set of unique terms, its language. These terms, along with the techniques which have been developed to determine appropriate illumination levels for various tasks, will be discussed in the final sections of this chapter.

In Chapter 10, we will discuss the characteristics of the various light sources commonly used today. These sources, which include incandescent, fluorescent and high intensity discharge, each have their own areas of application. Light sources alone would be of little value unless the illumination produced can be effectively controlled and distributed by the lighting fixtures into which the lamps are installed. The combination of the light source and fixture make up the complete lighting unit, called a *luminaire*. The various types of luminaires, along with their light distribution characteristics, or *photometrics*, will be discussed in Chapter 11.

Lighting terms, factors affecting human vision, light sources, luminaires and photometrics are combined in Chapter 12, where we will discuss the process of actually designing the building's illumination system.

THE EARLY DAYS

There has always been a need for artificial illumination systems. From the earliest days of man, when fire provided a means of heat and cooking as well as a source of light, to the widely used gas lighting systems of the 1800s, lighting has been a major factor in human development. However, it was not until the late 1800s, barely over a century ago, that the technology was available to produce artificial illumination as we know it today.

Although there were many attempts to produce an incandescent light source during the early and mid-1800s, it was Thomas Edison who developed the first economically viable lamp.

Edison was rather eccentric, tended to work long hours, and kept an erratic schedule. By the time Edison turned his attention to the incandescent lamp, he had already made significant technological contributions in the areas of telegraphy and the development of the phonograph. By the late 1870s his laboratory at Menlo Park, New Jersey, was the scene of his search for a practical incandescent light source. Here, along with a work force numbering around 100, Edison researched a wide variety of techniques for producing an incandescent lamp (Cox 1979).

The incandescent lamp has often been described as a "hot wire in a bottle" which is a pretty fair description. The basic concept involved the placement of a resistive element inside a glass jacket from which the air had been removed. The resistive element was connected to a source of electricity by means of wires which penetrated the glass envelope. The flow of current through the resistive element, later called a *filament*, increased the temperature of the element to the point that it began to glow or incandesce. It had already been established that the filament would fail in very a short length of time if burned in air, so the removal of air from the lamp was essential. The process of air removal in those days involved the use of a mercury pump, and required several hours. Edison had experimented with several types of filament material, including platinum.

By the fall of 1879, he considered carbon to be a potential source of filament material. He carbonized a wide variety of materials, produced filaments, installed them in glass envelopes, laboriously pumped out the air and sealed the envelopes. These test lamps were then connected to a power source and burning-life tests conducted.

On Sunday, October 19, 1879, Edison was ready to begin yet another burning-life test. On this occasion, the filament consisted of a piece of carbonized thread from his wife's sewing kit! Specifically, it was "Coats Co. Cord No. 29" (Cox 1979). The long process of assembling the lamp and evacuating the air consumed most of the day, and it was not until 8:00 P.M. that the lamp was ready for testing. Edison carefully observed the burning lamp for most of the night and the following morning, and found that the lamp was still glowing brightly. In fact, it continued to do so for 40 hours. Edison knew that his goal was within reach.

Understanding the importance of obtaining the public's interest, Edison decided on a demonstration of his new lamp on New Year's Eve, 1879. He knew that widespread public interest would make it much easier to obtain the financing he needed to continue his work.

As the fall of 1879 wore on, Edison installed the new light source in his laboratory, in several surrounding buildings and on the connecting paths. These brightly lighted structures were clearly visible from trains passing nearby, and soon the secret was out. The public was eventually invited to see Edison's laboratory on the snowy night of December 21, 1879. The roads were churned into mud by carriages, and the railroad had to add extra trains to carry the crowd that by the evening's end, totalled over 3,000 people.

This demonstration, that might be said to have marked the beginning of the electrical industry as we know it, caused immediate concern within the gas lighting companies, as gas stock plunged. In short order, electrical lighting in the form of incandescent lamps began to meet an ever-increasing percentage of the nation's overall illumination requirements. Factories found it possible to extend working hours well into the evening and, hence, were able to increase production. Downtown streets were lighted, thus making travel safer.

As discussed in Chapter 7, an entire new industry developed shortly after 1880—the electric utility industry. As these utilities grew and flourished, so did the application of electrical lighting. Mrs. Edison's humble sewing thread, coupled with Edison's vast creativity and vision, had forever changed our world.

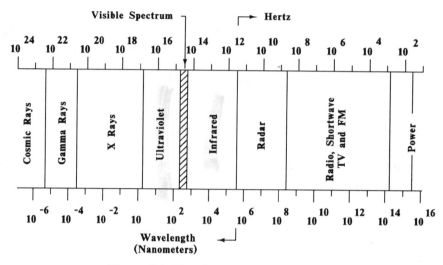

Figure 9.1 The electromagnetic spectrum.

THE NATURE OF LIGHT

There are two basic theories regarding the nature of light. One of these, the *quantum theory*, views light as consisting of a stream of individual photons. The second, or *electromagnetic theory*, views light as part of the electromagnetic spectrum. We will base our discussion on the electromagnetic approach because it best suits our work. Figure 9.1 shows the electromagnetic spectrum, and the relatively small part occupied by the portion responsible for human vision.

We will define *light* as the radiant energy that stimulates the retina, thereby producing a visual sensation.

The relationship between the wavelength and its frequency is given by

$$\lambda = \frac{c}{v} \quad m \tag{9.1}$$

where λ is the wavelength in meters, c the speed of light in meters per second, approximately $3 \cdot 10^8$ m/sec, and v the frequency in cycles per second. It is actually more convenient to deal with the wavelengths associated with the visible spectrum in terms of nanometers, $1 \cdot 10^{-9}$ m.

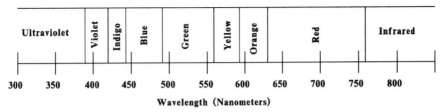

Figure 9.2 The visible portion of the electromagnetic spectrum.

Traditionally, the area of the electromagnetic spectrum to the right of the point indicated in Figure 9.1 is expressed in terms of *frequency*. This part of the electromagnetic spectrum includes radar, radio, television, and power waves. The portion of the spectrum to the left is specified in terms of its *wavelength* and includes the visible spectrum, as well as infrared, ultraviolet, gamma, and cosmic rays.

Figure 9.2 shows the visible portion of the electromagnetic spectrum. The visible spectrum extends from $0.38 \cdot 10^{-6}$ m to $0.76 \cdot 10^{-6}$ m, or from 380 nm to 760 nm (nanometers).

The human eye perceives the various wavelengths in the visible spectrum in terms of color. For example, the region near 380 nm is perceived as violet, while the region near 760 nm is perceived as red. Table 9.1 lists the seven pure spectral colors of the visible spectrum, along with their associated radiant energy wavelengths.

HUMAN VISION

The human eye is a truly incredible instrument. It is capable of adapting to a range of lighting levels, varying from almost total darkness to bright sunlight. At the same time, it can focus on objects as close as a few centimeters, or as far away as hundreds of meters. Our eyes provide us with binocular vision by combining the visual field of both eyes which our brain then interprets as a single image. All of this is accomplished without our even being aware of the diverse processes required to produce vision. In order to design effective illumination systems, we must have a basic understanding of human vision, and the way in which our eyes respond to lighting.

Let us now take a moment to review the major structural and neurological components of the eye. Figure 9.3 shows a somewhat simplified section of the human eye.

Table 9.1 Color Versus Wavelength.

COLOR	WAVELENGTH (nm)
Red	760-630
Orange	630-590
Yellow	590-560
Green	560-490
Blue	490-440
Indigo	440-420
Violet	420-380

The eye is basically spherical in shape, with a diameter of about one inch, and is composed of three discrete layers called *tunics*. The outermost tunic is composed of two components—the *sclera* and the *cornea*. The *sclera* is a tough membrane covering the posterior 5/6s of the eye that helps maintain the eye's spherical shape. The *cornea* comprises the front, or anterior 1/6 of the eye. The *cornea* has a fixed focus, and provides us with about 70% of the converging power of the eye.

The next layer, or middle tunic, consists of the *choroid*, the *ciliary body* and the *iris*. The *choroid* is very vascular, and provides the blood supply for much of the eye. The *ciliary body* plays an important part in altering the shape of the lens, and also provides the fluid, called *aqueous humor*, that fills the area between the cornea and the lens. Both the *aqueous humor* and the *vitreous humor*, that fills the remainder of the eye, help provide nutrients for the non-vascular components of the eye. The *zonules*, which are two fibrous sheets, extend from the ciliary body and join with the lens itself. The force of the muscles in the ciliary body affect the shape of the lens and, thereby, help focus the eye. The *iris* opens and closes to adjust the amount of light entering the eye.

The innermost layer or tunic is divided into two parts—the *pars optica*, or *retina*, and the *pars ceca*. The *pars ceca* is a thin layer at the front of the eye that extends from the point at which the choroid joins the ciliary body over the innermost surface of the ciliary body and iris. The *pars optica, or retina*, covers the remaining interior surface of the eye. It is the *retina* which contains the visual cells, or retinal receptors. Two important areas of the retina are the *macula*, and the *optic disc*. The *macula* is a small depression in the retina about 1.5 mm in diameter. The lowest part of the *macula* called the *fovea*, is about 0.4 mm in diameter, and contains

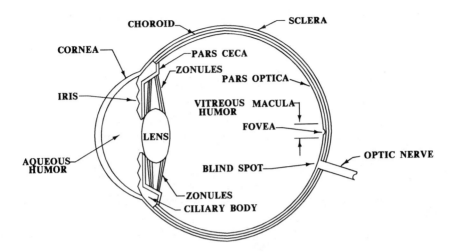

Figure 9.3 Cross section of the human eye.

the visual cells which are responsible for our sharpest vision and best color response. The *optic disc* is the point at which the optic nerve leaves the retina. It is through the *optic nerve* that the visual signals are transmitted to the brain.

These are the basic structural and neurological components of the eye. With this background, we can now delve a bit deeper into the actual process of seeing.

The Vision Process

Light enters the eye through the cornea. The quantity of light that enters is regulated by the iris. The lens focuses the light on the retina which contains light-sensitive cells. Human vision is essentially a *photochemical process* in which the incident radiant energy precipitates a chemical process that converts the radiant energy into electrical signals which are sent via the optic nerve to the brain, where they are interpreted as visual images. The process of focusing on objects is called *accommodation*, while the process of moving from darker to lighter environments or vice versa is called *light adaptation*.

There are two basic types of photo receptors or cells located in the retina. The first type are *rod cells*, which are cylindrical in shape, 0.002

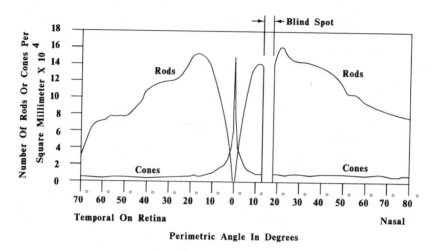

Figure 9.4 The distribution of rods and cones in the retina. The zero degree point represents the Fovea Centralis. (Reproduced with permission of Illuminating Engineering Society of North America.)

mm in diameter, and 0.07 mm long. *Cone cells* are conical in shape with a base diameter of 0.005 mm and a height of 0.07 mm.

Rod cells are extremely sensitive to light, but are not at all color sensitive. They are primarily responsible for our night vision, and since they are color insensitive, our night vision is basically monochromatic. Rods are, however, quite motion-sensitive, which is a great asset in night vision. Rod cells predominate in the retina, except in the area of the fovea, where cone cells predominate. *Cone cells* are most effective in daylight vision, and are concentrated in the very small area of the fovea itself. Cone cells are quite color sensitive, and are also responsible for our sharp vision. Figure 9.4 shows the distribution of rods and cones in the retina. Note that there are no rod cells in the fovea, and few cone cells outside the fovea.

Daylight vision, where cone cells play a predominant role, is called *photopic vision*, while night vision, where rod cells come into play, is called *scotopic vision*. The transition range between photopic and scotopic vision, where both cone cells and rod cells are operational, is called *mesopic vision* (Helms 1980).

During the shift from photopic to scotopic vision, another important change takes place. The maximum sensitivity of the eye shifts from 555 nm to about 510 nm, making blue and green colors appear more visible.

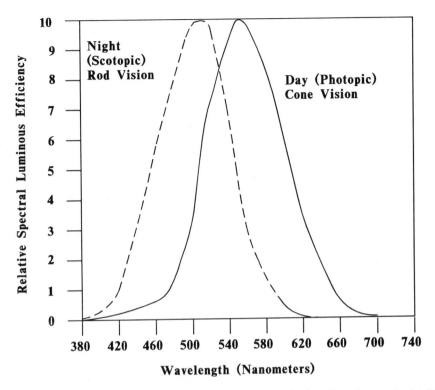

Figure 9.5 The relative spectral sensitivity for photopic (cone) and scotopic (rod) vision. (Reproduced with permission of Illuminating Engineering Society of North America.)

This is called the *Purkinje Shift* and is shown in Figure 9.5. The high nighttime visibility of green is one reason so many roadway signs are this color.

Color Vision

As mentioned previously, human vision is basically a photochemical process involving the rod and cone cells (IES Lighting Handbook 1984). The pigment associated with the rod cells is called *rhodopsin*, and as we have discussed, these cells are not color sensitive. There are three different pigments associated with cone cells, each having a different peak color

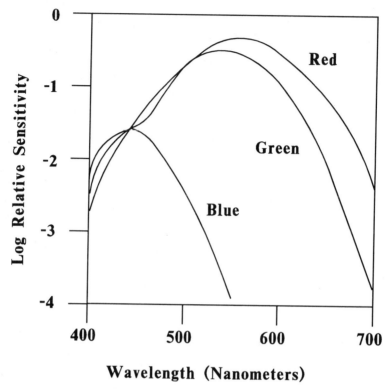

Figure 9.6 The relative spectral sensitivity curves of the three cone types - red, green, and blue. (Reproduced with permission of Illuminating Engineering Society of North America.)

sensitivity. These are the red, green and blue cone cells. Figure 9.6 shows a graph of the relative sensitivity of these three classes of cone cells. Human vision is then, actually, *trichromatic* in nature. Note that all three classes of cone cells have some degree of sensitivity throughout the visible spectrum.

Design Considerations

The nature of human vision requires that a number of factors be kept in mind during the design process. Several of these are as follows:

1. The photochemical process associated with human vision is not instantaneous. As a result, the eye and brain continue to perceive a stimulation for a short time, even after the stimulus ceases. This phenomena is called *afterimage*. *Afterimage* makes us perceive a rapid series of still images as moving. This is the principle on which the motion picture is based.

2. As we age, the eye loses some of its ability to focus on close objects. This is called *presbyopia* and is primarily due to increased rigidity of the lens material. Most people begin to experience some degree of *presbyopia* at about age 40.

3. Increased age also brings a decrease in the ability of the eye to adapt to dark and light environments. There is often also a decrease in the eye's sensitivity, especially at low levels of illumination. Finally, there is some degree of loss of the transmission of light due to increased opacity of the lens itself, and also a decrease in color sensitivity due to yellowing of the lens.

From the preceding discussion, you have probably realized that a number of factors must be carefully considered in the design of a lighting system, not the least of which is the age of the occupants who will be using the space. We must also consider the nature of the activities involved, which we normally refer to as the visual task. In an office area, the visual task might be reading typewritten pages or working at a computer terminal. In a warehouse, the visual task might be the reading of labels on stacked cartons. In a parking deck, the task might be the location of your car, followed by safe maneuvering of the car out of the deck. In some visual tasks, speed is crucial, while in others, true color rendition is important. In general, the requirements for effective task visibility can be broken down into four categories (IES Education Series ED-100 1985).

Factors Affecting Task Visibility

Contrast
The contrast between an object and the background against which it is viewed is crucial. For the same level of illuminance, light objects viewed against a dark background are more visible than the same object would be

viewed against a light background. In general, tasks with poor contrast require a higher level of illuminance than would be necessary for those with better contrast.

Size

The size of an object is also important in determining visibility. By size, we mean the visual size of the object, not its absolute size. For example, smaller objects can be made more visible by bringing them closer to the eye, thus increasing their visual size. We also know that increased visibility can be obtained by increasing the level of illuminance.

Time

An often overlooked factor in task visibility is time. Vision is not instantaneous, and in applications where, for example, rapidly moving machinery is involved, the time factor becomes quite important. Here again, increased levels of illuminance are often used to offset short periods of viewing time. For example, in tennis court lighting, relatively high levels of illuminance are required to help offset the short viewing time due to the speed of the ball.

Luminance

The last factor is luminance. The luminance of the task depends on the percentage of the incident light reflected back toward the eye. Visual tasks with lower levels of reflected light can be made to appear brighter by increasing the level of incident light.

PSYCHOLOGICAL ASPECTS OF LIGHTING

The psychological factors associated with the human response to lighting are undeniably an important consideration. Unfortunately, they are also virtually impossible to generalize, and downright dangerous to quantify. This is understandable when we take into consideration the wide range of personal preferences for colors, lamp types and levels of illumination. However, research has given us a few guidelines, several of which we will now review. We will relate these to two distinctly different visual environments—social areas and work areas (IES Education Series ED-100 1985).

1. People tend to prefer lower levels of illumination when relaxing than when working.
2. In social situations people generally prefer non-uniform lighting, while uniform lighting often seems preferable in work environments.
3. People tend to prefer warmer earth colors in social situations, while cooler greens and blues are more appropriate for the stimulating environment of the office.
4. In general, social spaces require lighting with very good color rendering qualities because people want to look their best. In office environments, color rendition may not be as important as the energy efficiency of the lighting system.

These psychological considerations merit consideration in the selection of room colors, as well as in the selection of light sources and levels of illumination. In fact, effective lighting design generally requires very close coordination between the designing architect and the electrical engineer.

Take time to examine your own personal feelings regarding these psychological issues. Do you have preferences which are not discussed above? Many engineers feel hopelessly inadequate in the area of psychological response to lighting, but these considerations are very much a part of effective lighting design. We will now turn our attention to the specialized terms associated with lighting.

INTRODUCTION TO LIGHTING TERMS

There are actually only a few key terms associated with illumination engineering, but these are quite specialized, therefore making it very important to clearly understand their meanings.

Prior to the introduction of electrical lighting, the predominant sources of artificial illumination were based on gas, oil and candles. In fact, the "standard" candle became one of the key factors in the definition of some of the fundamental terms. Before proceeding, let us take a moment to learn how some of the terms originated.

The *standard candle* is defined as a candle that emits a uniform illuminous intensity, in all directions, of one candlepower (IES Lighting Handbook 1984). Today, we refer to this intensity as one *candela* (*cd*). Figure 9.7 shows a standard candle located in the center of a sphere with a one foot radius, a unit sphere. For photometric purposes, we will assume that the interior surface of this sphere is completely black.

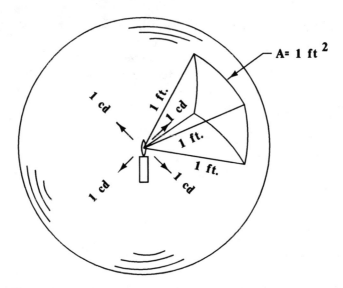

Figure 9.7 Unit sphere with standard candle source at center.

The total luminous flux emitted by a source is measured in lumens. The *lumen (lm)* is defined as the rate at which luminous flux falls on a 1 ft^2 surface of the unit sphere from a uniform source of one candela located at the center of the sphere. The surface area of a sphere is $A = 4\pi r^2$, so the area of the unit sphere is 4π ft^2. This means that the standard candle emits 4π or 12.57 lumens. It is important to understand that the candle is assumed to be a point source. This requirement is effectively met if the dimensions of the source are small compared with the other dimensions involved.

The unit of illuminance is the *footcandle (fc)*, which is defined as the ratio of the total lumens evenly distributed over a surface to the area of the surface in ft^2. Thus, the level of illuminance on the 1 ft^2 area shown in Figure 9.7 is 1 lumen/ft^2 or one footcandle.

From the above discussion, we can develop another important lighting concept. Figure 9.8 shows a section of two concentric spheres of radius 1 ft and 2 ft, respectively. The light source at the center of the sphere emits a uniform luminous intensity of 1 cd in all directions, and the total luminous flux rate is 12.57 (4π) lm. The luminous flux density on the 1 ft^2 area of the unit sphere is 1 lumen/ft^2 or 1 fc, as discussed above. The luminous flux density on the 2 ft radius sphere will be the total lumens, 12.57 (4π), divided by the surface area, 16π ft^2, or 1/4 fc.

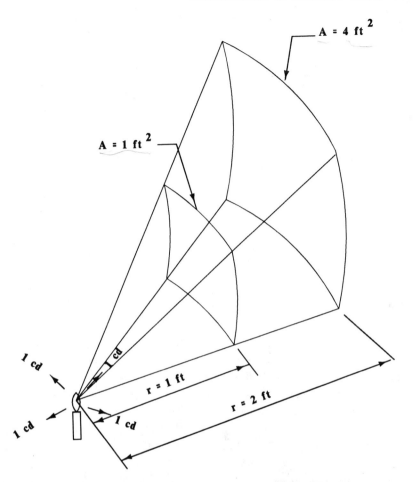

Figure 9.8 Section of spheres of radius 1 ft. and 2 ft.

If the sphere were 3 ft in radius, its area would be 36π ft^2, and the flux density would be 1/9 lumen/ft^2. From this, we see that the level of illumination produced by a source decreases with the square of the distance from the source to the area being illuminated. This is known as the *inverse square law of illumination*.

We should remember that the previous discussion assumed that all surfaces receiving illumination are perpendicular to the light falling on them. In actual application, this is rarely true, and this case will be discussed shortly.

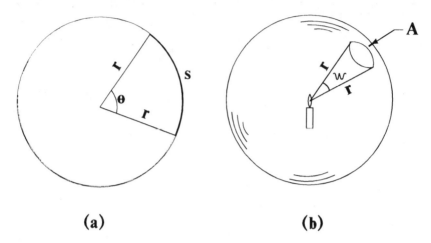

(a)   **(b)**

Figure 9.9 The (a) radian and (b) steradian.

Units

A few further words about units might now be in order. You will note that we have expressed distance in feet and illuminance in footcandles. Most building designs in the United States are presently based on the *English System* and, hence, we will follow this standard. Most of the rest of the world has converted to the *Systeme International* (*SI*) system. In this system, the applicable lighting units are described as length in meters (m) and illuminance levels in lux (lx). The units of luminous flux, lumen and lumen intensity, candela are the same for both the English and SI systems.

In the SI system, the level of illuminance, *lux*, is defined as the uniform distribution of one lumen over an area of 1 m^2 (10.76 ft^2). Thus, the same total luminous flux is spread over an area 10.76 times as large as before, making 1 *fc* = 10.76 lx.

Solid Angles

Before proceeding further, we should spend a few moments discussing solid angles. Figure 9.9(a) shows a two-dimensional angle, θ, associated with a circle of radius r, and arc length, s. In the special case where $s = r$, the angle θ is defined as 1 radian (rad). Since the circumference of the circle is $2\pi r$, there are 2π radians in the total circle. Figure 9.9(b) shows a sphere of radius r with a solid angle ω, and intercepted area on the surface of the sphere A. The solid angle ω is defined as the ratio of the area, A, to the square of the radius, r^2, and the units are expressed in steradians (sr).

In the case where $A = r^2$, $\omega = 1$ sr. If we consider the unit sphere of radius 1 ft with a surface area of 4π ft^2, we see that a point in space is surrounded by 4π sr. We will find the steradian quite useful in more rigorously defining the candela and lumen as well as other terms.

$$\omega = \frac{A}{r^2} \quad \textit{Steradians (sr)} \quad (9.2)$$

Lighting Terms

1 candel = 4π lums

With the above background, we are now in a position to more rigorously define the previous terms, plus additional ones (Murdoch 1985).

Luminous Energy (Q) is the term used to describe luminous power and is related to luminous flux. The units of Q are lumen-seconds.

Luminous Flux (φ) is the rate of flow of luminous energy and is given by:

$$\varphi = \frac{dQ}{dt} \quad \textit{Lumens (lu).} \quad (9.3)$$

Amount of Light Flux in One Sr

Luminous Intensity (I) is the solid angular density of luminous flux in a specific direction and is given by:

$$I = \frac{d\varphi}{d\omega} \quad \textit{Candelas (cd).} \quad (9.4)$$

Illumination (E) is the incident luminous flux (φ) per unit area of receiving surface and is given by:

$$E = \frac{d\varphi}{dA} \quad \textit{Foodcandles (fc).} \quad (9.5)$$

Note that these definitions of Intensity (I) and Illuminance (E) are quite consistent with our previous introductory remarks. Based on these definitions, we now introduce several additional terms:

Luminous Exitance (M) is the luminous flux emitted by a small area of surface divided by the area of the surface, and is given by:

$$M = \frac{d\varphi}{dA} \quad Lumens\,/ft^2. \tag{9.6}$$

Luminance (L) is the luminous flux per unit of projected area (A_θ) per solid angle ($d\omega$) leaving a point in a given direction and is given by:

$$L = \frac{d^2\varphi}{d\omega\,dA_\theta} \quad Candelas\,/ft^2. \tag{9.7}$$

Luminance is associated with the brightness in a given direction, while *luminous exitance* is associated with the density of luminous flux leaving an area. An example frequently used to illustrate these terms is a lamp shade. A lamp shade emitting 50 lumens per square foot of its surface area has a luminous exitance of 50 lu/ft^2. If the candlepower in a given direction from the same 1 ft^2 area is 40 candela, the luminance is 40 candelas/ft^2 in the same direction. At one time, it was quite common to refer to luminance in terms of Footlamberts (candelas/ft^2), but this term is no longer in widespread use.

The concept of projected area is important in the definition of luminance, because it is this area that is actually perceived by a viewer. Figure 9.10 shows a light-emitting surface of length L and width, W. If observed from the location at angle θ, measured from a line perpendicular to the surface, the area may appear less than its actual area, $L \cdot W$, depending on the value of θ.

In fact, the projected area appears as

$$A_p = (L \cdot W)\cos(\theta) \quad ft^2. \tag{9.8}$$

Note that in the special case where $\theta = 0°$, the area is seen as its true total area, $L \cdot W$. In the case where the surface is seen from its side or end, $\theta = 90°$, the projected area is zero.

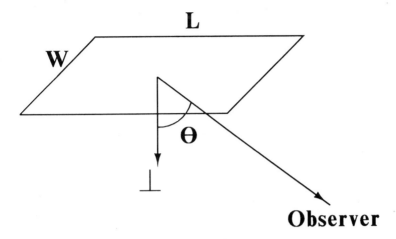

Figure 9.10 Variation of visual size with viewing angle.

Inverse Square Law of Illumination

Using the definitions of luminous intensity (I) and solid angle (ω), we can develop an alternative form for Equation 9.5 which will give us Illumination (E).

Recalling that

$$I \ = \ \frac{d\varphi}{d\omega} \tag{9.4}$$

and

$$d\omega \ = \ \frac{dA}{r^2} \tag{9.2}$$

so that

$$d\varphi \ = \ I \, d\omega$$

and

$$dA \ = \ r^2 \, d\omega.$$

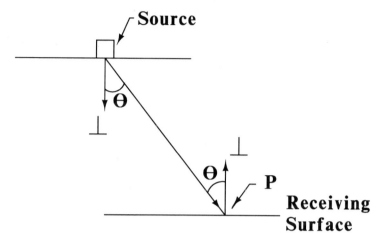

Figure 9.11 Inverse Square Law.

We can express the level of illumination at a point as:

$$E = \frac{d\varphi}{dA} = \frac{I\,d\omega}{r^2\,d\omega} = \frac{I}{r^2} \quad fc. \tag{9.9}$$

The level of illuminance given by Equation 9.9 is correct only if the source of luminous intensity is perpendicular to the surface being illuminated. If the source is at an angle, θ, with the receiving surface as shown in Figure 9.11, the resulting illumination will be given by the modified form of Equation 9.9

$$E = \frac{I}{d^2}\cos(\theta) \quad fc \tag{9.10}$$

where I is the luminous intensity (cd) of the source in the direction of the receiving surface, d the distance in feet from the source to the receiving surface, and θ the angle between the source and a line perpendicular to the receiving surface.

Example 9.1: A given light source emits a luminous intensity of 1500 candelas (cd) in the direction of a surface, as shown in Figure 9.11. The distance from the source to the surface is 15 feet, and the angle between the incident luminous intensity and the perpendicular to the surface is 60°.

What is the level of illuminance at point P?

$$E = \frac{I}{d^2} \cos(\theta) = \frac{1500}{15^2} \cos(60°) = 3.33 \quad fc$$

How Light Interacts with a Surface

When light strikes a surface, it can be *reflected, transmitted, absorbed*, or *refracted*. We will consider each of these in turn (IES Education Series ED-100 1985).

Reflection. Reflection is light which strikes a luminaire surface or a room surface and is reflected. There are several ways light can be reflected, which can be described as *specular, spread, diffuse*, or a *combination* of these. Figure 9.12 shows the characteristics of each of these types of reflection.

Specular Reflection. In specular reflection, shown in Figure 9.12(a), the angle of the incident light is equal to the angle of reflected light. Specular reflectance is exhibited by mirrors or highly polished metals. Several common luminaire materials also exhibit specular reflection, among these are polished or anodized (specular) aluminum and optically coated plastics.

Spread Reflection. Figure 9.12(b) shows the characteristics of spread reflection. Notice that the primary angle of the reflected light is equal to the angle of incident light, as in specular reflection. However, due to the nature of the surface, light is also distributed at angles centered around the primary angle of reflection. Some luminaire surfaces which exhibit spread reflection are etched or brushed aluminum.

Diffuse Reflection. A diffuse surface, shown in Figure 9.12(c), exhibits constant luminance (cd/ft^2) in all directions, and also obeys *Lambert's Law*. This law states that the intensity (cd) in any direction from a uniformly diffuse surface is given by:

$$r_\theta = i \cos(\theta) \quad cd \qquad (9.11)$$

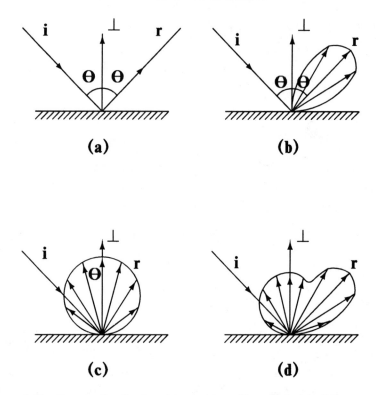

Figure 9.12 Types of reflection (a) specular, (b) spread, (c) diffuse, and (d) combination.

The resulting distribution is often simply called a *Lambert Cosine Distribution*. Notice that the distribution is quite independent of the angle of incidence. Of course, there are no perfect diffusers, but many surfaces do exhibit similar characteristics, among them are flat non-glossy painted surfaces.

Combination. Actually, most surfaces exhibit a combination of specular, spread and diffuse reflection. Figure 9.12(d) shows this kind of surface. The white paint used in many luminaires exhibits such a combination of characteristics.

We will now discuss the other ways light interacts with a surface— transmission, absorption, and refraction.

Transmission. Materials which permit the passage of light as well as visual images are called *transparent*. Material such as clear glass or plastic are examples of transparent materials. On the other hand, materials which allow light to pass through, but do not permit images to be seen are called *translucent*. An example of a translucent material might be milk white plastic. Materials which permit no light to pass through are referred to as *opaque*.

Absorption. Light that is not reflected from the surface, whether it be specular, spread diffuse or a combination of these, is *absorbed*. Absorption plays an important role in our perception of color. If, for example, a surface absorbs all visible radiant energy wavelengths except those in the range of 630 to 760 nm, which are reflected, the surface will appear red. Thus, the color of an object is due to the selective absorption/reflection phenomena.

Refraction. The bending of light rays as they travel from one medium to another of different density is referred to as *refraction*. This bending occurs due to a small slowing in the speed of light in denser materials. Figure 9.13 shows how refraction occurs in the case of a glass panel with air on either side.

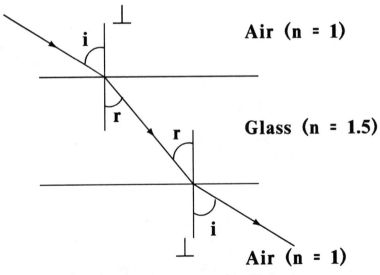

Figure 9.13 Illustration of Snell's Law.

Snell's Law gives us the relationship between the angle of incidence and refraction as a function of the angles involved and the index of refraction of the materials. Snell's law is given by:

$$n_1 \sin(i) = n_2 \sin(r) \tag{9.12}$$

where n_1 and n_2 equal the index of refraction of the first and second materials, respectively, and i and r the angles of incidence and refraction.

This principle is used extensively in the design of luminaire lenses and refractors in order to achieve the desired distribution and brightness control. In actual practice, the design of a luminaire might involve all the principles discussed—reflection, absorption, transmission, and refraction—to achieve the desired photometric distribution.

Glare and Visual Comfort

Most people are quite adaptable to a wide range of lighting environments, even quite poor ones. Perhaps you have noticed how you react when attempting to read in a poorly lighted environment, by turning the written material this way and that, in order to see more easily. This type of corrective measure becomes almost subconscious and most of us are unaware that we attempt to offset poor lighting in this manner. We also hear people complain about spaces that are "too bright" when they are actually suffering from excessive glare resulting from poorly shielded light sources, shiny room surfaces or shiny furnishings. Remembering that our eyes adapt quite well to levels of illumination, varying from several thousand *fc* on a sunny day to well under 50 *fc* in some interior environments, we can see that "too bright" does not refer to the level of illumination, but rather the level of visual discomfort.

We classify glare, according to its source, as either direct or reflected. Visual discomfort can also result from the interaction of light sources with the visual task itself. We will discuss each of these factors in some detail.

Direct Glare. Direct glare originates from a light source directly in the normal field of view. This source of glare can be either natural, such as a poorly shielded window, or artificial, such as a poorly designed or located luminaire. When we consider that our normal field of vision encompasses a large area, including much of the ceiling area and other room surfaces, it is not surprising that direct glare can be quite a problem.

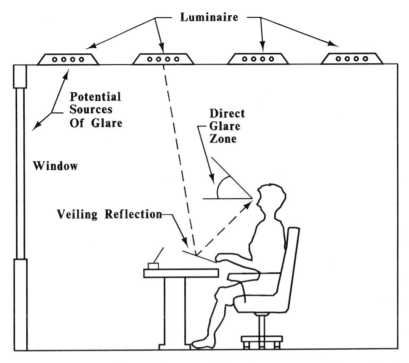

Figure 9.14 Typical office environment illustrating sources of glare and veiling reflections.

Figure 9.14 shows a cross-section of a typical room and the sources of objectionable brightness. The direct glare zone is considered to be the zone that is 45 degrees from the horizontal as shown in Figure 9.14. As you can imagine, one luminaire design goal is to reduce luminous intensity in this zone.

Reflected Glare. Reflected glare originates from reflective or shiny surfaces located within our field of view which reflect light from either natural or artificial sources. Sources of reflected glare may vary from polished picture frames to shiny desk surfaces, or even office equipment, such as computer keyboards. Whatever the source, reflected glare can become quite distracting, thus reducing productivity, as well as causing excessive eye fatigue. We will now discuss how other factors can affect the overall visual comfort of a space.

Table 9.2 Recommended Room Surface Reflectance Ranges for Room Surfaces, Furniture and Equipment. (Reprinted with Permission of Illuminating Engineering Society of North America.)

SURFACE	REFLECTANCE (%)
Ceiling	80 or more
Walls	40 - 70
Furniture	25 - 45
Office Machines and Equipment	25 - 45
Floors	20 - 40

Veiling Reflections.   Veiling reflections result from images of light sources, natural or artificial, reflected from the visual task itself. This veiling reflection tends to reduce the task contrast, thus making effective vision more difficult.

Light sources positioned directly above and slightly forward of the viewing task, as shown in Figure 9.14, can be a serious problem, especially if the task is shiny. Techniques have been developed to determine the degree of lost contrast caused by the location of light sources in this offending zone. The discussion of these techniques are beyond the scope of this work, however, those interested in learning more can refer to the *IES Lighting Handbook* referenced at the end of this chapter.

Room Surface Reflectances. Research has pointed out the importance of room surface reflectances to the overall visual comfort of a space. These surfaces include equipment and furniture located within the space. Since the electrical engineer has little control over the selection of room surface colors, it is very important that he coordinate very carefully with the architect or interior decorator in making choices about lighting. As with so many aspects of building design, the quality and effectiveness of the final product, in this case the illumination system, depends on effective planning and cooperation between several members of the design team.

Recommended room surface reflectance ranges for room surfaces, furniture and equipment are shown in Table 9.2. Reflectances can be expressed in either decimal or percentage form.

EVALUATION OF ILLUMINANCE REQUIREMENTS

One of the first tasks involved in designing the lighting system for a space is making a decision regarding the appropriate level of illuminance. During the 1950s and 1960s the trend seemed to be toward higher and higher levels of illumination. In many instances, designs were conceived without concern for energy consumption. The more conservative period, following the Arab Oil Embargo of 1973, changed that upward trend dramatically. Before 1973, it was common to have a single recommended level of illuminance for a given task, whereas today, we first select an appropriate illuminance range, followed by a process designed to narrow this range down to a specific target value. This newer technique permits us to take factors, such as the worker's age, task, room surface reflectance, and the importance of speed and accuracy into account. The overall result is that the design can be more accurately tailored to specific applications, thus conserving energy. In order to effectively utilize the new system, we must have information on the following items:

1. *The Visual Task*—The exact nature of the visual task. For example, reading black text on a white background is easier than similar text on a dark gray background.
2. *The Age of the Observer*—As discussed in the earlier sections of this chapter, several aspects of vision change with age, making it necessary to provide increased levels of illumination for older workers.
3. *The Importance of Speed and/or Accuracy for Visual Performance*—Here, we must consider whether the visual task is casual, important, or critical. For example, the task of assembling heart monitoring equipment would be considered critical because lives may depend on the finished equipment.
4. *The Reflectance of the Task*—This reflectance will determine the adaptation luminance produced by the lighting.

Careful consideration of all these factors helps us determine the appropriate level of illuminance for the given visual task.

Determining the Appropriate Lighting Level

The appropriate lighting level for a given application is determined as follows:

Type of Activity	Illuminance Category	Ranges of Illuminances		Reference Work-Plane
		Lux	Footcandles	
Public Spaces with Dark Surroundings	A	20-30-50	2-3-5	
Simple Orientation for Short Temporary Visits	B	50-75-100	5-7.5-10	General Lighting Throughout Spaces
Working Spaces Where Visual Tasks are only Occasionally Performed	C	100-150-200	10-15-20	
Performance of Visual Tasks of High Contrast or Large Size	D	200-300-500	20-30-50	
Performance of Visual Tasks of Medium Contrast or Small Size	E	500-750-1000	50-75-100	Illuminance of Task
Performance of Visual Tasks of Low Contrast or Very Small Size	F	1000-1500-2000	100-150-200	
Performance of Visual Tasks of Low Contrast and Very Small Size over a Prolonged Period	G	2000-3000-5000	200-300-500	Illuminance on Task, Obtained by a Combination of General and Local (Supplementary Lighting)
Performance of Very Prolonged and Exacting Visual Tasks	H	5000-7500-10000	500-750-1000	
Performance of Very Special Visual Tasks of Extremely Low Contrast and Small Size	I	10000-15000-20000	1000-1500-2000	

Figure 9.15 Illuminance categories and illuminance values for generic types of activities in interiors. (Reproduced with permission of Illuminating Engineering Society of North America.)

The first step is to determine the appropriate illuminance range as shown in Figure 9.15. Note that there are nine ranges designated "A" through "I." These ranges were established by the Illuminating Engineering Society of North America Application Committee. These nine ranges cover illuminance levels of 2 to 2,000 fc (20 to 20,000 lx). Illuminance ranges A through C represent tasks which normally require uniform, overall illumination of the space. Ranges D through F represent tasks which are generally isolated to specific areas and, hence, these levels of illumination should be restricted to the actual task area. Ranges G through I represent extremely difficult visual tasks. Such levels of illumination may not be economically feasible over even relatively small areas, so a combination of ambient lighting and task lighting (lighting located directly at the work area) is recommended.

To aid in the selection of the appropriate illuminance range, several tables have been prepared by the Illuminating Engineering Society of North America which describe a wide variety of specific lighting applications and suggested illuminance ranges. Figure 9.16 lists a small part of this table that covers lighting applications for reading, assembly and machine shop-type visual tasks.

Once the appropriate illuminance range is established, the next task is to select the specific target value within the range. This is accomplished by determining the sum of several weighting factors, as shown in Figure 9.17.

For categories A through C, we consider the occupant's age and the surface reflectances, and assign appropriate weighting factors. If the sum of these factors, including sign, is -2, we use the lowest of the three levels of illumination, if the total is +2, we use the higher. For all other values, -1, 0 or +1, we use the middle value of the range.

For categories D through I, we consider the worker's age, the importance of speed and accuracy and the reflectance of the task background. If the sum of these three factors is -3 or -2, we use the lowest of the three levels of illumination. If the sum is +2 or +3, we use the highest of the levels. For -1, 0 or +1, we use the middle value of the range.

Example 9.2:   An industrial facility is engaged in the assembly of components for lawn mowers. The visual task is considered to be moderately difficult. Most of the workers are fairly young, and vary in age from 20 to 45 years. The room surface reflectances are approximately 50%, and the reflectance of the task background is less than 30%.

Referring to Figure 9.16, we see that the appropriate illuminance category range for moderately difficult assembly tasks is E. From our previous discussion, we know that illuminance levels in the D through F ranges should be restricted to the specific task areas. Referring next to Figure 9.15, we see that the illuminance range is 50-75-100 *fc*. We will next turn our attention to the weighting factors, shown in Figure 9.17.

The age of the workers varies from 20 to 45 years, so the appropriate weighting factor would be 0. The speed and/or accuracy requirements of the task might be considered to be important, but not critical, so, again, the appropriate weighting factor would be 0. The relatively low task background reflectance, less than 30%, requires a weighting factor of +1. The sum of these (0+0+1 = +1) becomes our overall total.

Area/Activity	Illuminance Category	Area/Activity	Illuminance Category
Conference Rooms		Printed Tasks	
Conferencing	D	6 point type	E[3]
Critical seeing (refer to		8 & 10 point type	D[3]
individual task)		Glossy Magazines	D[13]
		Maps	E
Reading		Newsprint	D
Copied Tasks		Typed originals	D
Ditto Copy	E[3]	Typed 2nd carbon and later	E
Micro-fiche Reader	B[12,13]	Telephone Books	E
Mimeograph	D		
Photographs, moderate detail	E[13]	Assembly	
Thermal copy, poor copy	F[3]	Simple	D
Xerograph	D	Moderately difficult	E
Xerography, 3rd generation	E	Difficult	F
and greater		Very Difficult	G
		Exacting	H
Electronic Data Processing Tasks			
CRT Screens	B[12,13]	Machine Shops	
Impact Printer		Rough bench or machine work	D
Good Ribbon	D	Medium bench or machine work,	
Poor Ribbon	E	ordinary automatic machines,	
2nd Carbon & greater	E	rough grinding, medium	
Inkjet Printer	D	buffing and polishing	E
Keyboard Reading	D	Fine bench or machine work,	
Machine Rooms		fine automatic machines,	
Active Operations	D	medium grinding, fine	
Tape Storage	D	buffing and polishing	G
Machine Area	C	Extra-fine bench or machine	
Equipment Service	E[10]	work, grinding, fine work	H
Thermal Print	E		
Handwritten Tasks			
#3 Pencil & Softer leads	E[3]		
#4 Pencil & Harder leads	F[3]		

Area/Activity	Illuminance Category	Area/Activity	Illuminance Category
Ball-point pen	D[3]		
Felt-tip pen	D		
Handwritten carbon copies	E		
Non-photographically			
reproducible colors	F		
Chalkboards	E[3]		

[3] Task subject to veiling reflections, illuminance listed is not an ESI value. Currently, insufficient experience in the use of ESI target values precludes the direct use of Equivalent Sphere Illumination in the present consensus approach to recommend illuminance values. Equivalent Sphere Illumination may be used as a tool in determining the effectiveness of controlling veiling reflections and as a part of the evaluation of lighting systems.

[10] Only when actual equipment service is in process. May be achieved by a general lighting system or by localized or portable equipment.

[12] Veiling reflections may be produced on glass surfaces. It may be necessary to treat plus weighting factors as minus in order to obtain proper illuminance.

[13] Especially subject to veiling reflections. It may be necessary to shield the task or to reorient it.

Figure 9.16 Selected activities and recommended illuminance ranges. (Reproduced with permission of Illuminating Engineering Society of North America.)

a. For Illuminance Categories A through C

Room and Occupant Characteristics	Weighting Factor		
	-1	0	+1
Occupants Ages	Under 40	40-55	Over 55
Room Surface Reflectances*	Greater than 70 percent	30 to 70 percent	Less than 30 percent

b. For Illuminance Categories D through I

Task and Worker Characteristics	Weighting Factor		
	-1	0	+1
Workers Ages	Under 40	40-55	Over 55
Speed and/or Accuracy**	Not Important	Important	Critical
Reflectance of Task Background***	Greater than 70 percent	30 to 70 percent	Less than 30 percent

*   Average weighted surface reflectances, including wall, floor and ceiling reflectances, if they encompass a large portion of the task area or visual surround. For instance, in an elevator lobby, where the ceiling height is 7.6 meters (25 feet), neither the task nor the visual surround encompass the ceiling, so only the floor and wall reflectances would be considered.
**   In determining whether speed and/or accuracy is not important, important or critical, the following questions need to be answered: What are the time limitations? How important is it to perform the task rapidly? Will errors produce an unsafe condition or product? Will errors reduce productivity and be costly? For example, in reading for leisure there are no time limitations and it is not important to read rapidly. Errors will not be costly and will not be related to safety. Thus, speed and /or accuracy is not important. If, however, prescription notes are to be read by a pharmacist, accuracy is critical because errors could produce an unsafe condition and time is important for customer relations.
***   The task background is that portion of the task upon which the meaningful visual display is exhibited. For example, on this page the meaningful visual display includes each letter which combines with other letters to form words and phrases. The display medium, or task background, is the paper, which has a reflectance of approximately 85 percent.

Figure 9.17 Weighting factors to be considered in selecting specific illuminance with ranges of values for each category. (Reproduced with permission of Illuminating Engineering Society of North America.)

From this, we see that the appropriate choice within the illuminance range is the middle value, 75 *fc*. With the specific target level of 75 *fc*, we can then proceed with our design.

References

Boylan, Bernard R. 1987. *The Lighting Primer.* Iowa: Iowa State University Press.

Cox, James A. 1987. *A Century of Light*. New York: The Benjamin Company, Inc.

General Electric Company Lighting Business Group. 1978. *Light and Color*. Ohio: General Electric Company.

Helms, Ronald N. 1980. *Illumination Engineering for Energy Efficient Luminous Environments*. New Jersey: Prentice-Hall, Inc.

Illuminating Engineering Society of North America. 1985. *IES Education Series ED-100 - IES Lighting Fundamentals*. New York: Illuminating Engineering Society of North America.

Illuminating Engineering Society of North America. 1984. *IES Lighting Handbook -Reference Volume*. New York: Illuminating Engineering Society of North America.

Murdoch, Joseph B. 1985. *Illumination Engineering - From Edison's Lamp to the Laser*. New York: Macmillan Publishing Company.

10

Light Sources

In this chapter we will consider the various types of light sources commonly used in building systems. These sources can be broadly classified as incandescent, fluorescent, or high intensity discharge sources. The incandescent source is the oldest, and as discussed in the introductory section of Chapter 9, was introduced in 1879. The fluorescent lamp was introduced in 1938. There are actually three discrete types of high intensity discharge, or HID, sources. These are the mercury vapor lamp, first introduced in 1930, the metal halide lamp, introduced in 1960, and the high pressure sodium lamp, introduced in 1965.

Each of these light sources has undergone several major refinements, and further development and research continue even today. This research has led to lamps which are even more efficient and compact, and which produce light at higher levels of color rendition and luminous efficacy.

From a design standpoint, we will be concerned with several significant characteristics of each light source. Among these are:

- The construction and operation of the light source
- The requirement for starting aids, such as ballasts
- The quality of the light output, primarily in terms of the color rendering qualities
- The overall efficiency of the source in terms of its lumens-per-watt ratio
- The rated lamp life
- The depreciation in light output over the life of the lamp
- The various lamp wattages and configurations commonly in use

367

By careful consideration of the above characteristics, it is possible to select the appropriate light source for a given application. It should also be remembered that each source has advantages and disadvantages which must be given consideration. In the final analysis, the actual selection is often a carefully considered compromise, based on the design objective and the characteristics of the various available light sources. Let us now consider each of these light sources in turn.

INCANDESCENT

The incandescent light source was first made feasible by Edison's work in the 1870s. This source has undergone almost continuous evolution since, and the process promises to continue well into the future. The incandescent lamp has often been described as a "hot wire in a bottle" and this is a pretty fair description. The basic theory of operation is the passing of current through a resistive element called a filament. This current raises the temperature of the filament to the point that it begins to glow or incandesce. As discussed in Chapter 9, the search for appropriate filament materials was extensive and ranged over a wide variety of carbonized materials. The *bottle* served to both protect the relatively fragile filament assembly, and also to contain the necessary environment for optimal filament life. It was discovered early on that filaments operated in air for only a short time before failure. By evacuating the air from the lamp envelope, the life was significantly extended.

Figure 10.1 shows the major components of a modern day incandescent lamp. Today, the filament is generally made of tungsten, and is in either a coiled coil or triple coiled configuration to optimize lamp performance. The glass envelope is filled with an inert gas, containing argon, krypton, or helium, which reduces the evaporation of filament material, thus extending lamp life. This gas fill is accomplished during the manufacturing process by, first, evacuating the air from the lamp by means of the exhaust tube, shown at the bottom of Figure 10.1, followed by injection of the desired inert gas mixture, and the sealing of the tube end. Wires are installed in the lamp interior to provide the necessary degree of filament support, and also to provide a path for external electrical connection. In many lamps, the connection path contains a fused element to rapidly isolate the lamp in the event of an internal short. Actual connection of the lamp to the electrical system, as well as mechanical

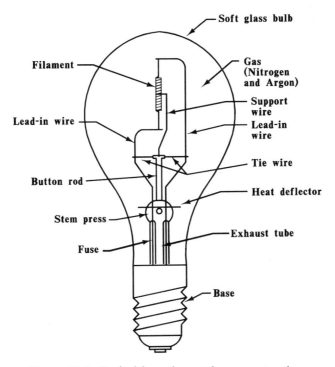

Figure 10.1 Typical incandescent lamp construction.

support, is provided by the lamp base. Although there are a wide variety of incandescent lamp bases, the medium base type, shown in Figure 10.1, is one of the most widely employed. The interior of the lamp is often frosted in order to more evenly distribute the generated light and also to reduce glare.

Incandescent Lamp Configurations

Incandescent lamps are available in a wide variety of configurations, some of which are shown in Figure 10.2. Each basic configuration is designated by a letter, or combination of letters, of the alphabet. For example, the *G* designation represents lamps that are basically globe-like in shape, and the *F* designation represents lamps which are shaped somewhat like flames. *T* lamps are tubular in shape, and *PS* lamps are pear-shaped with straight sides. Perhaps one of the most widely used lamp shapes is the *A* lamp, which is somewhat pear-shaped, but with an elongated neck.

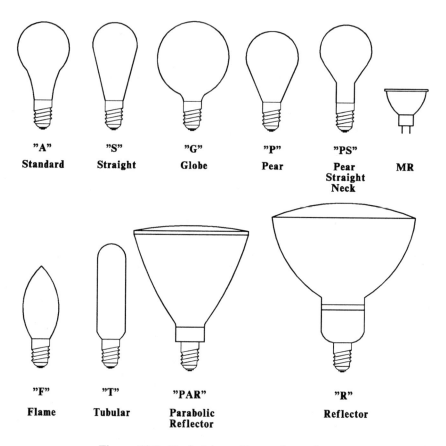

Figure 10.2 Typical incandescent lamp shapes.

A second general incandescent lamp group consists of lamps which contain their own reflectors. These are the *PAR*, parabolic reflector lamps, and the *R*, reflector lamps.

Incandescent Lamp Designations

Incandescent lamps, like the other light sources we will study, are designated in a standard manner. For example, consider the following lamp designation:

100 A 23/IF 120 V

The *100* refers to the lamp wattage, 100 watts, and the *A* to the lamp shape. The *23* represents the lamp diameter at its widest point in 1/8s of an inch. The lamp is then 23/8s, or slightly under three inches in diameter. The *IF* indicates that the lamp is frosted inside, and the *120 V* is the nominal lamp voltage.

Incandescent Lamp Rated Life

The *rated life* of an incandescent lamp is defined as the point in time during the burning-life at which 50% of the lamps have failed. This is generally in the 750 to 1000 hour range for most incandescent lamps. Figure 10.3 shows a typical incandescent lamp mortality curve. Note that

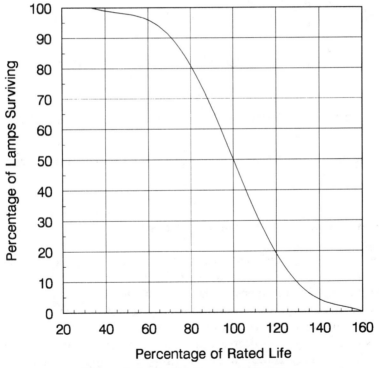

Figure 10.3 Typical lamp mortality curve.

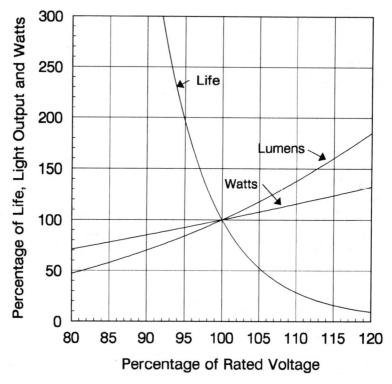

Figure 10.4 Effect of voltage on incandescent lamp life, lumen output and watts.

some lamps fail well before rated life, while others continue to burn well beyond the end of rated life.

Effects of Voltage on Incandescent Lamps

Incandescent lamps are quite sensitive to variations in applied voltage. In 1932, Barbow and Meyer developed techniques for quantifying the effect of applied voltage variations on the lumen output and power input. Later work extended these relationships to include lamp life. Figure 10.4 shows the effect of voltage variations on those three parameters.

Note that only a 5% decrease in applied voltage significantly increases life (approximately 200%), while decreasing light output to about 80% of the rated value. On the other hand, increasing the applied voltage by 5%

reduces life to about 55% of rated value, while increasing light output by only about 20%.

Light output, lamp life, and applied voltage are frequently traded off in various design applications. For example, in applications in which the burning-hours per year are very low, increased light output can be obtained by increased voltage, and conversely, when more burning-hours are involved, increased lamp life can be obtained, at the expense of light output, by reducing the applied voltage. In actuality, the relatively low overall efficacy of the incandescent lamp, in the range of 15-20 lumens/watt, and short lamp life, seriously restrict the use of incandescent sources in all but residential applications.

Lamp Lumen Depreciation

As with all light sources, the lumen output of the incandescent lamp decreases over the life of the lamp. This is due to factors such as the evaporation of the filament material, and the blackening of the interior of the lamp. Filament evaporation results in a decrease in the filament cross sectional area and, hence, greater resistance, lower current, lower operating temperature, and lower light output. The evaporated filament material is deposited on the interior surface of the lamp jacket, resulting in darkening and loss of light transmission. Figure 10.5 shows a typical lumen depreciation curve for an incandescent lamp. Note that the lumen depreciation over the life of the lamp is, essentially, a straight line.

Tungsten Halogen Cycle

The recognition that filament evaporation, and the attendant blackening of the lamp interior, significantly affect the life and lumen output of the lamp, prompted considerable research to find a means of reducing filament evaporation. It was discovered that the evaporated filament material is carried to the lamp wall by gas convection currents. It was found that by the addition of a small amount of halogen gas, filament evaporation could be reduced as follows: The tungsten evaporated from the filament combines with the halogen gas to form tungsten iodide, which is conveyed toward the lamp jacket by convective currents. If the lamp wall is maintained at

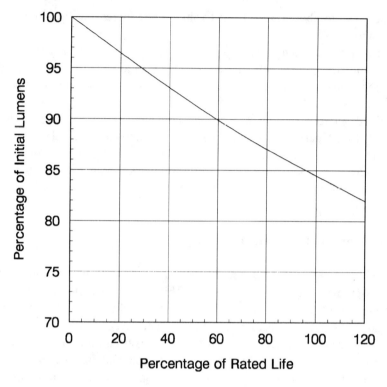

Figure 10.5 Typical incandescent lamp lumen depreciation curve.

a much higher temperature than in normal incandescent lamps tungsten iodide is not deposited on the bulb wall, but rather is carried back to the vicinity of the filament, where it is reduced to tungsten and iodine vapor. A significant portion of the tungsten is then redeposited on the filament. This process is repeated continuously during the life of the lamp and is referred to as the tungsten halogen cycle. Tungsten halogen lamps have improved light output, lumen depreciation characteristics and a longer lamp life. The *MR* lamp, shown in Figure 10.2, is a member of the tungsten halogen family.

Table 10.17 lists data on a representative sampling of incandescent lamps. In actuality, there are hundreds of different lamp types for a wide variety of both general and specialized applications.

Table 10.1 Typical Incandescent Lamp Performance Data.

Bulb Type	Watts	Rated Life (hours)	Initial Lumens	Initial Lu/W	Comments
A-19	75	750	1190	16	General Service Lamp
A-19	100	750	1750	18	General Service Lamp
A-23	150	750	2780	19	General Service Lamp
G-25	60	1500	660	11	Decorative Globe
G-40	100	2500	1280	13	Decorative Globe
PAR-38	75	2000	765	10	Parabolic Reflector
PAR-38	150	2000	2000	13	Parabolic Reflector
R-30	75	2000	900	12	Reflector Lamp
R-40	100	2000	1190	12	Reflector Lamp
R-40	150	2000	1900	13	Reflector Lamp
MR-11	20	3000	260	13	Tungsten Halogen
MR-16	50	3000	895	18	Tungsten Halogen
MR-16	75	3500	1300	17	Tungsten Halogen
T-3	75	2000	1350	18	Tungsten Halogen
T-4	100	2000	1600	16	Tungsten Halogen
T-4	150	2000	2800	19	Tungsten Halogen
T-4	150	2000	5000	20	Tungsten Halogen
T-4	500	2000	10450	21	Tungsten Halogen

Reprinted with permission of G.E. Lighting, Nela Park, Cleveland, Ohio.

Incandescent Lamp Color

Figure 10.6 shows the spectral distribution characteristics of a typical incandescent lamp. Note that the light output is essentially continuous over the 300 to 900 nanometer range. Note also that the relative power output is greatest in the warmer colors, such as orange and red. Most people find this color rendering quality quite desirable, and as a result, the spectral characteristics of other light sources are often compared with that of the

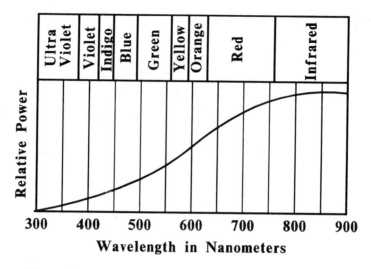

Figure 10.6 Typical incandescent lamp spectral distribution.

incandescent lamp. Under incandescent light, earth colors are enhanced, making this source popular for residential applications, and for other social spaces where people often prefer such colors.

FLUORESCENT LAMPS

Fluorescent lamps are low-pressure, gas-discharge sources which were first introduced in the 1930s. Fluorescent lamps are typically tubular in shape and vary in length from just a few inches to eight feet. The very first lamps, introduced in 1938, were the 15 W (18"), 20 W (24"), and 30 W (36") lamps. In 1939, the 40 W (48") lamp was introduced. During these early days, the luminous efficacy of the fluorescent lamp was about 35 lu/w. Today, this efficacy has doubled. Fluorescent lamps have made an enormous impact on offices, factories, schools, shopping centers and many other types of buildings. Recently, it was estimated that 70% of all light generated in North America is generated by fluorescent sources. The evolution of the fluorescent source has continued, and the recent development of greatly improved phosphors and new compact lamp designs promise to keep this versatile source in the forefront for many more years.

We will now consider the more important characteristics of the fluorescent source, including its construction, theory of operation, auxiliary equipment requirements, color rendition, lamp life and lumen maintenance.

Theory of Operation and Construction

The fluorescent lamp is a gas-discharge source. Figure 10.7 shows the major construction characteristics of a typical lamp. The lamp is cylindrical in shape, with an overall diameter varying from under 0.5" to about 1.5". At each end of the tube is an electrode, or "hot cathode," which is quite similar to the incandescent lamp tungsten filament.

Despite this similarity, the fluorescent cathode performs a different function from its incandescent counterpart. The incandescent filament is heated to incandescence by the passage of current, thus directly producing light. The fluorescent cathode's function is to produce electrons, and as a result, the cathode is coated with an emissive material. In addition, the fluorescent cathode is operated at a far lower temperature than its incandescent counterpart, therefore vastly extending its life. The fluorescent tube is filled with a gas consisting of mercury, as well as several inert gasses, such as Argon, Krypton, and Neon. The tube pressure is very low. A voltage is applied between the opposing ends of the tube, and electron flow is established between the cathodes. This arc occurs in the presence of mercury atoms. When the electrons collide with the mercury atoms there is a release of energy. This energy release occurs primarily in the non-visible portion of the ultraviolet band, at about 253.7 nanometers. The key to the successful light production of the fluorescent source is the conversion of this non-visible energy to visible energy by means of a coating of phosphor on the interior surface of the tube. This phosphor is sensitive to ultraviolet energy, and reacts by fluorescing or reradiating a significant portion of this incident energy in the visible spectrum. The phosphors are carefully formulated to have a peak sensitivity near 253.7 nanometers. In addition, it is possible to formulate phosphors which produce various colors, thus making a wide variety of fluorescent lamp colors possible. Fluorescent lamp phosphors are temperature sensitive, and their efficiency drops significantly at ambient temperatures above or below the design level.

We will now consider another important aspect of fluorescent lamp operation and that is the nature of an arc through gas. Once established, the arc itself has a very low impedance, and the resulting current flow

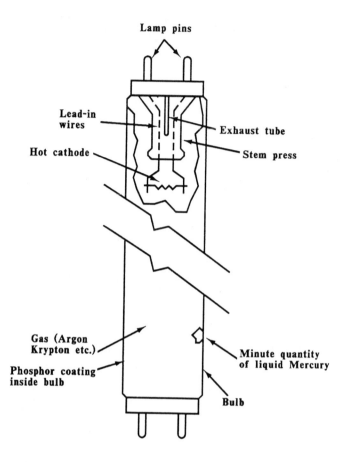

Figure 10.7 Typical fluorescent lamp construction.

would rapidly destroy the lamp unless limited by external means. The net result is that all fluorescent sources require a device called a *ballast* which performs the dual function of providing the correct voltage for starting and operating the lamp, and also of limiting the current flow to the appropriate value.

Despite the fact that all fluorescent lamps operate in essentially the same way, namely an arc passing through gas, the actual techniques employed to start and operate the lamps vary. Fluorescent lamps are available in several configurations, as shown in Figure 10.8. We will now consider several of these starting methods.

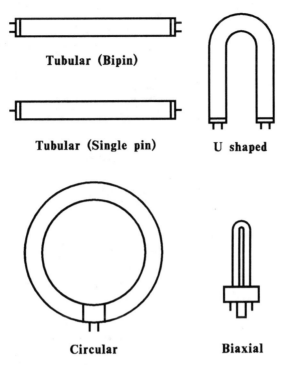

Tubular (Bipin)

Tubular (Single pin) **U shaped**

Circular **Biaxial**

Figure 10.8 Typical fluorescent lamp shapes.

Fluorescent Lamp Starting Techniques

There are three basic approaches to starting and operating a fluorescent lamp. These are the *preheat approach*, first employed in 1938, the *instant start* method, introduced in 1944, and the *rapid start* method, introduced in 1952. The rapid start system became so popular that it is probably the most widely used starting technique today. We will consider each of these techniques in some detail.

Preheat Method
In this approach, the cathodes are heated (preheated) for several seconds, prior to the application of a high voltage to start the lamp. The preheat method was the first starting technique developed, and is still in use today. Figure 10.9(a) shows the ballast circuit for the preheat method. Note the starting switch, which preheats the cathodes for several seconds, and then opens to allow application of starting voltage across the tube. In most

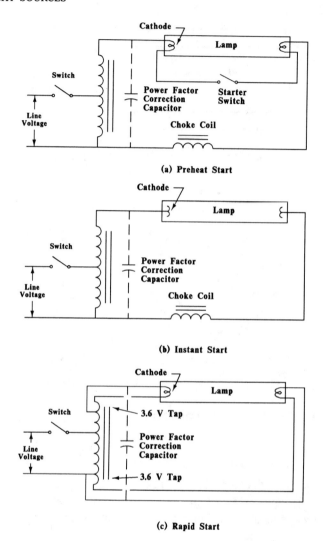

Figure 10.9 Fluorescent ballast circuits (a) preheat, (b) instant start, and (c) rapid start.

applications, the starter functions automatically, and is actually a current-sensitive switch which opens after several seconds. In some applications, the starter is a momentary contact switch, which is manually closed to preheat the lamp(s), then released to permit lamp starting. This starting technique is often employed in desk lamps. The remainder of the circuit, shown in Figure 10.9(a), essentially a transformer, provides the correct lamp voltage, and also acts as a current limiter, or choke, to limit the lamp current to the specified design level.

Instant Start Method

In the instant start method, there is no preheating of the cathode material, in fact, there is only one conductor and, hence, a single pin at each end of the lamp. Figure 10.9(b) shows a typical instant start system. Lamp starting is accomplished by the application of a relatively high voltage (400 - 1,000 volts) to the cathodes. The arc itself, once established, provides the necessary cathode heating. As you might imagine, this brute force starting is somewhat stressful on the cathodes. Instant start lamps are commonly called *slimline* lamps, and the higher voltage present at the lamp pins requires additional protection to lessen the hazard to personnel. This increased protection is normally provided by recessed spring-loaded lamp contacts.

Rapid Start Method

The rapid start method is perhaps the most widely used lamp starting method today. The cathodes are constantly heated during lamp operation by a relatively low voltage, in the range of 3.6 volts, supplied by separate windings within the ballast. The actual starting voltage requirements are similar to those of the preheat lamp, and lamp starting occurs in about one second. A typical circuit for rapid start lamp operation is shown in Figure 10.9(c).

Ballast Construction

The major components of a typical ballast are a transformer core and coil, and possibly a capacitor. The transformer supplies the appropriate starting and operating voltage, and also acts as a current-limiting choke. The capacitor is employed to raise the overall power factor of the ballast, offsetting the inductive nature of the choke. The entire ballast assembly is mounted in a metal enclosure that is filled with a potting material that encapsulates the core, coil and transformer. Because the transformer utilizes many laminations of magnetic steel, it is possible to experience some degree of ballast noise, based on the frequency of the applied voltages. The total potential ballast noise in a space will be a function of the ballast type, as well as the luminaire design and method of mounting. Ballasts are available which are designed to exhibit various degrees of sound reduction, and these are often given sound ratings, using the letters A through F, with A being the quietest. At the present time, there are no recognized national standards covering these sound ratings.

During recent years, we have seen the introduction of new energy-saving ballasts which can be used with standard lamps, or a new generation of lower wattage energy-saving lamps. Special ballast designs are available to permit fluorescent lamp starting at low temperatures (-20°F), for outdoor applications. All modern ballast designs include a thermal cutout device to automatically remove the ballast from operation if a predetermined ballast case temperature limit, 110°C, is exceeded. During the past several years, several manufacturers have also introduced new ballast designs which employ electronic circuits, rather than the traditional core and coil arrangements. These new ballasts generally offer very high power factor operation, and are also lighter in weight. One potential drawback of electronic ballasts is the possible increased level of current wave distortion. This harmonic distortion can create significant power system problems, which are addressed in the chapter on power quality. Properly installed and maintained, fluorescent ballasts should have a useful life of 12 to 15 years, although many ballasts may operate well beyond this time.

Fluorescent Lamp Rated Life

Fluorescent lamps have very long-rated life, often in the 7,500 to 20,000 hour range. Table 10.2 lists typical characteristics for a variety of fluorescent lamps. Note, also, that the luminous efficacy of modern fluorescent sources ranges from about 45 lu/w to well over 80 lu/w. It should be noted that the life of a fluorescent lamp is affected by the frequency of lamp starting because, each time the lamp is started, some amount of deterioration of the cathode occurs. The longest lamp life is obtained by continuous burning, but this is not normally an economically feasible possibility. The data shown in Table 10.2 is based upon three hours of operation per start.

Table 10.2 Typical Fluorescent Lamp Performance Data.

Bulb Type	Length (in.)	Watts	Description	Starting	Initial Lumens	Initial Lu/W	Rated Life	Hrs/ Start
T12	18	15	Cool White	Preheat	675	45	9000	3
T12	18	15	Warm White	Preheat	870	58	7500	3
T12	24	20	Cool White	Preheat	1250	63	9000	3
T12	24	20	Warm White	Preheat	1300	65	9000	3
T12	36	30	Cool White	Preheat	2200	73	7500	3
T12	36	30	Warm White	Preheat	2300	77	7500	3
T12	48	40	Cool White	Rapid Start	3150	79	20000	3
T12	48	40	Warm White	Rapid Start	3200	80	20000	3
T12	48	40	Deluxe Cool White	Rapid Start	2250	56	20000	3
T12	48	40	Cool White	Instant Start	3000	75	13000	12
T12	48	40	Warm White	Instant Start	3050	76	13000	12
T12	96	75	Cool White	Instant Start	6300	84	18000	12
T12	96	75	Warm White	Instant Start	6500	87	18000	12
T12	96	75	Deluxe Warm White	Instant Start	4350	58	18000	12
T12	48	60	Cool White	Rapid Start	4300	72	18000	12
T12	48	60	Warm White	Rapid Start	4300	72	18000	12
T12	96	110	Cool White	Rapid Start	9200	84	18000	12
T12	96	110	Warm White	Rapid Start	9200	84	18000	12
T12	96	100	Deluxe Warm White	Rapid Start	6550	66	18000	12
T12	48	110	Cool White	Rapid Start	6500	59	12500	12
T12	48	110	Warm White	Rapid Start	6300	58	12500	12
T12	96	215	Cool White	Rapid Start	14000	65	12500	12
T12	96	215	Warm White	Rapid Start	13600	63	12500	12
T4	5-5/16	7	Biaxial (SPx27)	Preheat	400	57	10000	3
T4	7-1/2	13	Biaxial (SPx27)	Preheat	850	65	10000	3
T5	10-1/2	18	Biaxial (SPx30)	Rapid Start	1250	69	20000	3
T5	12-27/32	24	Biaxial (SPx30)	Rapid Start	1800	75	10000	3
T5	16-1/2	36	Biaxial (SPx30)	Rapid Start	2850	79	10000	3
T5	22-1/2	40	Biaxial (SPx30)	Rapid Start	3150	79	20000	3

Reprinted with permission of G.E. Lighting, Cleveland, Ohio. Rated Life Values are given for three hours per start. Initial lumens are measured after 100 hours of operation.

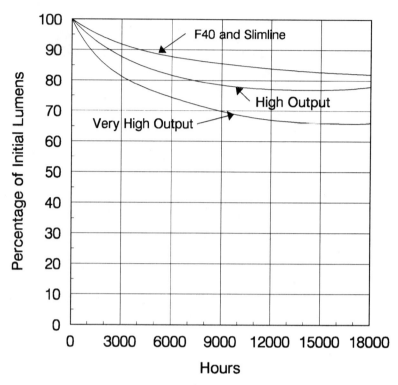

Figure 10.10 Typical fluorescent lamp lumen depreciation curves.

Fluorescent Lamp Lumen Depreciation

As with all other light sources, the light output of fluorescent lamps decreases as the lamp ages. To some extent, this lumen depreciation is a function of the level of current flowing through the tube, and its effect on the fluorescent phosphors. Lightly loaded lamps, such as the standard 40 W lamp, have the best lumen maintenance characteristics, while the heavier loaded lamps, such as the very high output (*VHO*) lamp, have the poorest characteristics. A graph of the lumen maintenance characteristics of three types of fluorescent lamps is shown in Figure 10.10.

Fluorescent Lamp Color

As mentioned earlier, the selection and formulation of fluorescent phosphors determines the spectral range of the fluorescent lamp. Figure 10.11 shows the spectral characteristics of three of the most commonly used fluorescent sources which are cool white, warm white and deluxe warm white. The cool or warm white lamps are commonly used in applications where good color rendering qualities are desirable, but overall lamp efficacy is also a major concern. These lamps have an initial luminous efficacy of approximately 80 lu/w. The spectral distribution of the cool white and warm white lamps is shown in Figure 10.11(a) and 10.11(b), respectively. Note that the warm white lamp is somewhat richer in the yellow, orange and red wavelengths, which results in its excellent color rendering properties. The deluxe warm white lamp is employed

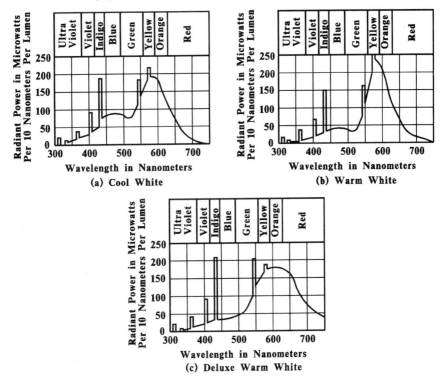

Figure 10.11 Typical fluorescent lamp spectral distribution (a) cool white, (b) warm white, and (c) deluxe warm white.

where color rendering characteristics are a major concern, and its lower luminous efficacy (56 lu/w) is acceptable. Figure 10.11(c) shows the spectral distribution for a typical deluxe warm white lamp. Note the shift in distribution toward the orange and red area of the spectrum.

MERCURY VAPOR

The mercury vapor lamp is the oldest of the high intensity discharge (HID) light sources, and was introduced in 1930. Like the other HID sources, metal halide and high pressure sodium, the mercury vapor lamp produces light by passing current through a gas contained in an arc tube. These arc tubes are quite compact, and are generally mounted inside an outer glass envelope. In this section, we will discuss the characteristics of the mercury vapor lamp, and in the succeeding sections, we will present a similar treatment of metal halide and high pressure sodium light sources.

Mercury Vapor Lamp Construction

Figure 10.12 shows the details of a typical mercury vapor lamp. As mentioned previously, this lamp produces light by passing an electric arc through a mixture of mercury and argon gas. The original arc is struck through the argon gas, and as the heat builds up, the mercury is vaporized. The arc tube contains the mercury and argon gas, as well as the starting and operating electrodes. Under normal operating conditions, the pressure in the arc tube is between two and four atmospheres. The arc tube itself is made of a fused silicon, while the outer jacket is typically made of a hard borosilicate glass. The purpose of the starting electrode is to establish a small arc between itself and the adjacent operating electrode. This arc initiates the heat buildup and ionization of the gas. The arc is then established between the operating electrodes, and the lamp gradually develops its full light output over a period of several minutes. Unfortunately, should an inadvertent power outage occur, the lamp must cool down for several minutes in order for the starting process to reoccur. The electrodes are, like fluorescent lamp cathodes, coated with an emissive material. The arc tube is supported within the outer lamp envelope by support wires, which help dampen vibration, and which also allow for the expansion and contraction of the assembly during operation. The outer

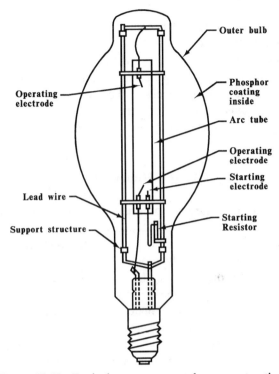

Outer bulb

Phosphor
coating
inside

Arc tube

Operating
electrode

Starting
electrode

Starting
Resistor

Operating
electrode

Lead wire

Support structure

Figure 10.12 Typical mercury vapor lamp construction.

jacket is normally filled with an inert gas, such as nitrogen, that reduces oxidation of the internal components.

As we will see in a later section, the color emitted by a mercury vapor lamp is very heavy in the blue-green portion of the visible spectrum. In addition, the lamp emits significant levels of ultraviolet energy. In order to improve the somewhat dismal color rendering qualities of this lamp, a phosphor coating is often added to the interior surface of the outer jacket. This phosphor coating functions in much the same manner as the phosphor in a fluorescent lamp, by reradiating the incident ultraviolet energy in the visible spectrum, thus dramatically improving the lamp color. This improved color rendering quality does not come without a price, however. In many luminaire designs the small size of the arc tube, and, hence, the light source, makes it possible to design reflectors based on an almost point source of light. The clear outer jacket tends not to trap excessive light in the luminaire. In the case of the phosphor coated lamp, the light emitting surface is no longer the small arc tube, but the entire phosphor-coated jacket. Accurate beam control becomes more difficult, and an increased level of light entrapment occurs.

Newer and more efficient members of the HID family have largely replaced the mercury vapor lamp during the past few years. The mercury vapor lamp, like the other members of the HID and fluorescent family of lamps, requires a ballast for proper starting and operation. These ballasts will be discussed in a later section of this chapter.

Mercury Vapor Lamp Life and Lumen Maintenance

Over the years, since its introduction, the design of the mercury vapor lamp has evolved considerably, and the lamp enjoys a rated life of as long as 24,000 hours. As mentioned previously, the electrodes are coated with an emissive material that is slowly eroded over the life of the lamp. The evaporated material is deposited on the interior surface of the arc tube, first as a white film, and then as a darker material. The lamp reaches the end of its life when all the emissive material is exhausted, and the applied voltage is insufficient to start the lamp.

The initial lamp lumens, listed for most mercury vapor lamps, are taken after 100 hours of operation. At this point, the lamp has cleaned up any initial impurities present in the arc tube. Figure 10.13 shows a typical lumen maintenance curve for a mercury vapor lamp.

Unfortunately, the mercury vapor lamp has somewhat poor lumen maintenance characteristics. This poor lumen maintenance, combined with the unfortunate color rendering properties, has resulted in decreased use of this light source during recent years, in favor of the metal halide and high pressure sodium lamps. The combination of long lamp life and poor lumen maintenance has resulted in a change in the definition of rated life for the mercury lamp. The lumen depreciation is so great that, at the actual point in time at which 50% of the lamps have failed, the lumen output is so low that continued use cannot be justified. In fact, at 24,000 hours, the rated life, approximately 65% of the lamps survive.

The luminous efficacy of the mercury vapor lamp is in the 37 to 63 lumens-per-watt range. The efficacy does not include ballast losses. Table 10.3 lists representative lamp data for the mercury vapor, as well as other HID sources. The rated lamp life values are based on 10 hours per start.

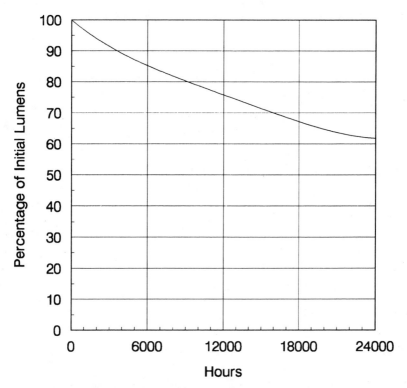

Figure 10.13 Typical mercury vapor lamp lumen depreciation.

METAL HALIDE

The metal halide lamp was introduced in 1960, and is very similar to the mercury vapor lamp, but offers several design improvements in the area of both color rendition and lumen maintenance. Figure 10.14 shows the construction details of a typical metal halide lamp. The arc tube is much smaller than that of a similar mercury vapor lamp. The tube contains metal halide in addition to mercury and argon. For lamps designed for horizontal burning, the arc tube is often curved in order to maintain the arc near the center of the tube. There are several combinations of halides commonly used. Common combinations are sodium, thallium, and indium iodides, sodium and scandium iodides and dysprosium and thallium iodides. The metal halide lamp requires a higher voltage to initiate the arc than the mercury vapor lamp.

Table 10.3 Typical High Intensity Discharge (HID) Lamp Performance Data.

Mercury Vapor

Watts	Bulb	Base	Lamp Type	Initial Lumens	Initial Lu/w	Rated Life
100	E17	Mogul	Clear	3700	37	18000
100	E23½	Mogul	Deluxe White	4200	42	24000
175	E28	Mogul	Clear	7950	45	24000
175	E28	Mogul	Deluxe White	8600	49	24000
250	E28	Mogul	Clear	11200	49	24000
250	E28	Mogul	Deluxe White	12100	48	24000
400	E37	Mogul	Clear	21000	53	24000
400	E37	Mogul	Deluxe White	22500	56	24000
1000	BT56	Mogul	Clear	57000	57	24000
1000	BT56	Mogul	Deluxe White	63000	63	24000

Metal Halide

Watts	Bulb	Base	Lamp Type	Initial Lumens	Initial Lu/w	Rated Life
175	E28	Mogul	Clear	14000	80	10000
175	E28	Mogul	Phosphor Coated	14000	80	10000
250	E28	Mogul	Clear	20500	82	10000
250	E28	Mogul	Phosphor Coated	20500	82	10000
400	E37	Mogul	Clear	36000	90	20000
400	E37	Mogul	Phosphor Coated	36000	90	20000
1000	BT56	Mogul	Clear	110000	110	12000
1000	BT56	Mogul	Phosphor Coated	105000	105	12000

High Pressure Sodium

Watts	Bulb	Base	Lamp Type	Initial Lumens	Initial Lu/W	Rated Life
35	E17	Medium	Clear	2250	64	24000
50	E17	Medium	Clear	4000	80	24000
70	E23½	Mogul	Clear	6400	91	24000
100	E23½	Mogul	Clear	9500	95	24000
150	E23½	Mogul	Clear	16000	107	24000
200	E18	Mogul	Clear	22000	110	24000
250	E18	Mogul	Clear	27500	110	24000
310	E18	Mogul	Clear	37000	119	24000
400	E18	Mogul	Clear	50000	125	24000
1000	E25	Mogul	Clear	140000	140	24000

Reprinted with permission of G.E. Lighting, Cleveland, Ohio. Rated Life Values are given for ten hours of operation per start. Initial lumens are measured after 100 hours of operation.

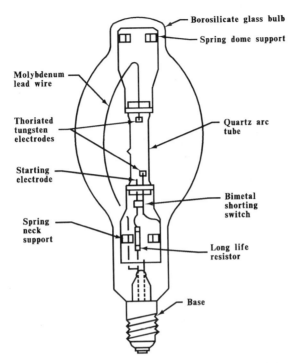

Figure 10.14 Typical metal halide lamp construction.

The metal halide lamp has a significantly higher luminous efficacy than the mercury vapor lamp, with typical efficacies in the 80 to 110 lumen-per-watt range. Note that this range of efficacies does not include ballast losses, which must be included for a more accurate total view.

The metal halide lamp also exhibits a unique characteristic during startup. As the lamp arc is initiated and stabilized over a period of several minutes, the lamp goes through a sequence of color changes before stabilization at its operating color. Fortunately, the metal halide increases the overall color rendering characteristics of the lamp. The color is often described as *white,* and as a result, this source was the first HID source to be widely employed in indoor applications in which color rendering was an important consideration. Another significant difference between the metal halide and mercury vapor lamps is the time required to restrike the arc following a power failure. The metal halide lamp requires up to 15 minutes for cooldown and arc restrike to occur.

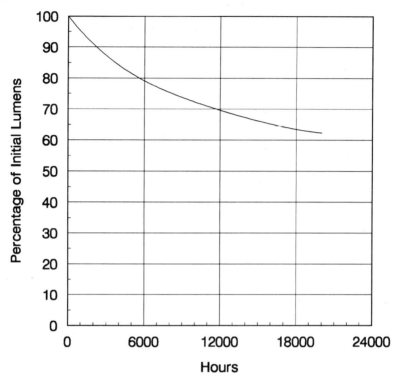

Figure 10.15 Typical metal halide lamp lumen depreciation curve.

Metal Halide Lamp Life and Lumen Maintenance

The metal halides in the arc tube have a negative effect on the electrode emissive material, resulting in a somewhat increased rate of evaporation of electrode material. As a result, the metal halide lamp has a shorter lamp life than the mercury vapor lamp. The lamp life is generally in the 18,000 hour range. Figure 10.15 shows a lumen maintenance curve for a typical metal halide lamp. Like the mercury vapor lamp, the initial lamp lumen value is established at the 100 hour burning point. Note that the lumen maintenance characteristics are somewhat dependent on the number of burning hours per start, with the best overall lumen maintenance occurring with continuous burning.

Like the other fluorescent and HID sources, the metal halide lamp requires a ballast for proper starting and operation. However, because of its higher starting voltage requirements, the metal halide lamp requires a ballast especially designed for its requirements, and in general, mercury vapor lamps and metal halide lamps do not operate efficiently on the same ballast.

HIGH PRESSURE SODIUM

The last member of the HID family we will consider is the high pressure sodium lamp, which was introduced in 1965. This lamp has undergone significant refinement, and is the most efficient of the HID family, with a luminous efficacy of 64 to 140 lumens-per-watt, not counting ballast losses. The lamp is often characterized by its distinctive color, described as warm white by some and, less charitably, as *plain yellow* by others. Figure 10.16 shows the construction details of a typical high pressure sodium, HPS, lamp.

Notice that the arc tube is longer and thinner than in either the mercury vapor or metal halide lamps. In this lamp, light is produced by passing current through a sodium vapor. By its very nature, sodium vapor is quite difficult to contain, and the development of a successful HPS lamp depended, in part, on the development of a polycrystaline alumin arc tube. The development of this material, and the necessary technology to seal the ends of the tube, was a formidable task. The outer envelope is made of a borosilicate glass, and is evacuated to lessen chemical deterioration of the arc tube components. The outer envelope also acts to isolate the arc tube from the effects of ambient temperature and drafts.

The narrow shape of the arc tube prevents the inclusion of a starting electrode, therefore, a very high-starting voltage is required to initiate the arc. This starting voltage is normally provided by a special electronic ballast, designed specifically for use with HPS lamps.

High Pressure Sodium Lamps and Lumen Maintenance

The lamp life and lumen maintenance characteristics of the HPS lamp have constantly been refined. Today, the rated life of the HPS lamp is in the 24,000 hour range, and the lamp exhibits very good lumen maintenance

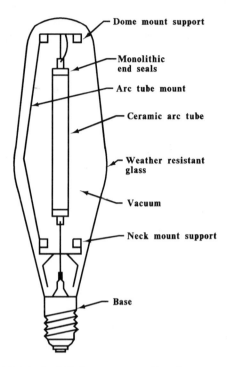

- Dome mount support
- Monolithic end seals
- Arc tube mount
- Ceramic arc tube
- Weather resistant glass
- Vacuum
- Neck mount support
- Base

Figure 10.16 Typical high pressure sodium lamp construction.

characteristics, as shown in Figure 10.17. As with the other members of the HID family, the initial lumen values are taken at 100 hours to allow for lamp stabilization. Table 10.3 lists representative data for several HPS lamp wattages.

The life of the HPS lamp is determined by several factors. As the lamp ages, the lamp voltage increases slightly. This is due to the blackening of the arc tube ends, which increases the temperature buildup at the ends of the tube. This increased temperature, in turn, results in higher arc tube pressure and, hence, in higher arc voltages. Due to this sequence, the HPS lamp *announces* the end of its life in a unique way. When the lamp voltage reaches a point at which the ballast can no longer maintain the arc, the lamp extinguishes. After a short cooldown period, the lamp arc restrikes, and the lamp operates until the arc voltage again reaches the point where it can no longer be supported by the ballast. Thus, the end of lamp life can generally be detected by the frequent cycling of the lamp.

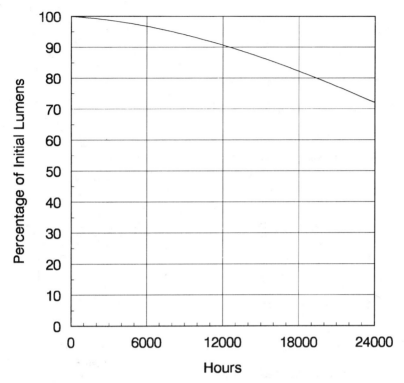

Figure 10.17 Typical high pressure sodium lamp lumen depreciation curve.

The lamp restrike period following a power interruption is decidedly superior to either the mercury vapor or metal halide lamp, with initial restrike occurring in less than one minute, and stabilized operation following in three to four minutes.

HID LAMP COLOR

As mentioned in earlier sections, each member of the HID family exhibits its own color characteristics. The clear mercury vapor lamp emits light at the characteristic lines of 404.7, 435.8, 546.1, 577.1 and 579 nanometers, as shown in Figure 10.18(a). This results in a blue-green light, which is not generally viewed as being particularly pleasing. The addition of a phosphor-coated jacket enhances the red portion of the light output. A

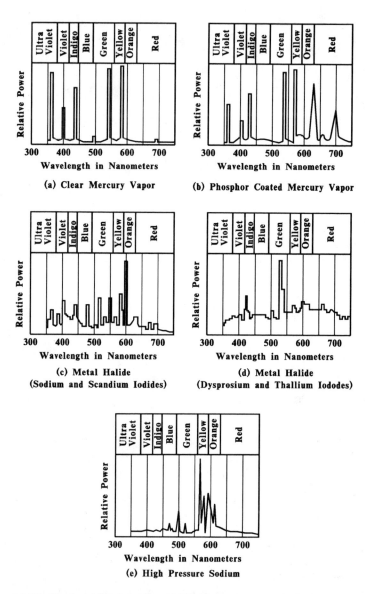

Figure 10.18 Typical high intensity discharge lamp spectral distribution (a) clear mercury vapor, (b) phosphor coated mercury vapor, (c) metal halide (sodium and scandium iodides), (d) metal halide (dysprosium and thallium iodides), and (e) high pressure sodium.

representative spectral distribution curve for a deluxe warm white mercury vapor lamp is shown in Figure 10.18(b).

Metal halide lamps exhibit spectral distribution curves, depending on the nature of the metal halide added. Figure 10.18(c) shows the spectral distribution curve for lamps with sodium and scandium iodides. Figure 10.18(d) shows a similar spectral distribution for lamps containing dysprosium and thallium iodides. Note the overall wide range of colors emitted by the metal halide lamp versus the mercury vapor lamps.

The high pressure sodium lamp emits light across the entire visible spectrum, with emphasis in the yellow and red wavelengths, as shown in Figure 10.18(e). A new generation of HPS lamps exhibits the potential for an even more evenly balanced spectral distribution. HPS lamps are widely used in outdoor or indoor applications where their tremendous lumen efficacy, excellent lamp life and lumen maintenance are more significant than the requirement for superior color rendering qualities.

HID BALLASTS

As discussed previously, HID lamps have a negative volt-ampere characteristic which requires that a current-limiting device be provided to prevent lamp destruction. The HID ballast fulfills this requirement and also provides the necessary voltage for proper lamp starting, and often contains capacitors for power factor correction. Each of the HID lamp types has special requirements which must be addressed in the ballast design.

There are several basic types of HID ballast design. Figure 10.19(a) shows the simplest ballast design, called a reactor, which basically acts as a choke to limit current flow. This design can be used only in applications where the available line voltage exceeds the lamp starting voltage requirements. In some designs, a capacitor is included to correct the power factor, which would otherwise be in the range of 50%. The reactor ballast without the capacitor is called a *normal* or *low power factor* design. Although the reactor design has clear economic advantages, this must be weighed against its disadvantages. For example, the starting current is about 50% higher than the steady state operating current, so conductors and circuit breakers must be appropriately oversized or, alternatively, fewer fixtures connected to each circuit. Also, the reactor provides very little regulation for variations in line voltage. A 3% change in line voltage can result in a change in lamp wattage of 6%.

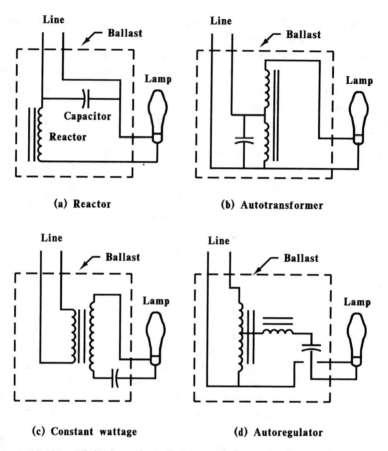

Figure 10.19 High intensity discharge ballast circuits (a) reactor, (b) autotransformer, (c) constant wattage, and (d) autoregulator.

Figure 10.19(b) shows a design that features both a reactor and an autotransformer. This design, called an autotransformer ballast, does not offer an improvement over the reactor ballast voltage regulation, but does permit application over a wider range of line voltages.

Figure 10.19(c) shows a constant wattage ballast suitable for mercury lamps. In this design, the primary and secondary are isolated, and current limitation is achieved by means of the series capacitor. This type of HID ballast also offers the advantage of very good regulation, with a line voltage change of ±13%, resulting in a lamp wattage change of only ±3%. The overall power factor of this design is normally in the range of 95%.

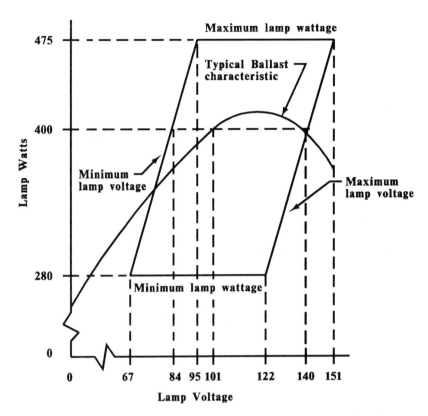

Figure 10.20 Typical 400 W high pressure sodium lamp operating locus.

Figure 10.19(d) shows an autoregulator ballast that contains design features of both the lag ballast, shown in Figure 10.19(b), and the constant wattage ballast, shown in Figure 10.19(c). The autoregulator ballast is less costly than the constant wattage ballast, but offers reduced voltage regulation. A 10% change in line voltage results in a 5% change in lamp wattage. Finally, there is no isolation between primary and secondary as in the constant wattage design.

As mentioned previously, the high pressure sodium lamp does not have a starting electrode and, therefore, requires a significantly higher voltage to start the lamp. This starting pulse is normally supplied by a special electronic circuit and, hence, ballasts for HPS lamps are especially designed. In fact, HPS lamp ballasts are designed around a locus of operating points, as shown in Figure 10.20 which represents a typical 400 W HPS lamp. This operating locus is necessitated by the fact that the HPS lamp wattage and arc voltage change during the life of the lamp. The proper operating range is a trapezoid, as shown, with voltage and wattage limits determining the sides of the trapezoid. A typical ballast operating scheme is shown as a curve. As this curve shows, the lamp wattage is initially somewhat below the rated value, but increases during most of the lamp life, with a decrease occurring only near the end of rated life.

References

Boylan, Bernard R., P.E. 1987. *The Lighting Primer.* Iowa: Iowa State University Press.

General Electric Lighting Business Group. 1978. *Light and Color.* General Electric Company.

General Electric Company. 1988. *Selection Guide for Quality Lighting.* General Electric Company.

Helms, Ronald N. 1980. *Illuminating Engineering for Energy Efficient Luminous Environments.* New Jersey: Prentice Hall, Inc.

IES Education Committee. 1988. *ED-100. IES Education Series.* New York: Illuminating Engineering Society of North America.

Illuminating Engineering Society of North America. 1987. *IES Lighting Handbook - Application Volume.* New York: Illuminating Society of North America.

Illuminating Engineering Society of North America. 1984. *IES Lighting Handbook - Reference Volume.* New York: Illuminating Society of North America.

Murdoch, Joseph B. 1985. *Illumination Engineering - From Edison's Lamp to the Laser.* New York: Macmillan Publishing Company.

11

Luminaires and Photometrics

In the previous chapter, we discussed the commonly used light sources. In this chapter, we will learn how we combine these sources, along with equipment designed to distribute the generated light, into a complete lighting system called a luminaire. In this sense, the luminaire actually consists of two discrete types of components—the lamps and the fixture. The lamps are the light sources themselves, and can be incandescent, fluorescent or HID. The fixtures can contain one or several lamps, depending on the design goals. The fixture provides the necessary physical protection for the light source(s), the required auxiliary equipment, such as the ballast, as well as the optical assembly, to effectively distribute the light generated by the lamps. The complete assembly of lamps and fixture becomes the luminaire.

As you might imagine, luminaires are available in literally hundreds of styles and designs, using a wide variety of light sources. Some luminaires are primarily decorative in nature, while others are designed with purely utilitarian goals in mind. The physical environment in which the luminaire is to be installed also plays a major role in the design. Some luminaires are designed for interior use and, therefore, need not be designed to include protection from the weather, while other luminaires are designed for exposure to extremes in humidity and temperatures, encountered in exterior applications. Construction cost also plays an important role in luminaire design. Luminaires are available which meet a range of economic requirements. Some buildings are designed to last a long time and, therefore, luminaire first cost might be less important than durability, while for other buildings, first cost might be a primary consideration. Hospitals might be an example of the former, and retail shopping centers the latter.

Many new engineers are often baffled by the vast number of lighting equipment catalogs, often numbering well over one hundred, available in engineering offices. Through experience, the engineer learns to balance design requirements, aesthetic goals and economic constraints in order to select the appropriate luminaires for a given project. This selection process often involves comparison of luminaires manufactured by different manufacturers, making standardized presentation of luminaire data very important. Luminaire data also includes information on light production, distribution and efficiency. This information is crucial for both luminaire selection, and also for use in the design process during which the appropriate number of luminaires for a space is determined.

In order to effectively communicate with manufacturers, contractors, architects, and other engineers, it is necessary to thoroughly understand luminaires and their light distribution characteristics, called photometric data. Fortunately, this information, as well as its associated terminology, is fairly well developed and also fairly standardized.

We will divide our discussion of luminaires and photometrics into two parts—interior luminaires and exterior luminaires.

INTERIOR LUMINAIRES

Luminaires for interior lighting are normally designated according to their light source, light distribution and method of mounting. Interior luminaires may employ any of the light sources discussed in Chapter 10, such as the incandescent, fluorescent or HID light sources. The luminaire's light distribution characteristics determine the standard CIE/IES category into which the luminaire falls. Figure 11.1 shows the six standard distribution categories. Note that determining the appropriate distribution category depends solely on the relative percentages of the luminaire's light emitted in the angles below 90°, and that emitted in the angles above 90°. These categories are direct, semi-direct, direct-indirect, semi-indirect, general diffuse, and indirect. The distribution varies from direct, in which 100% of the light is directed downward, to indirect, in which 100% of the light is directed upward. Note that both the general diffuse and direct-indirect

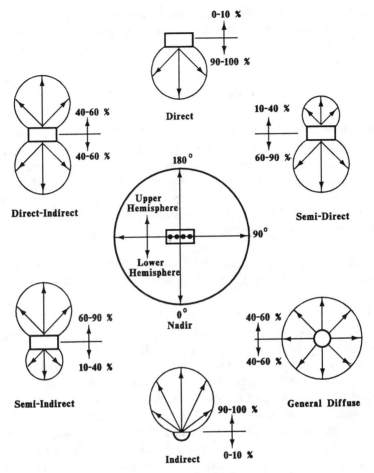

Figure 11.1 Luminaire distribution categories.

categories have the same distribution percentages, namely 40-60% upward and 40-60% downward. The basic difference between the two is in the percentage of light present in the 90° range. As shown in Figure 11.1, the direct-indirect luminaire emits very little light in the 90° range, while the general diffuse luminaire has a very significant level of light output in this range. The direct-indirect distribution is actually an IES classification and not one of the recognized CIE distributions. As we will see in Chapter 12, this classification is also employed to help determine how luminaire light output is affected by accumulated dirt, as well as how the overall lighting system is affected by the accumulation of dirt on the room walls, ceiling and floor. Figure 12.3 shows a few of the many types of available interior luminaire designs.

Fixture Construction and Mounting

As mentioned in earlier sections, luminaires are available for a wide variety of applications and environmental conditions. In large measure, the luminaire design goals determine the nature of construction, and often the mounting arrangements as well. One of the key aspects of luminaire design is the method of controlling the light leaving the lamps in such a way that as much luminous flux as possible reaches the work plane, while at the same time, reducing the direct glare component in the angles above 45°.

In some types of luminaires, this control is accomplished by means of a reflector mounted inside the luminaire. Figure 11.2 shows several commonly used reflector configurations. In general, these reflectors employ specular surfaces. If the reflector is semi-circular, with the source at the focal point, the light reflected from the reflector will pass through the focal point, as shown in Figure 11.2(a). If the reflector is parabolic in shape, with the source at the focal point, the reflected light will be parallel, as shown in Figure 11.2(b). In this same case, if the source is located between the focal point of the parabola and the surface, the light will tend to diverge, as shown in Figure 11.2(c). Finally, if the reflector is elliptical in shape and the light source is located at one focal point, the reflected light will pass through the other focal point. Each of these reflector designs has specific applications. For example, the reflector shown in Figure 11.2(b) might be appropriate for floodlight applications where very narrow distribution is required, while the configuration shown in Figure 11.2(c) might be applicable where broader floodlight distribution is desired.

The distribution produced by the configuration shown in Figure 11.2(d) is employed in recessed luminaires where a very narrow opening is desirable. In this case, the small opening would be located close to the second focal point. A common concern for each type of reflector is the desirability of having the light source closely approximate a point source. As a result, incandescent and HID sources are frequently employed with reflectors, rather than fluorescent lamps which are more of a linear light source.

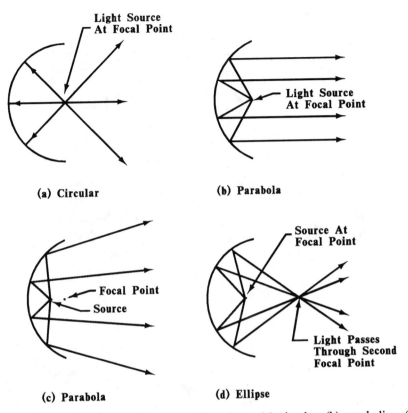

(a) Circular

(b) Parabola

(c) Parabola

(d) Ellipse

Figure 11.2 Luminaire reflector configurations (a) circular, (b) parabolic, (c) parabolic, and (d) elliptical.

Luminaire Diffusers and Louvers

Another means of controlling the light leaving a luminaire is by the placement of a diffuser or louver below the light source. Diffusers can be made of either glass or plastic. They can be flat sheets of clear or translucent material, or they can be molded with a prismatic pattern for light control. Translucent diffusers tend to spread the downward light in all directions, including that area within the direct glare zone between 45° and 90°. As a result, the visual comfort probability of such diffusers is quite low. Diffusers with molded prisms for light control can be designed to substantially reduce the light output in the direct glare zone, and direct a higher percentage downward toward the work plane. Plastic diffusers can

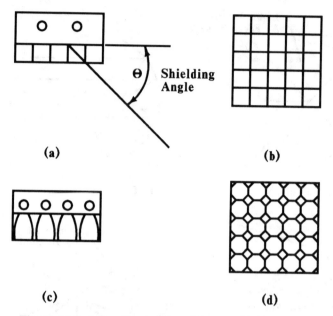

Figure 11.3 Louver materials and louver shielding angle.

be made of polystyrene or acrylic material. Polystyrene is less expensive, but is subject to yellowing due to exposure to ultraviolet energy emitted by the light source. As a result, most of today's diffusers are made of acrylic. Diffusers are available in several thicknesses, ranging from about 0.12 inches to under 0.10 inches.

Louvers have been available for many years and, today, seem to be enjoying renewed acceptance. The basic operating premise of the louver is shown in Figure 11.3(a). The louver is made up of a number of individual rectangular cells, with the depth of the cell chosen so as to provide the necessary degree of shielding. In the louver shown, all light within the shielding angle is blocked. Louvers can be made of metal or plastic. Early louvers first employed parabolic surfaces to redirect light downward toward the work plane. Unfortunately, the smaller the size of the rectangular or parabolic shape, often 0.5 inches, the greater the louver's top surface area blocks light leaving the lamp. This substantially lowered the efficiency of these early designs. Today's modern parabolic louvers are often several inches in size and are of significantly greater depth. The shielding of these louvers is very good and the resulting visual comfort the efficiency of these early designs. Today's modern parabolic louvers are often several inches in size and are of significantly greater depth. The shielding of these louvers is very good and the resulting visual comfort quite acceptable. As an added bonus, the overall luminaire efficiency is also high.

In addition to the square louver shapes, shown in Figures 11.3(b) and 11.3(c), there are many other louvers based upon other geometric patterns. These are generally selected on the basis of architectural design appeal and are employed only in relatively small areas. On the other hand, the newer generation of larger dimension parabolic louvers are suitable for general use.

INTERIOR LUMINAIRE PHOTOMETRICS

As discussed in the introduction to this chapter, it is important for the engineer to have complete information on the construction and performance of a luminaire. This information is used for two basic purposes. First, such information facilitates the luminaire selection process by presenting information on the size, construction details and appearance of the luminaire. This information is generally included in the manufacturer's sales literature. The second purpose is to accurately present data on the photometric performance of the luminaire. This information is used for both the comparison of similar luminaires during the selection process and also in the design process itself. Information on the photometric performance of a luminaire must be presented in a standardized way in order to facilitate rapid and accurate analysis of information on competing luminaires. In this section, we will discuss how the data from a photometric report is derived, presented and analyzed.

Photometric Measurements

A photometric report begins with a test to determine the luminous intensity or candlepower distribution curve for the luminaire. This test is performed by an instrument called a goniophotometer. The instrument is mounted in a special room designed to eliminate undesired reflectances. The luminaire to be tested is mounted in the instrument and candlepower intensity measurements are made at specific vertical angles to produce a

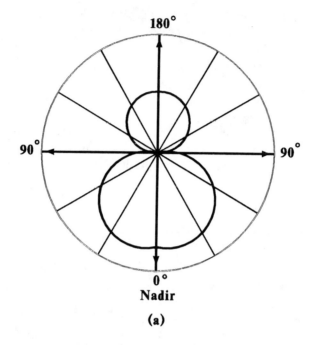

(a)

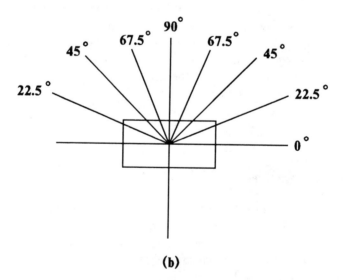

(b)

Figure 11.4 Photometric test measurement angles.

by a computer which also records the data. Figure 11.4(a) shows the general shape of the resulting curve. The luminous intensity is recorded on a polar plot, with 0° (Nadir) being straight down. Data is plotted for angles ranging from 0° to 180°. In many tests, the data is recorded in 10° increments, however, 5° or 2½° increments produce better results. In the case of luminaires with a symmetrical distribution, only the 0° to 180° data are shown.

The luminous intensity data are recorded at several vertical planes, as shown in Figure 11.4(b). In this figure, we see a plan view of a luminaire. The zero degree axis extends through the length of the luminaire, and the photometric tests measure a complete luminous intensity distribution curve through several such vertical planes. In the case shown in Figure 11.4(b), measurements are taken at 0, 22.5, 45.0, 67.5 and 90 degrees, respectively. This is commonly referred to as *5-plane photometry*. Note that there is only one 0 or 90 degree plane, while there are two 22.5, 45.0 and 67.5 degree planes. In some cases, only the 0, 45.0 and 90 degree planes are included.

The physical and environmental constraints for such a photometric test are extensive. If the light source is sensitive to temperature variations, the test room must be maintained at the correct temperature. In addition, special ballasts with *carefully determined* characteristics may be used. Finally, the exact lumen output of the test lamp must be known in order to produce exact, repeatable results. The setup for a typical photometric test can require a number of hours of work and might even make allowances for the luminaire to stabilize overnight in the test room, prior to the commencement of measurements. Test facilities such as these are both expensive to build and to maintain. It is also worth noting that generally the test facility also includes an environmental test room for temperature and humidity tests, and sound rooms to analyze luminaire and ballast noise characteristics. Many manufacturers utilize their test facilities on an almost around-the-clock basis to acquire data on new prototype luminaires, test sample luminaires from the production line, and also to test luminaires manufactured by competing companies.

Photometric Reports

It might be useful to pause for a moment to consider the nature of the information we would like to receive from a photometric report. First, we would like the report to contain basic information on the manufacturer, catalog number, date of test, testing organization, as well as any luminaire-specific information, such as adjustable lamp socket positions, etc. Secondly, we would like to have the report contain information which would be useful in evaluating the luminaire photometric performance and also data for use in the design process. Such information might include the following:

- Shielding information (lens or diffuser type)
- CIE/IES distribution class
- Lamp data
- Spacing criterion (guidance on maximum luminaire spacing)
- Luminaire zonal flux distribution (useful in evaluating luminaire efficiency)
- Luminaire efficiency
- Average zonal luminance in the direct glare zone
- A table of coefficient of utilization data (useful in evaluating the performance of the luminaire in a specific room configuration)

The luminaire zonal flux distribution and luminaire efficiency can be derived from the candlepower distribution data. The CIE/IES classification can be determined from the zonal flux distribution. The average zonal luminance values can be derived from the candlepower distribution data and projected luminaire area at various angles.

The Tables of Coefficients of Utilization (*CU*) data are based not only on the luminaire characteristics but also specific room geometries and room surface reflectances. The computation of *CU* data involves flux transfer theory. We will merely point out that this table of *CU* data helps the engineer predict the average level of work plane illumination provided by a given luminaire in a specific room configuration. We will reserve further discussion of *CU* tables until the design portion in Chapter 12.

Figure 11.5 shows the luminous intensity data included in a typical photometric report. The unit is a two-lamp open industrial luminaire which is 8 ft. in length and which utilizes two 8 ft. fluorescent lamps, rated 6300 lm. each. The reflector of the luminaire has slots permitting an upward component of light.

Let us now see how we translate the luminous intensity data into information on the luminaire zonal flux, luminaire efficiency and average zonal luminance.

Luminaire Test Data

Luminaire	1' X 8' Surface Or Pendant Mounted industrial fluorescent.
Lamps	2 - F96T12/CW (6300 Lu / 2310 fL)
Reflector	Slotted reflector, synthetic enameled.

Maximum Luminance (fL)

Angle	Along (0 Deg.)	Across (90 Deg.)
45	2449	2703
55	2336	2703
65	1887	2703
75	1239	1126
85	497	678

W = 13.5"
L = 96"
Lighted Area = 13.5" X 90.73"

Candela Distribution

Degrees	0 Deg.	22.5 Deg.	45 Deg.	67.5 Deg.	90 Deg.
5	2614	2617	2620	2624	2628
15	2534	2566	2598	2633	2647
25	2407	2489	2570	2611	2651
35	2157	2286	2415	2498	2580
45	1736	1936	2135	2198	2260
55	1322	1532	1742	1716	1690
65	883	1031	1179	1209	1239
75	455	595	735	650	565
85	115	141	166	162	157
95	39	71	103	104	104
105	146	137	127	173	218
115	313	284	255	233	210
125	477	449	420	391	361
135	640	625	609	576	543
145	793	727	660	703	745
155	902	860	817	777	737
165	963	951	939	917	894
175	1012	1011	1009	1007	1005

Figure 11.5 Typical photometric test results for industrial fluorescent luminaire.

Average Candlepower, Zonal Lumens and Luminaire Efficiency

The photometric data for the industrial fluorescent luminaire, shown in Figure 11.5, was developed by a standard 5-plane photometric test at 10 degree increments, so we must first determine the average mid-zone candlepower values. We compute these by means of Equation 11.1.

$$I_{avg} = \frac{I_0 + 2 \cdot I_{22.5} + 2 \cdot I_{45} + 2 \cdot I_{67.5} + I_{90}}{8}. \quad (11.1)$$

Here, I_0, $I_{22.5}$, I_{45}, $I_{67.5}$, and I_{90} represent the intensity at the five vertical planes shown in Figure 11.4. The average intensity must be computed for each of the mid-zone angles represented by the rows in Figure 11.5.

Note that the denominator of Equation 11.1 is 8 instead of 5 as you may have expected. This is because there are two planes at 22.5°, 45°, and 67.5° while there is only one plane at 0° and 90°. Note that the photometric data is recorded at mid-zone values of 5°, 15°, etc., so the vertical increment is 10°. For vertical angles of 5° and 15° the average intensity values, I_{avg} are as follows.

$$I_{avg(5)} = \frac{2614 + 2 \cdot 2617 + 2 \cdot 2620 + 2 \cdot 2624 + 2628}{8} = 2621 \; cd$$

$$I_{avg(15)} = \frac{2534 + 2 \cdot 2566 + 2 \cdot 2598 + 2 \cdot 2633 + 2647}{8} = 2597 \; cd$$

$$(11.2)$$

Table 11.1 shows the average intensity values for the various mid-zone angles.

Our next order of business is to determine the zonal lumens for each vertical increment which, in this case, is 10°. In order to determine these values, we will need to develop a technique for determining the lumens emitted in each of the vertical increments. Figure 11.6 shows a unit sphere which has been divided into several strips or zones.

We want to compute the unit sphere area associated with each increment of vertical angle. We will call these Zonal Constants. Since the total area of the unit sphere is 4π ft^2 we would expect the sum of all Zonal Constants to equal 4π. The Zonal Constant (ZC) for a strip whose mid-zone angle is θ is given by

$$ZC_\theta = 2\pi(\cos\theta_2 - \cos\theta_1) \qquad (11.3)$$

where θ_1 is the upper value of the vertical increment and θ_2 the lower value. In this case, the increment is 10°, so $\theta_1 = \theta + 5°$ and $\theta_2 = \theta - 5°$. For mid-zone angles of 5° and 15° the zonal constants will be

$$ZC_5 = 2\pi(\cos(0°) - \cos(10°)) = 2\pi \cdot 0.0152 = 0.0955$$
$$ZC_{15} = 2\pi(\cos(10°) - \cos(20°)) = 2\pi \cdot 0.0451 = 0.2835.$$

From our definition of luminous intensity, I, developed in Chapter 9, (Equation 9.4), we will recall that

$$I = \frac{d\varphi}{d\omega} \qquad candelas \; (cd) \qquad (9.4)$$

Table 11.1 Midzone Angles and Average Luminous Intensity Data for Industrial Fluorescent Luminaire.

ZONE DEGREES	MIDZONE ANGLE	AVERAGE INTENSITY
0-10	5	2621
10-20	15	2597
20-30	25	2550
30-40	35	2392
40-50	45	2067
50-60	55	1624
60-70	65	1120
70-80	75	623
80-90	85	151
90-100	95	87
100-110	105	155
110-120	115	258
120-130	125	420
130-140	135	600
140-150	145	715
150-160	155	818
160-170	165	934
170-180	175	1009

and

$$\omega = \frac{A}{r^2} \quad Steradians \ (sr)$$

where A is the surface area and r the radius of the sphere. In the case of a unit sphere

$$\omega = \frac{A}{1^2} = A \quad Steradians$$

so in this case

$$\varphi = I \cdot A \quad Lumens. \tag{11.4}$$

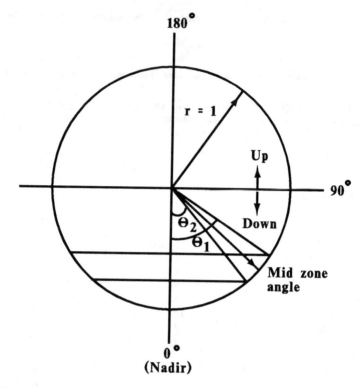

Figure 11.6 Unit sphere and angles associated with zonal constants.

Since our zonal constants (ZC) equal the incremental unit sphere areas (A) the final form of Equation 11.4 becomes

$$ZL_\theta = I_\theta \cdot ZC_\theta \quad Lumens. \tag{11.5}$$

Again, note that the sum of all zonal constants (0° to 180°) will total 4π.

Using Equations 11.3, 11.5, and the average intensity value shown in Table 11.1, we are in a position to compute the zonal lumens. For 5° and 15° the zonal lumens will be given by

$$ZL_5 = 2621 \cdot 0.0955 = 250 \quad Lumens$$
$$ZL_{15} = 2597 \cdot 0.2835 = 736 \quad Lumens.$$

Table 11.2   Zones, Zonal Constants, Average Luminous Intensity and Zonal Lumens for Industrial Fluorescent Luminaire.

Zone Degrees	Zonal Constants	Average Intensity	Zonal Lumens
5	0.0955	2621	250
15	0.2835	2597	736
25	0.4629	2550	1180
35	0.6282	2392	1503
45	0.7744	2067	1601
55	0.8972	1624	1457
65	0.9926	1120	1112
75	1.0579	623	659
85	1.0911	151	165
95	1.0911	87	95
105	1.0579	155	164
115	0.9926	258	256
125	0.8972	420	377
135	0.7744	600	465
145	0.6282	715	449
155	0.4629	818	379
165	0.2835	934	265
175	0.0955	1009	96
			11,208 Total

Table 11.2 shows the zones, zonal constants, average intensity and zonal lumens for each zone.

Our next task is to evaluate the overall luminaire efficiency, which is defined as the percentage of the total lamp lumens that actually leave the luminaire. Numerically, this is given by

$$\% \text{ Luminaire Efficiency} = \frac{\text{Lumens Leaving Luminaire}}{\text{Lamp Lumens}} \cdot 100. \quad (11.6)$$

In the case of this luminaire, the total lumens leaving the luminaire is simply the sum of the zonal lumens shown in Table 11.2. The total luminaire efficiency is then

$$\% \; Luminaire \; Efficiency \;\; = \;\; \frac{11,208}{12,600} \cdot 100 \;\; = \;\; 88.95\,\%.$$

Luminaire Brightness and
Coefficient of Utilization (*CU*) Data

Two additional areas addressed by a photometric report are luminaire brightness and coefficient of utilization data. Luminaire brightness is computed for the direct glare angles of 45, 55, 65, 75 and 85 degrees, and is expressed in Footlamberts (*fL*). The initial photometric report, shown in Figure 11.5, shows measured maximum brightness values for the angles of concern. We compute average brightness values (L_{avg}) using the candlepower distribution curve data, along with the projected luminaire area at the angle of viewing, PA_θ. The average brightness value in Footlamberts will be given by

$$L_{avg(\theta)} \;\; = \;\; \frac{I_\theta}{PA_\theta} \cdot 144\,\pi \quad fL. \tag{11.7}$$

Here PA_θ is expressed in square inches, and the 144π serves to convert the average luminance in candelas per square inch (cd/in^2) to Footlamberts (*fL*).

Average brightness values are normally computed for viewing angles along the luminaire axis (0°) and directly across the luminaire axis (90°). For the luminaire in Figure 11.5 the value of projected area at 0° would be the product of the lighted area width and length which is 1224.85 in^2. The projected area at other angles would be

$$PA_\theta \;\; = \;\; PA_0 \cdot \cos(\theta). \tag{11.8}$$

Using intensity data from Figure 11.5 and the projected areas, we can compute the average brightness values at 45 degrees for both along (0°) and across (90°) viewing angles. These values will be given by

$$Along: \;\; L_{avg(45)} \;\; = \;\; \frac{1736}{866.1} \cdot 144\,\pi \;\; = \;\; 907 \; fL$$

$$Across: \;\; L_{avg(45)} \;\; = \;\; \frac{2260}{866.1} \cdot 144\,\pi \;\; = \;\; 1180 \; fL.$$

Table 11.3 Zones, Midzone Angles, and Average Brightness Values for 45, 55, 65, 75, and 85 Degrees for Industrial Fluorescent Luminaire.

Angle	Average Brightness (Along - 0°)	Average Brightness (Across - 90°)
45	907	1180
55	851	1088
65	772	1083
75	649	806
85	487	665

Average brightness values for 45, 55, 65, 75, and 85 degrees are shown in Table 11.3.

With regard to the computations for projected area, *PA*, several points should be remembered. First, the luminaire projected area calculation becomes more complicated in the case of luminaires with luminous sides which effectively increase the projected lighted area. Secondly, the computations for luminaires which are stem- or pendant-mounted, and which have upward light components, may also require modified calculations for projected area. In these cases, the correct projected area calculation techniques may be found in Illuminating Engineering Society of North America standards.

Coefficient of Utilization Tables

As mentioned in earlier parts of this chapter, photometric tests are also utilized to produce tables of data referred to as coefficient of utilization (*CU*) tables. While the candela (candlepower) distribution curves are useful in performing inverse square law calculations, *CU* data help us predict the average level of illuminance over an area. In inverse square law calculations, we may not be as concerned with the complex room surface reflectances which affect the level of illumination, whereas in *CU* based calculations such reflectances are an integral part of the analysis. From this, you might suspect that these tables of *CU* data are not only luminaire specific but room configuration and room surface reflectance specific as well. For example, rooms which are long and narrow tend to have lower overall efficiencies than more rectangular spaces, due to the absorption of incident light by walls. Rooms with lower room surface reflectances also tend to absorb more light, and as a result, have lower overall efficiencies. The goal of the *CU* table is to facilitate the determination of the percentage of lamp lumens that successfully reach the work plane in a specific room geometry with specific surface reflectances. The development of such tables involves elements of flux transfer theory which are beyond the scope of this work. *CU* data is normally presented for several ceiling and wall reflectances as well as a number of room geometries. These room geometries are expressed in terms of a room cavity ratio (*RCR*) which is based on the ratio of total vertical room surface area to total floor surface area.

Coefficient of utilization values are always expressed in decimal form and fall between 0.0 and 1.0. In Chapter 12 we will discuss how we utilize candela distribution curves for inverse square law calculations, often called point-by-point calculations, and *CU* data for average level of illuminance calculations.

Figure 11.7 shows the additional data computed from the initial test results shown in Figure 11.5. In Figure 11.7 we see the results of zonal lumen, luminaire efficiency, luminaire brightness, and coefficient of utilization calculations. In addition to the information computed from the initial test data, the photometric report will provide guidance on the maximum luminaire spacing criteria, which is useful in avoiding overly dark areas between luminaires. Spacing criteria will be discussed in Chapter 12.

The combination of Figures 11.5 and 11.7 comprises a typical full photometric report. Such reports are available from the manufacturer of any luminaire, and are an important part of the selection and design process.

EXTERIOR LUMINAIRES

Like the luminaires used in interior lighting applications, those intended for exterior use are available in a wide variety of designs. Exterior luminaires can be broadly broken down into several categories, as shown in Figure 11.8. Some exterior luminaires are both decorative and functional in nature, such as the decorative sphere and traditional lantern-type luminaire

Zonal Flux	
Degrees	**Flux**
5	250
15	736
25	1180
35	1503
45	1601
55	1457
65	1112
75	659
85	165
95	95
105	164
115	256
125	377
135	465
145	449
155	379
165	265
175	96

Industrial Fluorescent Luminaire

Spacing Criterion: 1.5 X Mounting Height
CIE Type : Semi-Direct
IES Class : Wide Spread
Lamps: Two F96T12/CW - 6300 Lu Each
2310 fL

Zonal Summary

Zone	Lumens	% Lamp	% Fixt
0 To 30	2166	17.2	19.3
30 To 60	4560	36.2	40.7
60 To 90	1935	15.4	17.3
90 To 180	2546	20.2	22.7
	11207	89.0	100.0

Luminaire Brightness Data

	Average (fL)		Maximum (fL)		fL Ratios	
Degrees	**Along (0 Deg.)**	**Across (90 Deg.)**	**Along (0 Deg.)**	**Across (90 Deg.)**	**Along (0 Deg.)**	**Across (90 Deg.)**
45	907	1180	2449	2703	2.7	2.3
55	851	1088	2336	2703	2.7	2.5
65	772	1083	1887	2703	2.4	2.5
75	649	806	1239	1126	1.9	1.4
85	487	665	497	678	1.0	1.0

Coefficients Of Utilization

ρ_f	20 %														
ρ_{cc}	80 %				70 %				50 %			30 %		
ρ_w	70%	50%	30%	10%	70%	50%	30%	10%	50%	30%	10%	50%	30%	10%
1	93	89	85	82	88	85	81	79	77	75	72	70	69	67
2	84	78	72	67	80	74	69	65	68	64	60	62	59	56
3	77	68	62	56	73	65	59	54	60	55	51	55	51	48
4	70	60	53	47	67	58	51	46	53	48	43	49	44	41
5	64	53	46	40	61	51	44	39	47	41	37	43	38	34
6	59	47	40	34	56	46	38	33	42	36	31	39	34	30
7	54	42	35	30	51	41	34	29	38	32	27	35	30	26
8	50	38	31	25	47	37	30	25	34	28	23	31	26	22
9	46	34	27	22	44	33	26	21	30	24	20	28	23	19
10	43	31	24	19	40	30	23	19	27	22	18	25	20	17

Figure 11.7 Typical data computed from photometric test report of Figure 11.5.

shown in Figure 11.8(a). These luminaires are frequently found along walkways or paths in residential applications. The luminaires shown in Figure 11.8(b) are primarily functional in nature. The rectangular luminaire shown in this figure is often referred to as a "shoe-box" or "cut-off" luminaire. This luminaire is popular in area lighting applications such as shopping centers and other parking applications. The second luminaire shown in this figure is the traditional roadway type luminaire found on

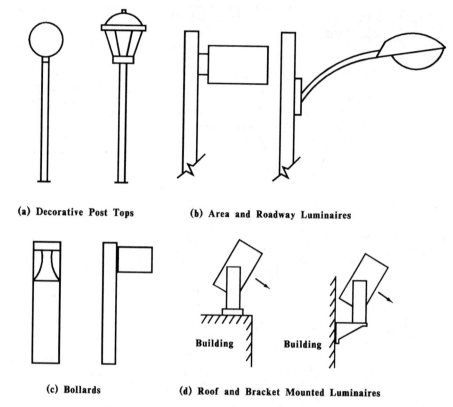

(a) Decorative Post Tops   (b) Area and Roadway Luminaires

(c) Bollards   (d) Roof and Bracket Mounted Luminaires

Figure 11.8 Typical exterior luminaire designs.

literally thousands of miles of roads throughout the country. Some luminaires are intended, primarily, for low level illumination of walkways and paths in both residential and commercial applications. Figure 11.8(c) shows a representative bollard type luminaire which employs a HID or incandescent lamp and special curved reflectors to distribute illumination with very low glare. These luminaires are available in a wide variety of shapes and are typically only three to four feet in height. The second luminaire is a variant of the area luminaire shown in Figure 11.8(b) and is also used on walkways or in plaza areas. It is also normally only three to four feet in height. Figure 11.8(d) shows floodlights designed for either roof mounting or bracket mounting on a vertical surface, such as a wall. This type of luminaire is ideal for providing area lighting for security purposes, or floodlighting architectural features of structures. Such floodlights are available with very precise beam control in order to minimize light falling on undesired areas.

Almost all exterior luminaires now employ one of the HID family of lamps. High pressure sodium is a favorite due to its availability in a variety of lamp wattages, ranging from 30 to 1000 watts, and its very long lamp life and excellent lumen maintenance. The use of incandescent sources has diminished greatly due to their inefficiency and short lamp life. Fluorescent sources have not found widespread application in exterior applications, except in signage applications, due to temperature sensitivity. Also, fluorescent sources, being linear, do not provide a point source of light. This makes reflector design for floodlighting or area lighting very difficult.

The design goals of the exterior lighting system largely determine the technique used in the design process. We are usually concerned with the specific footcandle level at various specific points, rather than simply the overall average level of illuminance. As a result, inverse square law calculations are normally used in exterior lighting design. This, in turn, determines the format of the photometric data which is in the form of candlepower distribution curves or tables.

Another major concern in exterior lighting is the shape of the lighting pattern projected by the luminaire on the work plane, and is often the flat plane of a parking area. The shape of this distribution pattern is important because areas are often rectangular in nature.

Luminaire distribution patterns which generally conform to the shape of the area, make the task of designing a relatively uniform lighting system, with minimal light spilling off the site, much easier. As a result of these concerns, exterior luminaires are often classified according to their light distribution patterns.

Luminaire Distribution Patterns

Five basic standard distribution patterns have been defined as shown in Figure 11.9. The patterns shown in this figure are the patterns projected by the luminaire on the surface below. The concentric lines are isofootcandle curves. The reader may be familiar with weather maps which use isobar lines which connect points of equal atmospheric pressure, or isotherm lines, showing similar curves for points of equal temperature. Isofootcandle patterns perform a similar function for luminaire

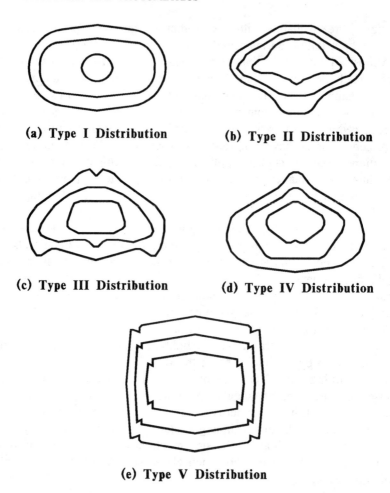

(a) Type I Distribution

(b) Type II Distribution

(c) Type III Distribution

(d) Type IV Distribution

(e) Type V Distribution

Figure 11.9 Exterior luminaire distribution patterns.

photometrics. An isofootcandle plot consists of a series of lines connecting points of equal footcandle levels. From the shape of these curves, it is possible to easily visualize the distribution pattern of a given luminaire. Let us examine each standard pattern. The Type I pattern is relatively long and narrow and, as such, is best suited for linear areas, such as roadways or narrow parking areas. The Type II pattern is asymmetrical. This pattern maximizes the distance between poles and is, like the Type I pattern, intended for linear spaces. The Type III pattern has a wider field

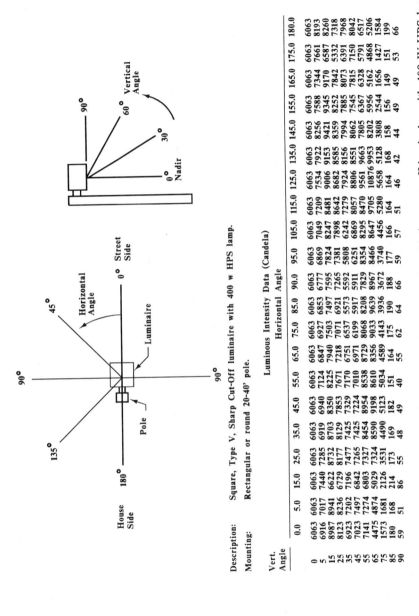

Description: Square, Type V, Sharp Cut-Off luminaire with 400 w HPS lamp.

Mounting: Rectangular or round 20-40' pole.

Luminous Intensity Data (Candela)

Horizontal Angle

Vert. Angle	0.0	5.0	15.0	25.0	35.0	45.0	55.0	65.0	75.0	85.0	90.0	95.0	105.0	115.0	125.0	135.0	145.0	155.0	165.0	175.0	180.0
0	6063	6063	6063	6063	6063	6063	6063	6063	6063	6063	6063	6063	6063	6063	6063	6063	6063	6063	6063	6063	6063
5	6916	7017	7440	7285	6919	6940	7124	6847	6927	6853	6777	6869	7049	7209	7534	7922	8256	7588	7344	7661	8193
15	8987	8941	6622	8732	8703	8350	8225	7940	7503	7497	7595	7824	8247	8481	9006	9153	9421	9345	9170	6587	8260
25	8123	8236	6729	8177	8129	7853	7671	7218	7071	6921	7265	7381	7898	8642	8682	8585	8359	8252	7842	5332	7318
35	6923	7202	7196	7477	7425	7329	7170	6751	6537	5573	5592	5808	6242	7279	7924	8156	7994	7885	8073	6391	7968
45	7023	7497	6842	7265	7425	7224	7010	6971	6199	5917	5911	6251	6869	8057	8806	8551	8062	7545	7815	7150	8042
55	7141	7274	6803	7327	8454	8954	8538	8729	8068	8208	7829	8354	8295	8470	9561	9663	7805	6367	6328	5791	6517
65	4475	4874	5029	7324	8590	9198	8610	8356	9033	9639	8967	8466	8647	9705	10876	9953	8202	5956	5162	4868	5206
75	1573	1681	2126	3531	4490	5123	5034	4589	4143	3936	3672	3740	4456	5280	5658	5128	3808	2544	1656	1427	1584
85	180	168	214	173	169	182	151	164	175	190	188	177	166	164	164	168	158	156	149	151	199
90	59	51	86	55	48	49	40	55	62	64	66	59	57	51	46	42	44	49	49	53	66

Figure 11.10 Typical exterior luminaire photometric data. Type V sharp cut-off luminaire with 400 W HPS lamp.

of light than the Type II and is often used in parking and area lighting. An example of this might be those areas where poles are located around the perimeter of a parking area. The Type IV distribution projects light well in front of the luminaire, and as such, is widely used in areas where the luminaire must be mounted along the perimeter of the space to be lighted.

The Type V pattern is, essentially, a square distribution pattern that is employed in larger areas where poles can be located within the area to be lighted. The square pattern also minimizes the degree of light spillage from rectangular areas. Most area lighting luminaires are available in any one of these distribution patterns. Because of their wide applicability, we will primarily concern ourselves with luminaires for area lighting.

Candela Distribution Data

Candlepower distribution data for area luminaires is presented in a fairly uniform manner. Figure 11.10 shows a typical candlepower (candela) distribution table for a Type V area luminaire, along with a diagram explaining the standard angular conventions.

Note that we are concerned with two angles. The vertical angle is measured from directly below the luminaire (0° Nadir) to 90°. The horizontal angle is measured from directly in front of the luminaire, 0°, to directly behind the luminaire, 180°. Since most of the photometry involved in area lighting has its origins in roadway lighting, the 0° angle is frequently referred to as "street side" and the 180° angle, as the "house side." The candlepower intensity table is then organized using rows to represent the vertical angles, and columns to represent horizontal angles. For example, the intensity at an angle of 65.0°, horizontally, and 35.0°, vertically, is 6751 cd.

In Chapter 12, we will see how area lighting systems are planned using candlepower distribution data, and also how the quality of the resulting design is evaluated.

References

General Electric Company Large Lamp Department. 1964. *General Lighting Design*. Ohio: General Electric Company.

General Electric Lighting. 1990. *Selection Guide for Quality Lighting*. Ohio: General Electric Company.

Helms, Ronald N. 1980. *Illumination Engineering for Energy Efficient Luminous Environments*. New Jersey: Prentice-Hall, Inc.

Illuminating Engineering Society of North America. 1983. *American National Standard Practice for Roadway Lighting, IES/ANSI RP-8-1983*. New York: Illuminating Engineering Society of North America.

Illuminating Engineering Society of North America. 1985. *Lighting for Parking Facilities*. New York: Illuminating Engineering Society of North America.

Illuminating Engineering Society of North America - Education Committee. 1988. *IES Education Series ED-100*. New York: Illuminating Engineering Society of North America.

Illuminating Engineering Society of North America. 1984. *IES Lighting Handbook - Reference Volume*. New York: Illuminating Engineering Society of North America.

Illuminating Engineering Society of North America. 1987. *IES Lighting Handbook - Application Volume*. New York: Illuminating Engineering Society of North America.

Journal of the Illuminating Engineering Society. 1974. *Lighting Roadway Safety Rest Areas*. New York: Illuminating Engineering Society of North America.

Murdoch, Joseph B. 1985. *Illuminating Engineering - From Edison's Lamp to the Laser*. New York: Macmillan Publishing Company.

→ Where do harmonics come from

- transformers
  - sat. of Iron core
  - exciting current →

- Electric Machines
- Electronic Power Sup.

- Arcing devices
  - Furnace
  - Arc Welder

→ Triplen harmonic : $3rd$, $6th$, $9th$, $12th$ ...
- They add directly : $I_{n3} = 3 \cdot I_{3rd}$
- $I_n = 3\sqrt{I_{9rd}^2 + I_{9th}^2 + I_{7th}^2}$ = RMS of current on Neutral.

→ Effects of harmonics

- Conductors
  - I^2R heat. Losses
  - RMS current increases due to harmonics
  - Higher losses → hotter conductors
  - skin effect , $f\uparrow$, $\uparrow R$

- transformers
  - copper losses
  - core losses

- Power electronic equipment
  - mis operation

- K-factor rated transformers
  Ex. $K = 4$ ∘ can operate with harmonics that cause $4 \times$ the normal eddy current loss in a std. transformer
  (k-

- C.B's, fuses, protection Relays
  - Variation in operation
  - high peak current may damage

Ex.) Line current in each phase

Harmonic No.	RMS current of Harmonic
1	100
3	57.7
5	33
7	17.2
9	5.8
11	4.7
13	2.7
15	2.3
17	1.6
19	1.4

$I_{rms} = \sqrt{I_1^2 + I_2^2 + I_3^2 + \cdots I_{10}^2}$ = $\boxed{121A}$

$THD = \sqrt{\dfrac{\sum\limits_{n=2}^{\infty} h_n^2}{h_1}} \times 100$

$THD = \sqrt{\dfrac{57.7^2 + 33^2 + 12.2^2 + 5.8^2 + 4.7^2 + 2.7^2 + 2.3^2 + 1.6^2 +}{100}}$

$= \boxed{68.1\%}$

K factor

$K \cdot factor = \sum\limits^{\infty} \left(\dfrac{I_n}{I_{rms}}\right)^2 h^2$

h	I_n/I_{rms}	$\left(\dfrac{I_n}{I_{rms}}\right)^2 \cdot h^2$
1	$100/121 = 0.83$	0.693
3	$57.7/121 = 0.477$	2.04
5	$33/121$	1.86
7		
9		
11		
13		
15		

$\boxed{K \cdot factor = 0.693 + 2.04 + 1.86 + \cdots + = 5.71}$
≈ 6 harmonic content

KVA of Required transf of $120/208$ v system
$S_{3\phi} = \sqrt{3}\,|V_L| \cdot |I_L| = \sqrt{3}\,(208) \cdot (121) = 43.6\,kV$

Triplen Harmonics → RMS floating on Neutral
$I_n = \sqrt{I_{3rd}^2 + I_{9th}^2 + I_{12}^2}$ $\times 3$
$= \sqrt{57.7^2 + 5.8^2 + 2.3^2}$ $\times 3 = 174.1$

12

Lighting Design

As discussed in Chapter 9, the building's illumination system is very important for two basic reasons. The first is *functional* in nature because the illumination system provides the light necessary for people to efficiently perform their tasks. The second reason is *aesthetic* in nature because the lighting system can also function as an architectural design tool in helping determine both the appearance and the mood of various interior spaces. Retail stores employ lighting to enhance the appearance of their products. Automobile dealerships employ lighting to add sparkle and color to the appearance of new cars. Restaurants use carefully designed illumination systems to create the mood of a space, often using warm earth colors and soft non-uniform lighting. On the other hand, office buildings use bright, cheerful colors and higher levels of illumination to promote a stimulating work environment.

Exterior lighting can also be both functional and decorative. It is well known that a carefully planned exterior lighting system can enhance security, as well as provide an attractive nighttime appearance. For example, shopping center parking areas which are well lighted promote increased customer traffic and, hence, sales. High quality exterior lighting also adds an increased sense of security, which is becoming increasingly important to many shoppers.

It should be clear from the preceding discussion that the design of an effective illumination system involves both engineering science and the art of design. The design process begins with an understanding of the lighting requirements necessary for effective human vision, as well as a knowledge of light sources, luminaires and photometrics. Based on this background and coupled with an understanding of the specific needs and design goals of the project, the design process can begin.

The process involves far more than simply the selection of luminaire types and illuminance levels. The ability of the selected luminaire to deliver the desired level of illuminance is directly affected by several factors. For example, the geometry of the room plays an important part in determining the amount of light actually reaching the work surface. The colors of a room's walls, ceiling and floor absorb or reflect light. We know that the illumination level produced by the lighting system will gradually decrease with time. This is due to several factors which include depreciation of light source output with age, reduction in room surface reflectances over time, and accumulation of dirt on the luminaire itself. These factors, and others, must be considered during the design process.

In this chapter, we will introduce the basic concepts associated with the lighting design process, as well as explain how we can assure that the system will retain its ability to provide effective illumination over the life of the building.

THE ZONAL CAVITY METHOD

The zonal cavity method is the currently recognized method of determining the average illuminance produced by an interior lighting system. As its name implies, it is based on the fact that a space can be divided into several sub-spaces or cavities. The zonal cavity method replaced earlier methods that were based on point-by-point techniques that used the inverse square law. Zonal cavity calculations are used to determine the coefficient of utilization, or CU, of the luminaire in the specific room configuration. The CU itself represents the fraction of the total lamp lumens that reach the work plane (IES Lighting Handbook 1984).

Figure 12.1 shows a cross section of a room and its three cavities, along with the room surface reflectances. The three room cavities are the *floor cavity*, the *room cavity* and the *ceiling cavity*. The *floor cavity* consists of the space between the floor and the work plane. The work plane height, h_{fc}, varies from application to application. For example, the work plane in an office might be 30" to 40", but for a corridor, the work plane might be the floor itself. The *room cavity* consists of the space between the work plane and the mounting height of luminaires. The *ceiling cavity* is the space between the bottom of the luminaire and the

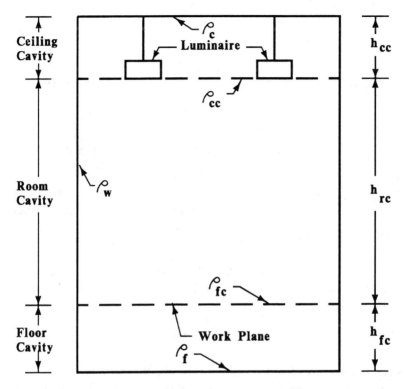

Figure 12.1 Typical room cross section showing cavities and reflectances.

ceiling itself. This cavity height, h_{cc}, varies according to the luminaire mounting arrangement. If the luminaires are mounted directly to the ceiling or recessed in the ceiling, as shown in Figure 12.2, the ceiling cavity height is zero. On the other hand, if the luminaires are suspended from the ceiling, also shown in Figure 12.2, the height of the ceiling cavity can vary from several inches to several feet. We must also realize that it is quite common for the ceiling, walls and floor to have different reflectance values, ρ, shown in Figure 12.1.

As mentioned previously, the geometry of a room plays an important part in how efficiently a given luminaire is able to deliver illumination to the work plane. The method of expressing the geometry of the space is by its *cavity ratio*, which is essentially the ratio of the vertical surface area to the horizontal surface area. For a rectangular space of length, L, width, W, and cavity height, h, the vertical surface area will be $2 \cdot h \cdot (L + W)$, and

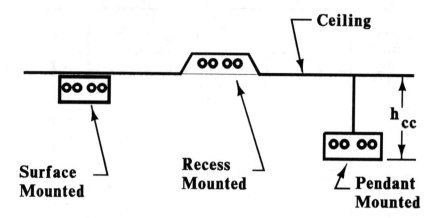

Figure 12.2 Typical luminaire mounting arrangements.

for the horizontal surfaces, $2 \cdot (L \cdot W)$. The ratio of this area is $h(L + W)/(L \cdot W)$. The cavity ratio is given by

$$CR = \frac{5 \cdot h \cdot (L + W)}{(L \cdot W)}. \tag{12.1}$$

The multiplier of five in Equation 12.1 resulted from the desire to have the cavity ratio for most common room sizes be greater than one. Long, narrow spaces tend to have high cavity ratios while rectangular spaces have lower cavity ratios. As we will see, the cavity ratio plays an important part in predicting the performance of a luminaire in a given space.

Example 12.1: What is the cavity ratio of a space with a length of 40 feet, a width of 20 feet, and a height of 9 feet?

$$CR = \frac{5 \cdot 9 \cdot (40 + 20)}{(40 \cdot 20)} = 3.38.$$

A space of 90 feet in length, width of 10 feet and a height of 12 feet will have a cavity ratio of:

$$CR = \frac{5 \cdot 12 \cdot (90 + 10)}{(90 \cdot 10)} = 6.67.$$

The reflectances of the various room surfaces also have a direct bearing on the overall ability of the lighting system to deliver illumination to the work plane. Referring once again to Figure 12.1, the reflectance values for the walls, ceiling and floor may be quite different due to the possible colors used for the surfaces. Typical reflectance values might be 20% for floors, 50% for walls, and 80% for ceilings.

Our ultimate goal is to reduce the ceiling, room and floor cavities to a single cavity—the room cavity. As a result, we are actually interested in the effective reflectance values, ρ_{eff}, of the ceiling and floor cavities at the boundary between them and the room cavity (Murdoch 1985). This effective reflectance value can be approximated by:

$$\rho_{eff} = \cfrac{1}{\cfrac{(1 + A_w/A_b)^2}{\rho_b + \rho_w(A_w/A_b)} - \cfrac{A_w}{A_b}} \qquad (12.2)$$

Here A_w represents the area of the wall surfaces and A_b the area of the base of the cavity. ρ_w is the wall reflectance and ρ_b the base reflectance. In the case of a ceiling cavity, A_b is the area of the ceiling itself and ρ_b is the reflectance of the ceiling. In the case of a floor cavity, A_b is the floor area, and ρ_b is the reflectance of the floor. Bear in mind that Equation 12.2 is a close approximation only.

Example 12.2: An office has the following dimensions and reflectance values:

Length = 60 ft	h_{fc} = 2.5 ft	ρ_c = .70
Width = 30 ft	h_{cc} = 3.0 ft	ρ_w = .50
		ρ_f = .20

What are the approximate effective reflectances for the ceiling cavity and floor cavity?

a) *Ceiling Cavity*

$$A_w = 2 \cdot 3.0 \cdot (60 + 30) = 540 \text{ ft}^2$$
$$A_b = 60 \cdot 30 = 1800 \text{ ft}^2$$
$$A_w/A_b = 540/1800 = 0.30$$

$$\rho_{eff} = \rho_{cc} = \cfrac{1}{\cfrac{1.3^2}{.7 + .5\,(.3)} - .3} = 0.59$$

b) *Floor Cavity*

$$A_w = 2 \cdot 2.5 \cdot (60 + 30) = 450 \text{ ft}^2$$
$$A_b = 60 \cdot 30 = 1800 \text{ ft}^2$$
$$A_w/A_b = 450/1800 = 0.25$$

$$\rho_{eff} = \rho_{fc} = \cfrac{1}{\cfrac{1.25^2}{.2 + .5\,(.25)} - .25} = 0.22$$

Photometric data for interior lighting application is normally presented in the form of either candlepower distribution curves and tables or in the form of coefficients of utilization tables, known as *CU* tables. Figure 12.3 shows several representative *CU* tables. The coefficient of utilization data presented in these tables is based on the Zonal Cavity Method, as discussed in previous sections of this chapter. The numerical value of the *CU* represents the fraction of the total lamp lumens that will reach the work plane in a specific room configuration with specific surface reflectances. Take a few moments to become familiar with the luminaire types in this figure. The total maintained level of illuminance in a room is then be given by:

$$E = \frac{\phi \cdot CU \cdot LLF}{A} \qquad Footcandles \qquad (12.3)$$

where E is the maintained work plane illuminance level in footcandles, and A is the room area in square feet. The light loss factor, LLF, represents the light lost due to several factors, such as dirt accumulation, lamp depreciation and room surface depreciation. These and other light loss factors will be discussed shortly. ϕ represents the total lamp lumens generated by all luminaires in the space. Let us now see how we determine the coefficient of utilization for a specific room configuration with known surface reflectances.

Figure 12.3 Representative luminaire photometric data.

Coefficients of Utilization for 20 Per Cent Effective Floor Cavity Reflectance (ρ_{FC} = 20)

Luminaire 1 — Fluorescent unit with flat prismatic lens, 4 lamp 610 mm (2') wide. Maint. Cat. V, SC = 1.4/1.2

ρ_{CC}	80			70			50			30			10			0
ρ_W	50	30	10	50	30	10	50	30	10	50	30	10	50	30	10	0
RCR																
 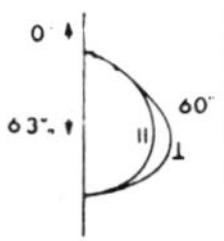 0	.75	.75	.75	.73	.73	.73	.70	.70	.70	.67	.67	.67	.64	.64	.64	.63
1	.67	.65	.63	.66	.64	.62	.63	.62	.60	.61	.60	.58	.59	.58	.57	.55
2	.60	.57	.54	.59	.56	.53	.57	.54	.52	.55	.53	.51	.53	.51	.50	.49
3	.54	.50	.47	.53	.49	.46	.52	.48	.45	.50	.47	.45	.48	.46	.44	.43
4	.49	.44	.40	.48	.44	.40	.47	.43	.40	.45	.42	.39	.44	.41	.39	.37
5	.44	.39	.35	.43	.38	.35	.42	.38	.34	.41	.37	.34	.40	.36	.34	.33
6	.40	.34	.31	.39	.34	.31	.38	.34	.30	.37	.33	.30	.36	.32	.30	.29
7	.36	.30	.27	.35	.30	.27	.34	.30	.27	.33	.29	.26	.32	.29	.26	.25
8	.32	.27	.23	.32	.27	.23	.31	.26	.23	.30	.26	.23	.29	.26	.23	.22
9	.29	.24	.20	.28	.23	.20	.28	.23	.20	.27	.23	.20	.26	.23	.20	.19
10	.26	.21	.18	.26	.21	.18	.25	.21	.18	.24	.20	.18	.24	.20	.18	.16

Luminaire 2 — 4 lamp, 610 mm (2') wide troffer with 45° white metal louver. Maint. Cat. IV, SC = 0.9

ρ_{CC}	80			70			50			30			10			0
ρ_W	50	30	10	50	30	10	50	30	10	50	30	10	50	30	10	0
RCR																
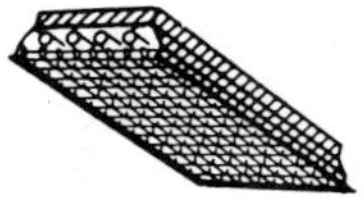 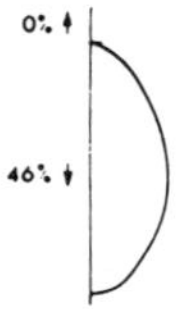 0	.55	.55	.55	.54	.54	.54	.51	.51	.51	.49	.49	.49	.47	.47	.47	.46
1	.50	.48	.47	.49	.47	.46	.47	.46	.45	.45	.44	.43	.43	.43	.42	.41
2	.45	.43	.41	.44	.42	.40	.43	.41	.39	.41	.40	.38	.40	.39	.37	.37
3	.41	.38	.36	.40	.38	.35	.39	.37	.35	.38	.36	.34	.37	.35	.34	.33
4	.37	.34	.32	.37	.34	.31	.36	.33	.31	.35	.32	.31	.34	.32	.30	.29
5	.34	.30	.28	.33	.30	.28	.32	.30	.27	.32	.29	.27	.31	.29	.27	.26
6	.31	.28	.25	.31	.27	.25	.30	.27	.25	.29	.27	.25	.29	.26	.24	.24
7	.29	.25	.23	.28	.25	.23	.28	.25	.22	.27	.24	.22	.26	.24	.22	.21
8	.26	.23	.20	.26	.23	.20	.25	.22	.20	.25	.22	.20	.24	.22	.20	.20
9	.24	.20	.18	.24	.20	.18	.23	.20	.18	.23	.20	.18	.22	.20	.18	.17
10	.22	.19	.16	.22	.19	.16	.21	.18	.16	.21	.18	.16	.20	.18	.16	.15

Luminaire 3 — 2 lamp prismatic wraparound. Maint. Cat. V, SC = 1.5/1.2

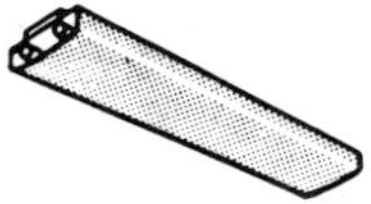

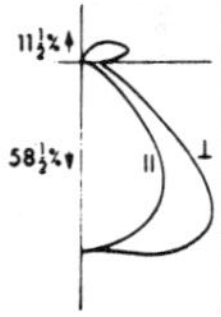

ρ_{CC}	80			70			50			30			10			0
ρ_W	50	30	10	50	30	10	50	30	10	50	30	10	50	30	10	0
RCR																
0	.81	.81	.81	.78	.78	.78	.72	.72	.72	.66	.66	.66	.61	.61	.61	.59
1	.71	.69	.66	.69	.66	.64	.64	.62	.60	.59	.58	.56	.55	.54	.53	.50
2	.64	.59	.56	.61	.58	.54	.57	.54	.51	.53	.51	.49	.49	.48	.46	.44
3	.57	.52	.48	.55	.50	.47	.51	.48	.45	.48	.45	.42	.45	.42	.40	.38
4	.51	.46	.41	.49	.44	.41	.46	.42	.39	.43	.40	.37	.41	.38	.35	.34
5	.46	.40	.36	.44	.39	.35	.41	.37	.34	.39	.35	.32	.38	.33	.31	.29
6	.41	.35	.31	.40	.35	.31	.38	.33	.30	.35	.31	.28	.33	.30	.27	.26
7	.37	.31	.27	.36	.31	.27	.34	.29	.26	.31	.28	.25	.30	.27	.24	.23
8	.33	.28	.24	.32	.27	.23	.30	.26	.22	.29	.25	.22	.27	.24	.21	.19
9	.30	.24	.20	.29	.24	.20	.27	.23	.20	.26	.22	.19	.24	.21	.18	.17
10	.27	.22	.18	.26	.21	.18	.25	.20	.17	.23	.19	.16	.22	.18	.16	.15

Figure 12.3 Representative luminaire photometric data. (Reproduced with permission of Illuminating Engineering Society of North America.) *(Continued)*.

Figure 12.3 Representative luminaire photometric data. (Reproduced with permission of Illuminating Engineering Society of North America.) *(Continued).*

Coefficients of Utilization for 20 Per Cent Effective Floor Cavity Reflectance ($\rho_{FC} = 20$)

Typical Luminaire	Maint. Cat.	SC	RCR	ρcc 80 (50,30,10)			70 (50,30,10)			50 (50,30,10)			30 (50,30,10)			10 (50,30,10)			0
4. Porcelain-enameled reflector with 35°CW shielding (22½%↑, 65%↓)	II	1.3	0	99	99	99	94	94	94	85	85	85	77	77	77	69	69	69	65
			1	88	85	82	84	81	78	76	74	72	69	67	66	62	61	60	57
			2	78	73	69	74	70	66	68	64	61	62	59	56	56	54	52	49
			3	70	63	58	67	61	57	61	56	53	56	52	49	51	48	46	43
			4	62	55	50	60	53	49	55	50	46	50	46	43	46	43	40	37
			5	55	48	43	53	47	42	49	44	39	45	41	37	41	38	35	32
			6	50	43	38	48	41	37	44	39	35	41	36	33	37	34	31	29
			7	45	38	33	43	37	32	40	34	30	37	32	29	34	30	27	25
			8	40	34	29	39	32	28	36	30	27	33	28	25	31	27	24	22
			9	36	30	25	35	29	24	32	27	23	30	25	23	28	24	21	19
			10	33	27	22	32	28	22	29	24	20	27	23	19	25	21	18	17
5. Pendant diffusing sphere with incandescent lamp (35%↑, 45%↓)	V	1.5	0	87	87	87	81	81	81	70	70	70	59	59	59	49	49	49	45
			1	71	67	63	66	62	59	56	53	50	47	45	42	38	37	35	31
			2	60	54	49	56	50	45	47	43	39	39	36	33	32	29	27	23
			3	52	45	39	48	42	37	41	36	31	34	30	26	27	24	22	18
			4	46	38	33	42	36	30	36	30	26	30	26	22	24	21	18	15
			5	40	33	27	37	30	25	31	26	22	26	22	18	21	18	15	12
			6	36	28	23	33	26	21	28	23	19	23	19	16	19	16	13	10
			7	32	25	20	29	23	18	25	20	16	21	16	13	17	13	11	09
			8	29	22	17	26	20	16	23	17	14	19	15	12	15	12	09	07
			9	26	19	15	23	17	14	20	15	12	17	13	10	14	11	08	06
			10	23	17	13	22	16	12	19	14	10	15	12	09	13	09	07	05
6. Recessed baffled downlight, 140 mm (5 ½") diameter aperture—150-PAR/FL lamp (0%↑, 68½%↓)	IV	0.5	0	82	82	82	80	80	80	76	76	76	73	73	73	70	70	70	69
			1	78	77	76	77	76	75	74	74	73	72	71	71	69	69	69	68
			2	76	74	73	74	73	72	73	71	70	71	70	69	68	68	67	67
			3	74	72	70	73	71	70	71	70	68	70	69	68	68	67	67	66
			4	72	70	68	71	69	68	70	68	67	69	67	66	67	66	66	65
			5	70	68	66	69	67	66	68	67	65	67	66	65	66	65	65	64
			6	69	66	65	68	66	65	67	66	64	66	65	64	65	64	64	63
			7	67	65	63	66	65	63	66	64	63	65	64	63	64	63	63	62
			8	66	64	62	65	63	62	65	63	62	64	63	62	64	62	62	61
			9	65	63	61	64	63	61	64	62	61	63	62	61	63	62	61	60
			10	63	61	60	63	61	60	63	61	60	62	61	60	62	61	60	59

434

Figure 12.3 — Representative luminaire photometric data

Coefficients of Utilization for 20 Per Cent Effective Floor Cavity Reflectance ($\rho_{fc} = 20$)

Typical Luminaire	Maint. Cat.	SC	% Lamp Lumens
7 — "High bay" narrow distribution ventilated reflector with clear HID lamp	III	0.7	1¼↑ / 77%↑
8 — "High bay" intermediate distribution ventilated reflector with clear HID lamp	III	1.0	1↑ / 76%↑
9 — "High bay" wide distribution ventilated reflector with clear HID lamp	III	1.5	¾↑ / 77½%↑

Luminaire 7

RCR	ρcc 80 (50)	(30)	(10)	70 (50)	(30)	(10)	50 (50)	(30)	(10)	30 (50)	(30)	(10)	10 (50)	(30)	(10)	0
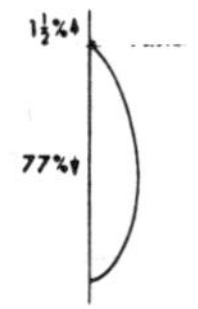 0	.93	.93	.93	.90	.90	.90	.86	.86	.86	.82	.82	.82	.78	.78	.78	.77
1	.87	.85	.83	.83	.81	.80	.81	.80	.79	.78	.77	.76	.75	.75	.74	.72
2	.81	.79	.76	.77	.75	.73	.77	.75	.73	.75	.73	.72	.72	.71	.70	.69
3	.77	.73	.69	.72	.70	.67	.73	.71	.69	.71	.69	.67	.70	.68	.65	.65
4	.73	.69	.66	.68	.65	.64	.70	.67	.64	.68	.66	.64	.67	.65	.63	.62
5	.69	.65	.62	.64	.61	.61	.66	.63	.61	.65	.62	.60	.64	.61	.59	.58
6	.65	.61	.58	.61	.58	.57	.63	.60	.58	.62	.59	.57	.61	.58	.56	.55
7	.62	.57	.54	.57	.54	.51	.60	.56	.54	.59	.56	.53	.58	.55	.53	.52
8	.58	.54	.51	.54	.51	.51	.57	.53	.51	.56	.53	.51	.55	.52	.50	.49
9	.55	.51	.48	.51	.48	.46	.54	.50	.48	.53	.50	.48	.53	.50	.47	.47
10	.53	.49	.46	.48	.46	.46	.52	.48	.46	.51	.48	.45	.50	.47	.45	.44

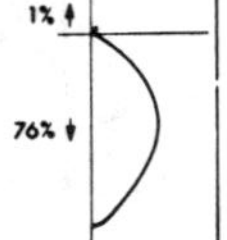

Luminaire 8

RCR	ρcc 80 (50)	(30)	(10)	70 (50)	(30)	(10)	50 (50)	(30)	(10)	30 (50)	(30)	(10)	10 (50)	(30)	(10)	0
0	.91	.91	.91	.89	.89	.89	.85	.85	.85	.81	.81	.81	.78	.78	.78	.76
1	.84	.82	.80	.82	.80	.78	.79	.77	.76	.76	.74	.73	.75	.72	.71	.69
2	.77	.73	.70	.76	.72	.70	.73	.70	.68	.70	.68	.66	.70	.68	.65	.63
3	.71	.66	.63	.69	.65	.62	.67	.64	.61	.65	.62	.60	.64	.61	.59	.57
4	.65	.60	.56	.64	.59	.56	.62	.58	.55	.60	.57	.54	.59	.56	.54	.52
5	.59	.54	.50	.59	.54	.50	.57	.53	.50	.56	.52	.49	.54	.51	.48	.47
6	.54	.49	.45	.54	.49	.45	.52	.48	.45	.51	.47	.44	.50	.47	.44	.42
7	.50	.44	.40	.49	.44	.40	.48	.43	.40	.47	.43	.39	.46	.42	.39	.38
8	.45	.40	.36	.45	.40	.36	.44	.39	.36	.43	.39	.35	.42	.38	.35	.34
9	.41	.36	.32	.41	.36	.32	.40	.35	.32	.39	.35	.32	.38	.35	.31	.30
10	.38	.33	.29	.37	.32	.29	.37	.32	.29	.36	.32	.28	.35	.31	.28	.27

Luminaire 9

RCR	ρcc 80 (50)	(30)	(10)	70 (50)	(30)	(10)	50 (50)	(30)	(10)	30 (50)	(30)	(10)	10 (50)	(30)	(10)	0
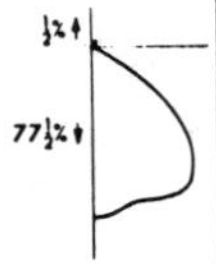 0	.93	.93	.93	.91	.91	.91	.87	.87	.87	.83	.83	.83	.79	.79	.79	.78
1	.85	.82	.80	.83	.81	.79	.79	.78	.76	.79	.78	.76	.76	.75	.74	.70
2	.77	.73	.70	.76	.72	.69	.73	.70	.67	.70	.68	.66	.68	.66	.64	.63
3	.70	.65	.61	.68	.64	.60	.66	.62	.57	.64	.61	.58	.62	.59	.57	.56
4	.63	.58	.53	.60	.57	.53	.60	.56	.50	.58	.55	.52	.57	.54	.51	.49
5	.57	.51	.47	.56	.51	.47	.55	.50	.46	.53	.49	.46	.52	.48	.45	.44
6	.51	.45	.41	.51	.45	.41	.50	.44	.40	.49	.43	.40	.47	.43	.40	.38
7	.46	.40	.35	.45	.39	.35	.44	.39	.35	.43	.38	.35	.42	.38	.35	.33
8	.41	.35	.31	.41	.35	.31	.39	.34	.31	.38	.34	.30	.38	.34	.30	.29
9	.37	.30	.27	.36	.30	.27	.36	.31	.27	.35	.30	.27	.34	.30	.26	.25
10	.33	.27	.24	.33	.27	.23	.32	.27	.23	.31	.27	.23	.31	.26	.23	.22

Figure 12.3 Representative luminaire photometric data. (Reproduced with permission of Illuminating Engineering Society of North America.)

435

Once the appropriate luminaire is selected and its photometric data obtained, the first step is the calculation of the three cavity ratios. Next we determine the effective reflectance of the floor and ceiling cavities. With this information, we are now ready to use the table of CU data using the computed room cavity ratios, along with wall reflectances, and effective ceiling cavity and floor cavity reflectances. It is usually necessary to interpolate the table data to arrive at the desired CU. Let us look at an example.

Example 12.3: A room has the following dimensions and reflectances. Compute the coefficient of utilization (CU) if Luminaire 1 is used.

$$\begin{array}{lll}
\text{Length} = 40 \text{ ft} & h_{fc} = 2.5 \text{ ft } \rho_c & = .70 \\
\text{Width } = 20 \text{ ft} & h_{cc} = 0.0 \text{ ft } \rho_w & = .30 \\
& h_{rc} = 6.0 \text{ ft } \rho_f & = .20
\end{array}$$

a.) *Ceiling Cavity Ratio*

$$CCR = 0 \text{ since } h_{cc} = 0$$

b.) *Room Cavity Ratio*

$$RCR = \frac{5 \cdot 6 \cdot (40 + 20)}{40 \cdot 20} = 2.25$$

c.) *Floor Cavity Ratio*

$$FCR = \frac{5 \cdot 2.5 \cdot (40 + 20)}{40 \cdot 20} = 0.94$$

d.) *Compute CU*

From the CU data for Luminaire 1

RCR	CU
2	.56
3	.49

$$CU = .56 - .25(.56 - .49) = 0.54$$

This *CU* is based on a floor cavity reflectance of 20% so we must now check our actual ρ_{fc} using Equation 12.2, with $A_w = 300$ ft^2, $A_b = 800$ ft^2, and $A_w/A_b = 0.38$

$$\rho_{fc} = \cfrac{1}{\cfrac{1.38^2}{.20 + .30 \cdot (.38)} - .38} = 0.18$$

Our actual floor cavity reflectance is somewhat less than 20%. Using Table 12.1 and interpolating between effective reflectances of 10% and 20%, we arrive at a multiplier of 0.99. Our final value for the *CU* is then $0.99 \cdot 0.54 = 0.53$. This means that 53% of the total lumens generated by all lamps in the room eventually reach the work plane.

Example 12.4: Using the luminaires and room size of Example 12.3, compute the number of luminaires required to provide an initial footcandle level of 75 *fc*. Each luminaire is equipped with 4, 40 W lamps rated 3150 lumens each.

In order to compute the initial footcandle level we will set the light loss factor, *LLF*, of Equation 12.3 to 1.0, thus neglecting all factors which tend to reduce the average illumination level.

By rearranging Equation 12.3 and solving for the required flux, ϕ

$$\phi = \frac{E \cdot A}{CU \cdot LLF}.$$

Using the room area, desired illuminance level and coefficient of utilization

$$\phi = \frac{75 \cdot 800}{.53 \cdot 1.0} = 113,208 \ \ Lumens.$$

The required number of luminaires

$$F = \frac{Total \ Lumens \ Required}{Lumens \ per \ Luminaire}$$

$$= \frac{113,208}{4 \cdot 3150} = 9 \ \ Luminaires.$$

Table 12.1 Multiplying Factors for other than 20% Effective Floor Cavity Reflectance. (Reprinted with Permission of Illuminating Engineering Society of North America.)

% Effective Ceiling Cavity Reflectance	80			70			50		
% Wall Reflectance	50	30	10	50	30	10	50	30	10
For 30% Effective Floor Cavity Reflectance (20% = 1.00)									
Room Cavity Ratio									
1	1.082	1.075	1.068	1.070	1.064	1.059	1.049	1.044	1.040
2	1.066	1.055	1.047	1.057	1.048	1.039	1.041	1.033	1.027
3	1.054	1.042	1.033	1.048	1.037	1.028	1.034	1.027	1.020
4	1.045	1.033	1.024	1.040	1.029	1.021	1.030	1.022	1.015
5	1.038	1.026	1.018	1.034	1.024	1.015	1.027	1.018	1.012
6	1.033	1.021	1.014	1.030	1.020	1.012	1.024	1.015	1.009
7	1.029	1.018	1.011	1.026	1.017	1.009	1.022	1.013	1.007
8	1.026	1.015	1.009	1.024	1.015	1.007	1.020	1.012	1.006
9	1.024	1.014	1.007	1.022	1.014	1.006	1.019	1.011	1.005
10	1.022	1.012	1.006	1.020	1.012	1.005	1.017	1.010	1.004
For 10% Effective Floor Cavity Reflectance (20% = 1.00)									
Room Cavity Ratio									
1	.929	.935	.940	.939	.943	.948	.956	.960	.963
2	.942	.950	.958	.949	.957	.963	.962	.968	.974
3	.951	.961	.969	.957	.966	.973	.967	.975	.981
4	.958	.969	.978	.963	.973	.980	.972	.980	.986
5	.964	.976	.983	.968	.978	.985	.975	.983	.989
6	.969	.980	.986	.972	.982	.989	.977	.985	.992
7	.973	.983	.991	.975	.985	.991	.979	.987	.994
8	.976	.986	.993	.977	.987	.993	.981	.988	.995
9	.978	.987	.994	.979	.989	.994	.983	.990	.996
10	.980	.989	.995	.981	.990	.995	.984	.991	.997
For 0% Effective Floor Cavity Reflectance (20% = 1.00)									
Room Cavity Ratio									
1	.870	.879	.886	.884	.893	.901	.916	.923	.929
2	.887	.903	.919	.902	.916	.928	.926	.938	.949
3	.904	.915	.942	.918	.934	.947	.936	.950	.964
4	.919	.941	.958	.930	.948	.961	.945	.961	.974
5	.931	.953	.969	.939	.958	.970	.951	.967	.980
6	.940	.961	.976	.945	.965	.977	.955	.972	.985
7	.947	.967	.981	.950	.970	.982	.959	.975	.988
8	.953	.971	.985	.955	.975	.986	.963	.978	.991
9	.958	.975	.988	.959	.980	.989	.966	.980	.993
10	.962	.979	.991	.963	.983	.992	.969	.982	.995

It is important to remember that we have ignored the light loss factor, *LLF*, in the preceding examples. In reality, this is a very important factor and must be considered in the actual design process. We will now address the light loss factor itself.

LIGHT LOSS FACTORS

From the moment a new lighting system is first energized, a number of factors combine to steadily decrease the illuminance level. It is convenient to group these factors into the two categories of *recoverable* and *unrecoverable* factors. The product of the various factors, all ranging from 0 to 1, is referred to as the light loss factor, or *LLF* (IES Lighting Handbook 1984). Since we normally wish to compute the average maintained level of illuminance, the light loss factor must be included. Let us now discuss these light loss factors in detail.

Recoverable Light Loss Factors

Lamp Burnout Factor (*LBO*)

As discussed in earlier chapters, all light sources have a specified average burning life. The actual definition of this average-rated life is the burning time, in hours, at which 50% of a sample group of lamps remains burning. Different light sources have differently rated lives. Statistically, we can expect some lamps to fail well before their rated life while others continue to burn many hours beyond it.

There are several approaches to the replacement of burned out lamps. The lamp burnout factor permits us to include the effect of the lamp replacement strategy in our design. For example, some building owners prefer to replace lamps only as they fail, while others employ partial or total replacement on a periodic basis. If lamps are replaced upon burnout, the lamp burnout factor, *LBO*, is equal to 1.0. If the replacement strategy is to replace all lamps when some fixed percentage of the lamps have failed, the *LBO* factor will be less than 1.0. For example, if the strategy is to replace all lamps when 10% have failed, the *LBO* factor will be 0.90.

Lamp Lumen Depreciation (*LLD*)

The lumen output of all light sources gradually decreases during the life of

Table 12.2 Typical Rated Lamp Life and Lamp Lumen Depreciation Data for Various Light Sources.

LAMP TYPE	TYPICAL RATED LIFE (hours)	TYPICAL LAMP LUMEN DEPRECIATION FACTOR
Incandescent		
General Service	1,000	0.80
Halogen	2,000	0.90
Fluorescent		
Light Loading	20,000	0.84
Medium Loading	12,000	0.79
Heavy Loading	10,000	0.72
High Intensity Discharge (HID)		
Mercury Vapor	24,000	0.83
Metal Halide	20,000	0.80
High Pressure Sodium	24,000	0.90

the lamp. The lamp lumen depreciation factor permits us to include an allowance for this depreciation in our design, again, to assure that our system does not fall below the desired maintained level of illuminance. Table 12.2 lists representative lamp lumen depreciation factors for several light sources. For fluorescent and HID sources, the lamp lumen depreciation is determined by dividing the lamp's lumen output at 100 hours by its output at 70% of its rated life. For incandescent sources, the *LLD* is established by dividing the initial light output by the light output at 100% rated life.

Luminaire Dirt Depreciation (*LDD*)

The luminaire dirt depreciation factor represents the decrease in the maintained level of illuminance due to the accumulation of dirt on the luminaire itself. Obviously the depreciation at any given point in time depends on both the degree of dirtiness of the environment, in which the luminaire is installed, and the design of the luminaire itself. This depreciation can have a serious effect on the overall efficiency of the lighting system because the system gradually delivers less and less total illumination, while continuing to consume the same electrical energy. Happily, the luminaire can be restored to original efficiency by periodic cleaning. Some owners perform luminaire cleaning at 12-, 18- or 24-month increments, while others perform cleaning in conjunction with their group lamp replacement program.

For the purpose of determining luminaire dirt depreciation, six luminaire maintenance categories have been established. These categories are shown in Table 12.3. The design of a luminaire, including the nature of its lens or louver, as well as openings for uplighting, determine the category into which it falls. In order to account for the varying dirt conditions, each luminaire maintenance category also considers five degrees of dirtiness. A description of these levels of dirtiness is shown in Table 12.4. The five levels vary from the location of luminaires in very clean areas, such as hospitals, to very dirty areas, such as industrial foundries. Between these two extremes are clean areas, such as high grade offices, medium areas, such as warehouses, and dirty areas, such as manufacturing plants.

Through years of experience and research, it has been found that the decrease in luminaire light output due to luminaire dirt accumulation follows an exponential decay, that can be expressed by

$$LDD = e^{-At^b} \tag{12.4}$$

where the constant A depends on both the degree of dirtiness and the luminaire maintenance category, while the constant b depends solely on the luminaire maintenance category. These constants are, essentially, shaping coefficients for the luminaire dirt depreciation curves. The t in Equation 12.4 represents the time in decimal years between luminaire cleanings. A table of the A and B coefficients is contained in Table 12.5, making manual calculations by calculator a straightforward matter. For many applications, however, sufficient accuracy may be obtained by referring to curves derived from Equation 12.4 and the data in Table 12.5. A family of six graphs is shown in Figure 12.4.

Example 12.5: A given lighting system contains luminaires which fall into Maintenance Category IV. These luminaires are installed in an environment classified as clean, and luminaire cleaning occurs at two-year intervals. Determine the LDD factor.

Table 12.3 Luminaire Maintenance Categories. (Reprinted with Permission of Illuminating Engineering Society of North America.)

Maintenance Category	Top Enclosure		Bottom Enclosure	
I	1.	None	1.	None
II	1.	None	1.	None
	2.	Transparent with 15% or more uplight through apertures.	2.	Louvers or baffles
	3.	Translucent with 15% or more uplight through apertures.		
	4.	Opaque with 15% or more uplight through apertures		
III	1.	Transparent with less than 15% upward light through apertures	1.	None
	2.	Translucent with less than 15% upward light through apertures	2.	Louvers or baffles
	3.	Opaque with less than 15% uplight through apertures		
IV	1.	Transparent unapertured	1.	None
	2.	Translucent unapertured	2.	Louvers
	3.	Opaque unapertured		
V	1.	Transparent unapertured	1.	Transparent unapertured
	2.	Translucent unapertured	2.	Translucent unapertured
	3.	Opaque unapertured		
VI	1.	None	1.	Transparent unapertured
	2.	Transparent unapertured	2.	Translucent unapertured
	3.	Translucent unapertured	3.	Opaque unapertured
	4.	Opaque unapertured		

Table 12.4 Degrees of Dirtiness. (Reprinted with Permission of Illuminating Engineering Society of North America.)

	VERY CLEAN	CLEAN	MEDIUM	DIRTY	VERY DIRTY
Generated Dirt	None	Very Little	Noticeable but not heavy	Accumulates rapidly	Constant Accumulation
Ambient Dirt	None (or none enters area)	Some (almost none enters)	Some enters area	Large amount enters area	Almost none excluded
Removal or Filtration	Excellent	Better than average	Poorer than average	Only fans or blowers, if any	None
Adhesion	None	Slight	Enough to be visible after some months	High-probably due to oil, humidity or static	High
Examples	High grade of offices, not near production laboratories; cleanrooms	Offices in older buildings or near production; light assembly; inspection	Mill offices; paper processing; light machining	Heat treating; high speed printing; rubber processing	Similar to Dirty but luminaires within immediate area of contamination

443

Table 12.5 Table of IES Coefficients for Six Maintenance Categories. (Reprinted with permission of Illuminating Engineering Society of North America.)

Luminaire Maintenance Category	b	A				
		Very Clean	Clean	Medium	Dirty	Very Dirty
I	0.69	0.038	0.071	0.111	0.162	0.301
II	0.62	0.033	0.068	0.102	0.147	0.188
III	0.70	0.079	0.106	0.143	0.184	0.236
IV	0.72	0.070	0.131	0.216	0.314	0.452
V	0.53	0.078	0.128	0.190	0.249	0.321
VI	0.88	0.076	0.145	0.218	0.284	0.396

From Table 12-5

$$A = 0.131 \quad \text{and} \quad b = 0.72$$

$$LDD = e^{-0.131\,(2^{0.72})} = 0.81$$

Room Surface Dirt Depreciation (*RSDD*)

The final recoverable luminaire dirt depreciation factor is room surface dirt depreciation. This factor takes into consideration the reduction in illumination over time due to the degradation of the room surface reflectance values which are caused by dirt accumulation.

It has been found that this factor also follows an exponential decrease and closely matches the curve for luminaires in Maintenance Category V (Murdoch 1984). The percent dirt depreciation is given by this modified form of Equation 12.4.

$$\% \text{ Expected Dirt Depreciation} = 100\,(1 - e^{-At^b}). \qquad (12.5)$$

For convenient reference, Equation 12.5 can be represented by a family of curves as shown in Figure 12.5.

The effect of room surface dirt depreciation is influenced, to some extent, by the nature of the luminaire's distribution characteristics. These

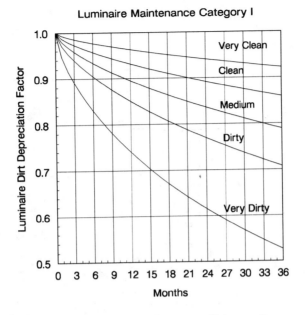

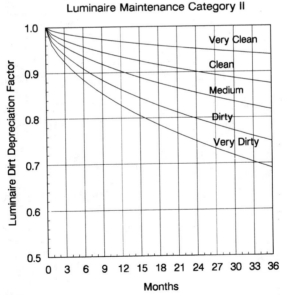

Figure 12.4 Luminaire dirt depreciation curves for various luminaire maintenance categories and degrees of dirtiness. *(Continued).*

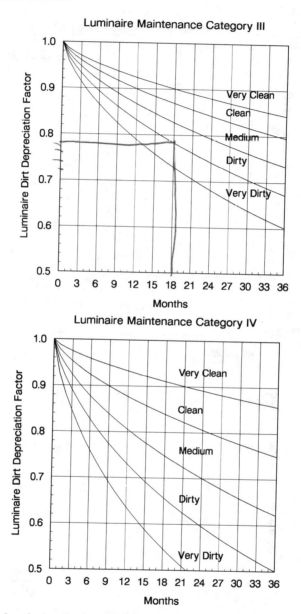

Figure 12.4 Luminaire dirt depreciation curves for various luminaire maintenance categories and degrees of dirtiness. *(Continued).*

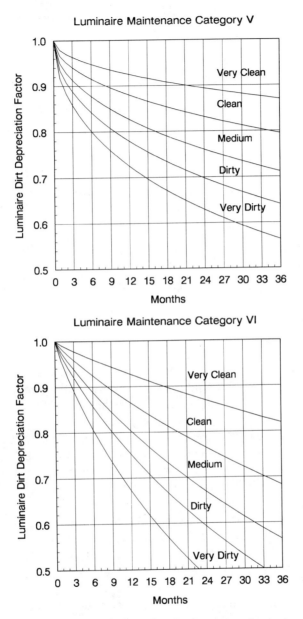

Figure 12.4 Luminaire dirt depreciation curves for various luminaire maintenance categories and degrees of dirtiness.

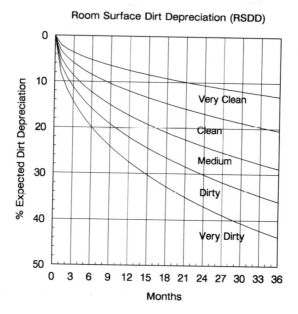

Figure 12.5 Percent expected room surface dirt depreciation curves for various degrees of dirtiness.

characteristics have been broken down into six broad categories, depending on the percentage of the total lumens emitted by the luminaire directed upward or downward. Table 12.6 shows these classifications and their associated upward and downward lighting components.

With the value of % expected dirt depreciation, the room's cavity ratio and the luminaire distribution classification, we use Table 12.7, interpolating if necessary. An example will demonstrate how we utilize the above data to arrive at a specific *RSDD* factor.

Example 12.6: A room is to be illuminated using direct luminaires. The room is considered to fall within the medium cleanliness classification, with room surfaces cleaned at 24-month increments. If the room cavity ratio is 3.5, determine the correct *RSDD* factor.

We will first need the percent expected dirt depreciation. Using Equation 12.5 and the appropriate coefficient for *A* and *b*:

$$\% \, Expected \, Dirt \, Depreciation \;=\; 100 \, (1 \,-\, e^{-0.190 \, (2^{0.53})}) \;=\; 24\%.$$

Luminaire Distribution Type

Room Cavity Ratio	Direct				Semi-Direct				Direct-Indirect				Semi-Indirect				Indirect			
Per Cent Expected Dirt Depreciation	10	20	30	40	10	20	30	40	10	20	30	40	10	20	30	40	10	20	30	40
1	.98	.96	.94	.92	.97	.92	.89	.84	.94	.87	.80	.76	.94	.87	.80	.73	.90	.80	.70	.60
2	.98	.96	.94	.92	.96	.92	.88	.83	.94	.87	.80	.75	.94	.87	.79	.72	.90	.80	.69	.59
3	.98	.95	.93	.90	.96	.91	.87	.82	.94	.86	.79	.74	.94	.86	.78	.71	.90	.79	.68	.58
4	.97	.95	.92	.90	.95	.90	.85	.80	.94	.86	.79	.73	.94	.86	.78	.70	.89	.78	.67	.56
5	.97	.94	.91	.89	.94	.90	.84	.79	.93	.86	.78	.72	.93	.86	.77	.69	.89	.78	.66	.55
6	.97	.94	.91	.88	.94	.89	.83	.78	.93	.85	.78	.71	.93	.85	.76	.68	.89	.77	.66	.54
7	.97	.94	.90	.87	.93	.88	.82	.77	.93	.84	.77	.70	.93	.84	.76	.68	.89	.76	.65	.53
8	.96	.93	.89	.86	.93	.87	.81	.75	.93	.84	.76	.69	.93	.84	.76	.68	.88	.76	.64	.52
9	.96	.92	.88	.85	.93	.87	.80	.74	.93	.84	.76	.68	.93	.84	.75	.67	.88	.75	.63	.51
10	.96	.92	.87	.83	.93	.86	.79	.72	.93	.84	.75	.67	.92	.83	.75	.67	.88	.75	.62	.50

Table 12.7 Room Surface Dirt Depreciation Factors. (Reprinted with Permission of Illuminating Engineering Society of North America.)

449

Using the *RCR* of 3.5, the direct luminaire classification, and the % expected dirt depreciation, we use Table 12.7 and locate the following data:

DIRECT

RCR	20%	30%
3.0	.95	.93
4.0	.95	.92

By interpolation, the correct *RSDD* is 0.94.

Unrecoverable Light Loss Factors

The recoverable light loss factors discussed previously resulted from factors which could be reset to nominal values by lamp replacement, luminaire cleaning or room surface cleaning. The factors we will now address cannot be so easily corrected and, once established, are considered to be permanent in nature.

These unrecoverable factors relate to the effects of ambient temperature, applied voltage, the efficiency of the ballast and luminaire surface depreciation. We will discuss these factors individually.

Luminaire Ambient Temperature (*LAT*)

While incandescent and HID light sources are relatively insensitive to variations in ambient temperature, fluorescent sources are not so fortunate. As discussed in Chapter 10, the temperature of the fluorescent lamp jacket or tube must be maintained in close proximity to optimal values or lamp output will suffer.

In most practical cases, this factor is set to unity. If the engineer feels that the ambient temperature in a given application will be significantly above or below the lamp's optimal temperature, the lamp manufacturer should be consulted for assistance and technical support. Fluorescent ballasts are available for low temperature operation, but the fluorescent lamps may still require an additional outer jacket to help raise the lamp's temperature. Operation of fluorescent systems at high temperatures not only affects light output, but also might shorten ballast life, or even result in the nuisance operation of the ballast thermal cutout.

Luminaire Voltage (*LV*)

As we have seen in Chapter 10, the incandescent lamp is quite sensitive to voltage variations where a 1% voltage decrease results in a 3% decrease in light output. In the case of fluorescent lamps, a voltage decrease of 1% will result in a decrease in light output of about 0.4%. Newer fluorescent electronic ballasts may not be as voltage-sensitive, and operation over a wider voltage range may be possible. HID ballasts vary widely in their voltage tolerance depending on their design. For most applications, the luminaire voltage factor is assumed to be unity, however, the engineer should be guided by the characteristics of the specific ballast to be used.

Ballast Factor (*BF*)

Luminaires are tested and photometric reports are derived using laboratory ballasts which closely match the optimal requirements of the lamp. Those special ballasts are quite expensive, and actual ballasts purchased for normal installation may not be quite as efficient. The ballast factor affords us the opportunity to take these differences into account. As incandescent lamps have no ballast, this factor is set to unity. Many engineers use a factor of 0.95 for fluorescent ballasts. There is little ballast factor data available for HID ballasts, therefore, in the present work, we will use a unity factor. Again, the engineer should be guided by information on the specific ballast to be used.

Luminaire Surface Depreciation (*LSD*)

The luminaire surface depreciation factor allows us to take into account any degradation of the luminaire's reflective surfaces over time. This degradation can result from the yellowing of paint or damage caused by a corrosive atmosphere. In most common applications, the luminaire surface depreciation factor is set to unity.

In summary the light loss factor, *LLF*, is the product of eight different factors. Four of these, the lamp burnout factor, *LBO*, the lamp lumen depreciation factor, *LLD*, the luminaire dirt depreciation factor, *LDD*, and the room surface dirt depreciation factor, *RSDD*, are considered to be recoverable factors. The remaining four factors, the luminaire ambient temperature factor, *LAT*, the luminaire voltage factor, *LV*, the ballast factor, *BF*, and the luminaire surface depreciation factor, *LSD*, are considered to be unrecoverable factors. The light loss factor is then given by:

$$LLF = LBO \cdot LLD \cdot LDD \cdot RSDD \cdot LAT \cdot LV \cdot BF \cdot LSD. \qquad (12.6)$$

LUMINAIRE SPACING CRITERION

In most applications, the uniformity of illumination on the work plane is an important consideration. The photometric characteristics of the luminaire, as well as the luminaire mounting height and spacing, determine this uniformity. If the luminaires are spaced too far apart, dark areas between luminaires may result. In practice, some degree of overlap in the distribution of luminaires is necessary in order to assure acceptable uniformity. Also, many luminaires have distribution patterns which are not symmetrical in nature. This results in different spacing requirements, depending on whether the luminaires are to be mounted end-to-end or side-by-side. Spacing criteria (SC) data is developed and presented as part of the photometric data for all luminaires. The SC data helps us determine the maximum distance between luminaires, and still maintain acceptable lighting uniformity.

There are two basic approaches to the determination of a spacing criterion. A layout for these approaches is shown in Figure 12.6. The basic idea is fairly simple. In Figure 12.6 (a) two luminaires, A and B, are mounted at a distance, S, apart. The highest level of illumination will

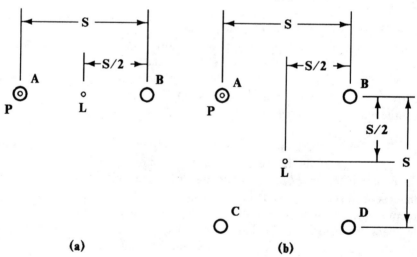

Figure 12.6 Layouts for determination of luminarire spacing criterion.

probably be directly under a luminaire, such as at point P. The lowest level of illumination will probably occur directly between the two luminaires at point L.

If we consider four luminaires, A, B, C, and D, arranged as shown in Figure 12.6 (b), the lowest level of illumination will probably occur at point L, directly in the center of all four luminaires, with the highest level occurring directly under one luminaire. The spacing criterion is determined by the distance, S, which will produce the same level of illumination at point L, as that directly under luminaire A, B, C, or D, due to that luminaire only.

If the luminaires in question have an asymmetrical distribution, the test is performed using the layout shown in Figure 12.6(a), twice. The first test places the luminaires perpendicular to a line between A and B. The second test rotates the luminaires so that they are parallel to a line through A and B. This results in two discrete spacing criterion values—one, which is perpendicular, gives us the distance between rows, while the second, which is parallel, represents the distance between units in a row. Fluorescent luminaires often have distributions which are asymmetrical in nature.

If the luminaires have symmetrical distributions, the layout in Figure 12.6(b) can be used, and the spacing criterion is determined based on four luminaires.

The representative photometric data shown in Figure 12.3, contains the spacing criterion, or SC, information for each luminaire. When two values appear, the first represents the maximum distance between rows and the second, the maximum distance between elements of a row. The ratios shown are the ratios of the distance between luminaires, S, to the mounting height above the work plane. Thus the SC is given by

$$SC = \frac{S}{MH}.$$ (12.7)

INTERIOR LIGHTING DESIGN EXAMPLES (CU) METHOD

In this section, we will examine two different lighting design problems, one of which is in an office area, and the other, in an industrial area.

Office Area Example

Example 12.7: An office area has the following dimensions and reflectances:

Length = 50	$h_{cc} = 0$	$\rho_c = .80$
Width = 30	$h_{rc} = 8'6''$	$\rho_w = .50$
	$h_{fc} = 2'6''$	$\rho_f = .20$

Use Luminaire Type 1. The environment is considered to be clean, and the luminaire employs 4 - 3150 lumen fluorescent lamps. The luminaires are cleaned at two-year increments, and lamps are replaced on burnout. The design calls for a maintained illumination level of 70 fc.

The *RCR* is given by:

$$RCR = \frac{5 \cdot 8.5 \cdot (80)}{1500} = 2.27.$$

Using the appropriate table in Figure 12.3

RCR	CU
2.0	.60
3.0	.54

by interpolation

$$CU = .60 - .27 \cdot (.60 - .54) = 0.58$$

floor cavity reflectance

$$A_w = 2 \cdot 2.5 \cdot 80 = 400 \ ft^2$$

$$A_b = 50 \cdot 30 = 1500 \ ft^2$$

$$A_w / A_b = 0.27$$

$$\rho_{fc} = \frac{1}{\dfrac{1.27^2}{.20 + .5 \cdot (0.27)} - 0.27} = .22$$

referring to Table 12.1

$$20\% = 1.00$$
$$30\% = 1.06$$

by interpolation, our multiplier is 1.01 and

$$CU = 1.01 \cdot 0.58 = 0.59.$$

Now determine the light loss factor:

1. Lamp Burnout $LBO = 1.00$ (Lamps replaced on burnout)
2. Lamp Lumen
   Depreciation $LLD = 0.84$ (Table 12.2)
3. Luminaire Dirt
   Depreciation $LDD = 0.83$ (Figure 12.4)
4. Room Surface Dirt
   Depreciation $RSDD = 0.97$ (Figure 12.5, Table 12.7)
        (% Expected = 16%)
5. Luminaire Ambient
   Temperature $LAT = 1.00$
6. Luminaire Voltage $LV = 1.00$
7. Ballast Factor $BF = 0.95$ (Fluorescent System)
8. Luminaire Surface
   Depreciation $LSD = 1.00$

The total light loss factor is then

$$LLF = 1.0 \cdot 0.84 \cdot 0.83 \cdot 0.97 \cdot 1.0 \cdot 1.0 \cdot 0.95 \cdot 1.0 = 0.64.$$

The number of luminaires required will be

$$Luminaires = \frac{70 \cdot 1500}{3150 \cdot 4 \cdot .59 \cdot .64} = 22.1.$$

Layout

There are several ways to arrange the luminaires in this area. Two possible arrangements are shown in Figure 12.7. The 2' × 4' luminaire selected is designed to be recess-mounted in a suspended tile ceiling, using either 2'

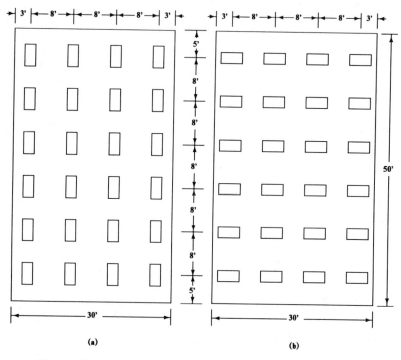

Figure 12.7 Luminaire layouts for office area - Example 12.7.

× 4' tiles or 2' × 2' tiles (with two tiles removed for fixture mounting). The nature of this ceiling system requires row-to-row spacing in increments of 2 feet. With 2' × 2' or 2' × 4' tile, the distance between elements of a row must occur in 2 ft. or 4 ft. increments, respectively. The spacing criteria for Luminaire 1 is 1.4/1.2. So, the maximum spacing between rows is 11.9 feet, and between units in a row, 10.2 feet. The arrangements shown in both Figure 12.7 (a) and (b) meet this criterion.

Industrial Area Example

Example 12.8: An industrial plant has the following dimensions:

Length = 150 ft.	h_{cc} = 5 ft.
Width = 100 ft.	h_{rc} = 18 ft.
	h_{fc} = 3 ft.

Due to floor mounted equipment and equipment suspended from the roof, we estimate the ceiling cavity and floor cavity reflectances to be 50% and 20%, respectively. The wall reflectance is 30%, therefore:

$$\rho_{cc} = .50$$
$$\rho_{w} = .30$$
$$\rho_{fc} = .20$$

We will use luminaire Type 9 with a 400 W HPS source rated at 50,000 lumens. The Maintenance Category is III, and the luminaire is classified as direct. The atmospheric conditions are dirty and cleaning occurs at 12-month increments. We are to design for a maintained footcandle level of 40 *fc*.

RCR
Our *RCR* is given by:

$$RCR = \frac{5 \cdot 18 \cdot (150 + 100)}{150 \cdot 100} = 1.50.$$

CU
Determine *CU* for Luminaire 9 from Figure 12.3

RCR	CU
1.0	.78
2.0	.70

by interpolation

$$CU = 0.78 - 0.5(.08) = 0.74.$$

The Light Loss Factor

1. Lamp Burnout Factor LBO = 1.00
2. Lamp Lumen Depreciation LLD = 0.90
3. Luminaire Dirt Depreciation LDD = 0.83
4. Room Surface Dirt Depreciation $RSDD$ = 0.96
5. Luminaire Ambient Temperature LAT = 1.00

6. Luminaire Voltage $\qquad$ LV $\quad = 1.00$
7. Ballast Factor $\qquad$ BF $\quad = 1.00$
8. Luminaire Surface Depreciation $\qquad$ LSD $\quad = 1.00$

The LLF will then be:

$$LLF \; = \; 1.0 \cdot 0.90 \cdot 0.83 \cdot 0.96 \cdot 1.0 \cdot 1.0 \cdot 1.0 \cdot 1.0 \; = \; 0.72.$$

The number of luminaires required will be:

$$Luminaires \; = \; \frac{40 \cdot (150 \cdot 100)}{50,000 \cdot 0.74 \cdot 0.72} \; = \; 22.5.$$

One possible layout of this area is shown in Figure 12.8.

Notice that the luminaires are spaced on increments of 25 feet with the first luminaire 12.5 feet from the wall. The spacing criterion for Luminaire 9 is $SC = 1.5$ or in our case, 27 feet. Thus, our spacing of 25 feet falls within acceptable limits. In order to produce a symmetrical arrangement of luminaires, it was necessary to utilize 24 luminaires instead of the number originally computed, so our actual level of illuminance will be given by

$$E \; = \; \frac{24.0}{22.5} \cdot 40 \; = \; 43 \; Footcandles.$$

INTERIOR LIGHTING DESIGN—
(POINT-BY-POINT METHOD)

In some applications, it may be important to know the level of illuminance at specific points, instead of simply the overall average for the space. In such applications, the coefficient of utilization method is not satisfactory so we must utilize inverse square law techniques, commonly referred to as point-by-point calculations. Examples of applications which might require point-by-point calculations are manufacturing areas, where we want to know the level of illuminance at a specific work surface, or in display area lighting, where we wish to know the level of illuminance at a specific display location.

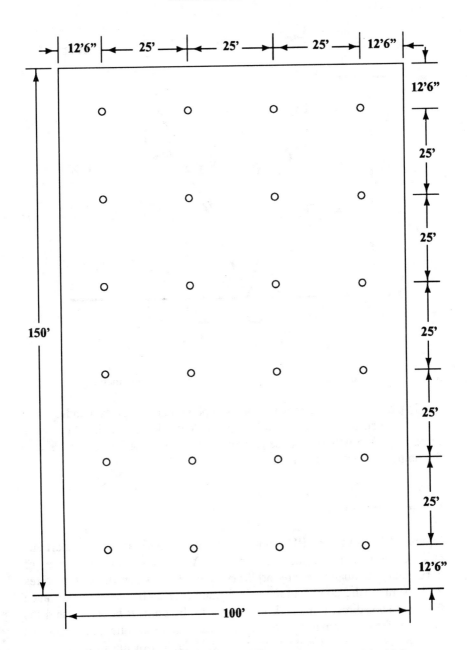

Figure 12.8 Luminaire layout for industrial area - Example 12.8.

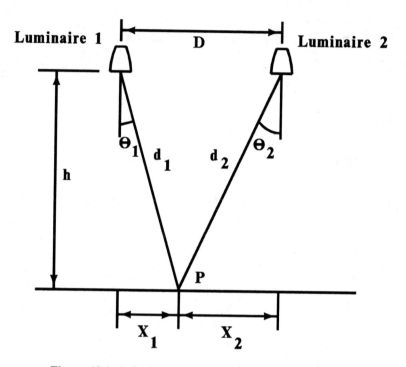

Figure 12.9 Point-by-point example for two luminaires.

In Chapter 11, the two basic forms of photometric reports, candlepower (candela) distribution curves and tables of Coefficients of Utilization, were presented. We will use the candlepower (candela) distribution curve data for point-by-point calculations.

General Methodology

Figure 12.9 shows two HID luminaires located at a distance of D apart and mounted h feet above the work plane. We can see that the level of horizontal illuminance at any arbitrary point, P, between these luminaires is the sum of the contribution of both luminaires. If we let the distance from the spot directly under Luminaire 1 to the point P be X_1 ft., and the distance from Luminaire 2 be X_2 ft., we can determine the necessary distances and angles for point-by-point calculations of the form

$$E = \frac{cd}{d^2} \cos(\theta) \quad fc. \qquad (12.8)$$

The distances and angles from the two luminaires to point P will be

$$
\begin{aligned}
d_1 &= \sqrt{X_1^2 + h^2} \\
d_2 &= \sqrt{X_2^2 + h^2} \\
\theta_1 &= \arctan\left[\frac{X_1}{h}\right] \\
\theta_2 &= \arctan\left[\frac{X_2}{h}\right].
\end{aligned}
\qquad (12.9)
$$

From the angles θ_1 and θ_2, the appropriate candlepower, $cd(\theta_1)$ and $cd(\theta_2)$, can be determined from the candlepower distribution curve. The total illuminance level at Point P, E_p, will be given by

$$E_p = \frac{cd(\theta_1)}{d_1^2} \cos(\theta_1) + \frac{cd(\theta_2)}{d_2^2} \cos(\theta_2) \quad fc. \qquad (12.10)$$

Example 12.9: Given two 400 W HPS industrial luminaires, as shown in Figure 12.9, and the candlepower (candela) distribution data, shown in Figure 12.10, find the level of illumination at point p. Assume $h = 25.0$ ft., $X_1 = 10$ ft. and $X_2 = 20$ ft. Using Equations 12.9, we find that

$$
\begin{aligned}
d_1 &= \sqrt{10^2 + 25^2} &= 26.93 \ ft. \\
d_2 &= \sqrt{20^2 + 25^2} &= 32.02 \ ft. \\
\theta_1 &= \arctan\left[\frac{10}{25}\right] &= 21.80° \\
\theta_2 &= \arctan\left[\frac{20}{25}\right] &= 38.66°.
\end{aligned}
$$

By linear interpolation of the candela distribution data

Description: **Medium distribution industrial HID**

Lamp: **400 Watt High Pressure Sodium rated 50,000 Lumens**

Spacing: **1.35 X Mounting height**

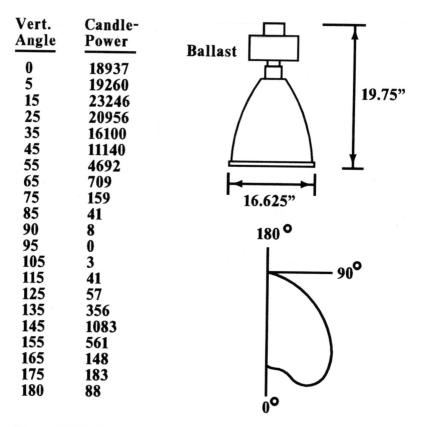

Vert. Angle	Candle-Power
0	18937
5	19260
15	23246
25	20956
35	16100
45	11140
55	4692
65	709
75	159
85	41
90	8
95	0
105	3
115	41
125	57
135	356
145	1083
155	561
165	148
175	183
180	88

Ballast

19.75"

16.625"

180°

90°

0°

Figure 12.10 Photometric data for 400 W HPS industrial luminaire.

$$cd(\theta_1) = cd(21.80°) = 21,689 \ cd$$
$$cd(\theta_2) = cd(38.66°) = 14,285 \ cd.$$

The illuminance level at point P will be given by

$$E_p = \left[\frac{21,689}{725} \right] 0.928 + \left[\frac{14,285}{1025} \right] 0.781 = 38.6 \ fc.$$

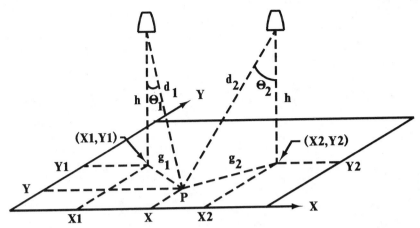

Figure 12.11 Generalized point-by-point example for two luminaires over rectangular work plane.

When more than one or two luminaires are involved, or the level of illuminance must be computed for a number of locations, the method outlined above is best implemented by computer. Also, we used linear interpolation to find the value of $cd(\theta_1)$ and $cd(\theta_2)$. In general, such candlepower distributions are curves, and straight-line interpolation between data points might introduce some degree of error.

Let us now see how we might generalize the techniques above for computer-generated solution. We again assume that we have two luminaires, but this time they are mounted over an area of length, L, and width, W, as shown in Figure 12.11. Let the L dimension be the X axis, and the W dimension be the Y axis. We will locate the luminaires at (X1, Y1) and (X2, Y2). The luminaire mounting height is again, h. We would like to determine the level of illuminance at incremental points (x, y) throughout the area. We will let these increments be the rows and columns of an array, E.

We will let the distance across the plane from directly under each luminaire to point P be called g.

The distances g_1 and g_2, distances d_1 and d_2 and angles θ_1 and θ_2 will then be given by

$$g_1(x,y) = \sqrt{(X1-x)^2 + (Y1-y)^2}$$

$$g_2(x,y) = \sqrt{(X2-x)^2 + (Y2-y)^2}$$

$$d_1(x,y) = \sqrt{g_1(x,y)^2 + h^2}$$

$$d_2(x,y) = \sqrt{g_2(x,y)^2 + h^2} \qquad (12.11)$$

$$\theta_1(x,y) = \arctan\left[\frac{g_1(x,y)}{h}\right]$$

$$\theta_2(x,y) = \arctan\left[\frac{g_2(x,y)}{h}\right].$$

The values of $cd(\theta_1)$ and $cd(\theta_2)$ are determined either by a table lookup algorithm and interpolation, or by a curve fit of the candlepower distribution data.

Using the above methods we can now compute the illuminance level as a function of x and y by

$$E(x,y) = \frac{cd(\theta_1(x,y))}{d_1(x,y)^2} \cos(\theta_1(x,y))$$
$$+ \frac{cd(\theta_2(x,y))}{d_2(x,y)} \cos(\theta_2(x,y)) \quad fc. \qquad (12.12)$$

The results are normally stored in an array, which is often then plotted to scale, yielding a graphical depiction of the overall area lighting. The above technique has the virtue of being easily expandable to any number of luminaires.

Example 12.10: Let us locate the same two luminaires, as in Example 12.9, over a flat plane, as shown in Figure 12.11. The luminaires are mounted at a height of 25 ft. above the work plane, which has a width of 40 ft. and a length of 60 ft. We will locate luminaires 1 and 2 above the plane at (x, y), points (15, 20) and (45, 20), respectively. We wish to compute the illuminance level at increments of 5 ft. along the X axis and 5 ft. along the Y axis.

The candlepower distribution data was curve-fitted with a polynomial curve fitting algorithm and the resulting equation for intensity (cd) was expressed as a function of θ.

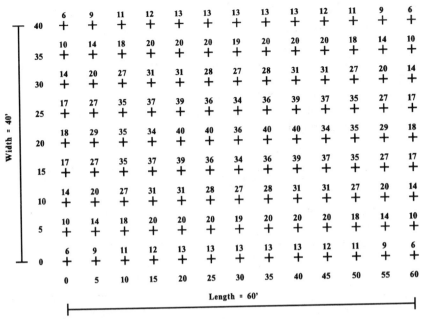

Figure 12.12 Results of point-by-point calculations - Example 12.10.

Using Equation 12.11 and the curve fitted equation for intensity, Equation 12.12 yields the array of illuminance values which have been plotted to scale in Figure 12.12. Similar results can be obtained by straight-line interpolation of the luminaire candlepower distribution data.

We must remember that the luminaire in this example has a symmetrical distribution. Had this not been true, the intensity in a given direction would have been a function of both the vertical angle, θ, and a horizontal angle, β, as well. Although symmetrical distribution is fairly common in interior luminaires, especially HID fixtures, we will see, in the next section, that exterior luminaires often have distributions which are asymmetrical in nature.

In summary, point-by point calculations are invaluable in applications where it is important to know the exact level of illuminance at a specific point, rather than overall averages. Such computations can become quite tedious if done manually and computer assistance is a necessity for systems involving large numbers of luminaires, such as sports lighting applications. We will now turn our attention to exterior lighting design.

AREA LIGHTING
POINT-BY-POINT DESIGN

The term *area lighting* actually applies to a variety of applications, including security lighting, roadway lighting and parking area lighting. There are a wide variety of different luminaire designs available for each application. In addition to area lighting, we have specialized applications, such as sports lighting and architectural lighting. In this section, we will focus on area lighting because of its universal application.

In planning an area lighting system, we must consider several important points. Among these are

- The goals of the design, including economic considerations
- The desired level of illuminance
- The location of the luminaires (including site constraints)
- The choice of luminaire type and light source

Once completed, the design must be evaluated to determine whether the desired level of illumination and uniformity have been achieved.

Clearly understanding the design goals is just as crucial for area lighting as for interior applications. The goals of area lighting vary widely. In some applications, only minimal lighting for security purposes is desired, while in others, area lighting is both functional and decorative in nature. Understanding the client's needs and desires requires careful communication before the design process begins. This coordination often involves the architect, especially if the area lighting is to have decorative qualities. If the building's owner is particularly knowledgeable or has developed several other projects, he may have specific area lighting criteria already in mind. This is particularly true of major retail stores who build many similar buildings. In such cases, the engineer must balance both the owner's requirements and the architect's desires along with his own engineering judgment in order to arrive at a design which pleases all concerned. This is often a very involved process, and effective professional communication between all parties will determine the success or failure of the endeavor.

The next step in the planning process is the determination of the desired level of illuminance. Fortunately, guidelines are available from recognized sources, such as the Illuminating Engineering Society of North America. Table 12.8 shows typical recommended maintained horizontal

Table 12.8 Recommended Illuminance Levels for Parking Area Lighting. (Reprinted with permission of Illuminating Engineering Society of North America.)

Open Parking Facilities General Parking and Pedestrian Area			Vehicle Use Areas	
Level of Activity	Footcandles[1]	Uniformity Ratio[2]	Footcandles[1]	Uniformity Ratio[2]
High	0.9	4:1	2.0	3:1
Medium	0.6	4:1	1.0	3:1
Low	0.2	4:1	0.5	4:1
Covered Parking Facilities				
Area	Day[1] Footcandles	Night[1] Footcandles	Uniformity Ratio[2]	
General Parking and Pedestrian Area	5.0	5.0	4:1	
Ramps and Corners	10.0	5.0	4:1	
Entrance Areas	5.0	5.0	4:1	
Stairways[3]	10-15-20 Footcandles			

[1] Minimum on Pavement. In covered parking areas, it is the sum of electric lighting and daylighting.
[2] Average to minimum.
[3] See Figure 2.2, IES Lighting Handbook, 1981 Application Volume.

footcandle levels for parking facilities. Notice that the recommended minimum levels of illuminance vary from 0.5 fc to 5.0 fc, depending on the application.

The location of luminaires is often a more serious problem than one might at first imagine. In many applications, pole locations might be possible only between parking aisles, while in other applications, it might be desirable to locate poles only along the side of the area to be illuminated. In some cases, adjacent buildings provide a mounting location, thereby saving the cost of poles, foundations and trenching. Increasingly, many state and local governments have enacted restrictions on pole heights, and in some cases, even on luminaire design and light source type. In actuality, the final selection of luminaire locations might well represent a compromise between the best theoretical solution and the realities posed by the above limitations.   With the variety of luminaire distribution patterns available today, almost any area can be effectively lighted.   We will discuss general pole location criteria later on in this chapter.

As mentioned in Chapter 11, luminaires are available to meet almost any design and economic requirements. In general, we wish to utilize the smallest number of luminaire locations that will meet the design goals, because pole, foundation and trenching costs are a significant part of the installation budget. Since most area lighting luminaires are somewhat difficult to maintain, due to their location, long-lived light sources with good lumen maintenance characteristics are very important. As a result, HID sources are almost universally used in area lighting applications. The long life, the very high efficiency and the excellent lumen maintenance characteristics of the HPS source has made it very popular with many engineers for area lighting applications.   Determining the appropriate luminaire design is often a joint venture between the engineer and the architect, since area lighting poles and luminaires, or even building-mounted luminaires, become an architectural feature of the building. The construction budget requirements may also play a role in determining the overall luminaire design and quality level. Weighing the economic factors early in the design process will help avoid last-minute compromises if the project is found to have exceeded its budget. Such budget-cutting changes late in the design are often less effective than thoughtful planning around a known budget earlier in the planning process, when overall design objectives are established.

Area Lighting Design

Once the luminaire type, light source, mounting location and level of illuminance are known, the actual design process can begin. By its very nature, area lighting design is normally accomplished by inverse square law calculations because we are generally interested in the level of illumination at specific points, as well as the overall average illuminance. The general computational technique is very similar to that employed in the interior lighting design, point-by-point example just presented.   One very significant difference, however, is the asymmetrical nature of most area luminaires. As discussed in Chapter 11, there are five basic distribution patterns, each with its own applications. The candlepower distribution data typically shows that the candlepower in a given direction is a function of both the vertical angle and the horizontal angle. It is necessary, then, to interpolate between candlepower values for both vertical and horizontal planes. As in the case of interior point-by-point calculations, it is necessary to establish a work plane height, as well as a coordinate system with incremental steps in both X and Y directions. In many applications, the work plane is taken to be the surface of the parking area or roadway. The luminaire height above the work plane must include the height of not only the pole, but also any concrete or metal base involved.

The computationally cumbersome nature of area lighting as well as other point-by-point calculations, make them well suited for computer application. Fortunately, a number of excellent area lighting programs are available. Let's look at a simple area lighting design problem and its solution by such a program.

Example 12.11: We are given a rectangular parking area, as shown in Figure 12.13. The owner has asked that the area be illuminated to an initial average footcandle level of 2.5 *fc*. Due to the location of parking aisles, we are restricted to the spaces between rows as shown.

In order to minimize the number of pole locations, two poles were placed as shown in Figure 12.13. The 400 W HPS luminaire selected, has a Type V (square) distribution and is pole-mounted 40 ft. above the ground. The luminaire photometric data is shown in Figure 11.10. A coordinate system was established with the X axis along the 150 ft. dimension, and the Y axis along the 100 ft. dimension. It was decided that computations should be carried out at 10 ft. increments in both X and Y directions. The resulting point by point computations are summarized in Figure 12.14.

Lighting Terms

At first glance, it is somewhat difficult to tell from the array of numbers whether the design objectives have been met and also if the overall

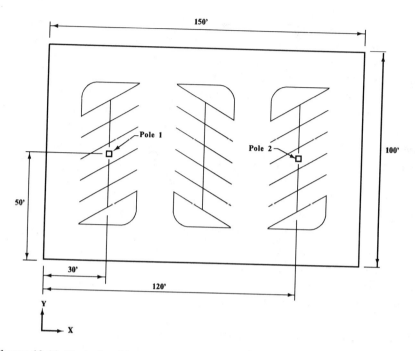

Figure 12.13 Typical parking area configuration and pole placement - Example 12.11.

uniformity is adequate. In order to evaluate these factors, we must establish several guidelines. These guidelines require that we add several new terms to our lighting vocabulary.

Maximum Footcandle Level -	The highest footcandle level at any point on the area.
Minimum Footcandle Level -	The lowest footcandle level at any point on the area.
Average Footcandle Level -	The numerical average of all the computed points.
Maximum to Minimum Ratio -	The ratio of the highest level of illuminance to the lowest level of illuminance.
Average to Minimum Ratio -	The ratio of the average level of illuminance to the minimum level of illuminance.

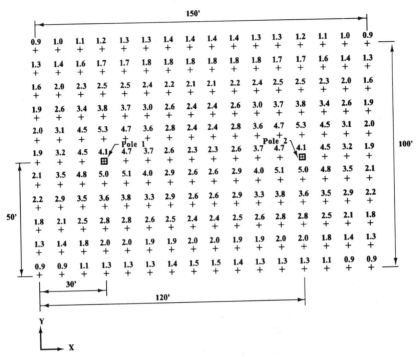

Figure 12.14 Results of area lighting point-by-point calculations - Example 12.11.

Experience has shown that maximum to minimum ratios of up to 6 may be acceptable while average to minimum ratios of up to 3 are fairly common.

Using this criteria, let us now evaluate the previous layout. An inspection of Figure 12.14 reveals that the maximum level is 5.3 *fc* and the minimum is 0.9 *fc*. There are 176 data points. The average of these is 2.5 *fc*. The corresponding ratios, using data accurate to six decimal places, are then

$$
\begin{aligned}
\text{Maximum} &= 5.3 \\
\text{Minimum} &= 0.9 \\
\text{Max/Min} &= 6.2 \\
\text{Avg/Min} &= 2.9
\end{aligned}
$$

The average to minimum ratio is generally acceptable, and the maximum to minimum ratio is slightly high. It might be possible to utilize perimeter poles instead of poles located within the area, or even an increased number of poles to improve this ratio. Obviously, such decisions would have economic consequences which must be considered along with the engineering goals.

Luminaire Placement Techniques

As with many areas of design, generalizations regarding pole placement techniques can sometimes be misleading. Despite this, there are several guidelines used by many design engineers, and in closing, it might be appropriate to review several of these.

- Do not space poles over four times the luminaire mounting height.
- In order to minimize glare, luminaires should be aimed 30 degrees or more below the horizontal.
- Luminaires should be aimed approximately two-thirds of the distance across the area to be lighted.
- Aim luminaires so that the edge of one beam overlaps the aiming point of the adjacent beam.
- Shadow can be minimized by increasing the number of directions from which light approaches.
- Select a beam spread or pattern which approximates the area to be lighted.

Note that these guidelines include both fixed luminaires, whose aiming points cannot be altered, as well as traditional floodlights which can be aimed at the time of installation. In reality, guidance on maximum pole spacing should be determined not only by the desired level of illuminance, but also by the light distribution characteristics of the luminaire, as well as the various illuminance ratios discussed in the previous section.

The above general guidelines might serve as a basis for rough preliminary design, which would then be followed by computer analysis and further refinement. Actual area lighting design is often a very iterative process, which begins with a clear understanding of the design goals, consideration of specific site constraints, selection of illuminance levels, light sources and luminaires, followed by a preliminary layout. A computer analysis of this preliminary layout often reveals design weaknesses, such as poorly lighted areas or unacceptable uniformity ratios.

At this point, more informed design decisions can be made regarding the luminaire placement, mounting height, distribution patterns, or even the number of poles. After refinement, another computer analysis will determine if further design is required. This cycle may occur several times. It may be seen from this discussion that computer analysis does not replace, in any way, the human judgment and experience that are required to produce an effective area lighting design. It does, however, provide information for informed decision making which increases the likelihood of a better overall design.

COMPUTER TECHNIQUES

As noted in previous sections, lighting design lends itself quite well to computer application. The required photometric data can be readily stored and retrieved. The various cavity ratios and reflectances can be determined, as well as the lamp lumen depreciation factors, luminaire dirt depreciation and room surface dirt depreciation. For many years, the widespread application of software for lighting design was hampered by the lack of availability of appropriate computer programs. Even using the available programs, it was often necessary to manually enter photometric data, and this could become quite time consuming. Beginning several years ago, several of the major manufacturers of lighting equipment started a development program aimed at making state-of-the-art lighting software available to the engineering community. Those software packages were made available, often, at little or no cost. Early software normally contained photometric data on the products manufactured by one manufacturer only, and entry of similar data by other manufacturers was not generally possible.

Since the mid-1980s, great strides have been made in manufacturer-developed software. Modern lighting software is quite sophisticated, and many packages feature highly interactive data entry and readily accessible, on-line help routines to supply a wide variety of information, such as lamp and ballast data. A number of these routines are capable of providing a recommended luminaire layout. These routines are also capable of rapidly computing illuminance levels for vertical surfaces, such as for warehouses, where the vertical illumination on the face of storage racks is very important. Such calculations would be very time consuming and tedious if performed by hand.

Recent advances in the standardized presentation of photometric data have now made it possible for many manufacturer-developed lighting analysis packages to import data from other manufacturers. This facilitates a comparison of the competing equipment alternatives. Several of the available software packages offer lighting economic analysis capability, thus allowing consideration of both the engineering and the financial aspects of a design. This software continues to undergo continuous development and modification, sometimes making it difficult for engineers to maintain current versions of programs. Beginning in the early 1990s, several manufacturers began to offer software and photometric data updates for registered users by modem, making it far easier to maintain current data. Since these rapid advances will, in all likelihood, continue, the 1990s should be an exciting time for the lighting designer.

References

Boylan, Bernard R. 1987. *The Lighting Primer*. Iowa: Iowa State University Press.

Cox, James A. 1979. *A Century of Light*. New York: The Benjamin Company, Inc.

General Electric Company Large Lamp Department. 1964. *General Lighting Design*. Ohio: General Electric Company.

General Electric Company Lighting Business Group. 1978. *Light and Color*. Ohio: General Electric Company.

General Electric Lighting. 1990. *Selection Guide for Quality Lighting*. Ohio: General Electric Company.

Helms, Ronald N. 1980. *Illumination Engineering for Energy Efficient Luminous Environments*. New Jersey: Prentice-Hall, Inc.

Illuminating Engineering Society of North America - Education Committee. 1988. *IES Education Series ED-100*. New York: Illuminating Engineering Society of North America.

Illuminating Engineering Society of North America. 1984. *IES Lighting Handbook -Reference Volume*. New York: Illuminating Engineering Society of North America.

Illuminating Engineering Society of North America. 1987. *IES Lighting Handbook -Application Volume*. New York: Illuminating Engineering Society of North America.

Murdoch, Joseph B. 1985. *Illumination Engineering - From Edison's Lamp to the Laser*. New York: Macmillan Publishing Company.

Sylvania. 1988. *Large Lamp Catalog*. Massachusetts: GTE Products Corporation.

Index

O   week 1   =>   chapter 1, chapter 2

week 2
M   297 -   309   ✓

W   309 -   319 + handout ✓

F   319 - 332 ✓

week 3

M :   HOLIDAY

W :   59 - 77 ✓

F :   95 - 109

WEEK 4

M :   109 - 120   13 Sept

W :   122 - 129

F :   129 - 145
EXAM #1

WEEK 5

M :   EXAM 1

W :   147 - 155 : + westishouse h.o.

F :   154 - 171